Springer Series in Astrophysics and Cosmology

The series covers all areas of astrophysics and cosmology, including theory, observations, and instrumentation. It publishes monographs and edited volumes. All books are authored or edited by leading experts in the field and are primarily intended for researchers and graduate students.

Cosimo Bambi · Jiachen Jiang
Editors

High-Resolution X-ray Spectroscopy

Instrumentation, Data Analysis, and Science

Springer

Editors
Cosimo Bambi
Department of Physics
Fudan University
Shanghai, China

Jiachen Jiang
Institute of Astronomy
University of Cambridge
Cambridge, UK

ISSN 2731-734X ISSN 2731-7358 (electronic)
Springer Series in Astrophysics and Cosmology
ISBN 978-981-99-4408-8 ISBN 978-981-99-4409-5 (eBook)
https://doi.org/10.1007/978-981-99-4409-5

This Springer imprint is published by the registered company Springer Nature Singapore Pte Ltd.
The registered company address is: 152 Beach Road, #21-01/04 Gateway East, Singapore 189721, Singapore

Preface

In 1895, Wilhelm Conrad Roentgen discovered X-rays, which are a type of electromagnetic radiation with very short wavelengths that can penetrate many materials, including human tissue. In the following years, X-rays were used primarily for medical imaging. In the 1940s and 1950s, scientists began using X-rays to study the properties of materials and the structure of crystals. However, it wasn't until the advent of space exploration that X-ray astronomy became possible.

In 1962, the first X-ray telescope was launched into space by the United States. This telescope, called Uhuru, was designed to study X-rays emitted by celestial objects, including stars and galaxies. Over the next few decades, several more X-ray telescopes were launched, including the Einstein Observatory, which was launched in 1978 and was the first telescope to produce detailed X-ray images of celestial objects. X-ray astronomy has provided important insights into many areas of astrophysics, including the study of black holes, the evolution of galaxies, and the nature of dark matter. Today, X-ray astronomy continues to be an active area of research, with new discoveries being made all the time.

The application of X-ray reflection grating instruments, which are used on both the XMM-Newton and Chandra X-ray observatories, opened the window of high-resolution X-ray spectroscopy, compared to previous CCD-resolution spectral data, at the end of the 1990s. These instruments use diffraction gratings to spread out the incoming X-rays into a spectrum, which allows scientists to analyze the composition and motion of the source of the X-rays.

One of the most significant discoveries made using X-ray reflection grating instruments on XMM-Newton and Chandra is the detection of supermassive black holes at the centers of galaxies. By studying the X-ray emissions from the accretion disks around these black holes, scientists have been able to learn about the properties of these mysterious objects and the way they affect their surroundings. Another important discovery made using these instruments is the study of hot gas in galaxy clusters. By analyzing the X-ray emissions from these clusters, scientists have been able to learn about the distribution and dynamics of the hot gas, which provides important clues about the structure and evolution of the Universe.

Overall, X-ray reflection grating instruments have proven invaluable tools for studying the high-energy Universe and have led to numerous groundbreaking discoveries in astrophysics and cosmology. However, the study of high-resolution, high-signal-to-noise X-ray data has been limited to the soft X-ray band until the availability of X-ray micro-calorimeters. X-ray micro-calorimeters are a type of X-ray detector that operates based on the principle of calorimetry. X-ray micro-calorimeters are highly sensitive and can measure the energy of individual X-ray photons with very high precision.

The Hitomi X-ray mission, also known as ASTRO-H, was a Japanese X-ray observatory featured with X-ray micro-calorimeters that was launched in 2016. Unfortunately, the mission was lost due to an attitude control system malfunction shortly after launch, but the data it collected during its short operational time has provided valuable insights into a number of astrophysical phenomena, such as much improved measurements of the velocity distribution in the Perseus Cluster. When this book is prepared, XRISM, the recovery mission of Hitomi, will soon be launched. The re-definition process of New Athena, ESA's next flagship mission, is undergoing.

This book has been crafted with a dual purpose in mind; firstly, to serve as a comprehensive guide for beginners looking to learn the skills required for high-resolution X-ray spectroscopy; and secondly, to provide a summary of some of the most significant discoveries made using high-resolution X-ray spectroscopy across a range of topics, from Galactic objects to galaxies. However, the true value of this book lies in its emphasis on the importance of future high-resolution X-ray spectral data. With ongoing advancements in X-ray telescope technology, the future of X-ray astronomy looks brighter than ever before. The wealth of data that will be collected through high-resolution X-ray spectroscopy in the coming years will be truly transformative, providing unprecedented insights into the inner workings of some of the most extreme objects in the Universe, including black holes, neutron stars, and galaxy clusters. This book serves as a reminder of the critical role that high-resolution X-ray spectroscopy will play in reshaping our understanding of the Universe and the physical processes that govern it.

Shanghai, China
Cambridge, UK
May 2023

Cosimo Bambi
Jiachen Jiang

Contents

Part I
Instrumentation and Data Analysis

Chapter 1
A Brief Survey of the History, Present and Future of Cosmic High-Resolution X-ray Spectroscopy

Andrew C. Fabian

High resolution X-ray Spectroscopy (HRXS) is a vital component of a cosmic X-ray astronomer's toolbox. It enables precise measurements of the abundances, ionisation, temperature, density, velocity and velocity dispersion of hot X-ray emitting matter and the composition, column density and velocity of cold X-ray absorbing gas and dust along the line of sight, Images and timing analyses of course provide much but spectroscopy supplies many of the hard numbers needed in our quest to understand the Universe [10].

HRXS is also a tool of Solar physicists [see e.g. 6] and laboratory physicists [see e.g. 18] where high photon fluxes are often available to enable very high resolution of 1000s to be achieved with small apertures. Cosmic X-ray astronomy must however deal with what the sources themselves provide. In the 2–10 keV band the flux of one of the X-ray brightest steady sources in the Sky—the Crab Nebula—is only about $4\,\mathrm{ph\,cm^{-2}\,s^{-1}}$. The brightest extragalactic sources are about 30 times fainter, so large effective collecting areas $>100\,\mathrm{cm}^2$ are needed to obtain even low resolution X-ray spectra in reasonable exposure times. Many high resolution X-ray spectra of cosmic objects obtained so far are therefore photon starved.

The information yield of a spectrum depends on the Spectral Resolution or Resolving Power $R = E/\Delta E = \lambda/\Delta\lambda$. Early X-ray astronomers relied on proportional counters for which $R \sim 6E_6^{0.5}$, where E_6 is the photon energy in units of 6 keV (the square root reflects the Poisson nature of the measurement of energy via the size of the electron cloud produced by photoelectric absorption of the incident X-ray photon, modified by the Fano factor). This was sufficient to see the iron K emission line in galaxy clusters [20, 21] demonstrating that the X-ray source was hot gas filling the cluster and not inverse Compton emission which would have given a featureless spectrum. Proportional Counters continued to be used in EXOSAT, Ginga, ROSAT, RXTE and other observatories. Solid State spectrometers such as the single pixel SSS on the Einstein Observatory launched in 1978 had a better resolution of about

A. C. Fabian (✉)
Institute of Astronomy, Madingley Road, Cambridge CB3 0HA, UK
e-mail: acf@ast.cam.ac.uk

C. Bambi and J. Jiang (eds.), *High-Resolution X-ray Spectroscopy*, Springer Series in Astrophysics and Cosmology, https://doi.org/10.1007/978-981-99-4409-5_1

$R \sim 40E_6^{0.5}$. Multipixel CCDs arrived (from MIT) with ASCA, launched in 1993, followed by the Chandra ACIS and XMM pn and MOS. In the meantime, Gas Scintillation Proportional Counters on EXOSAT and ASCA had an intermediate value of R.

Greater spectral resolution was achieved first with Bragg Crystal Spectrometers which enabled the first detection of the OVIII line [23] from a rocket flight, followed in 1979 by an instrument on the Einstein Observatory (EO see e.g. spectrum of Puppis A by [22]). Next came diffraction Gratings that are still operating on Chandra (LETG [3], MEG and HEG [4]) and XMM (RGS [7]). Since gratings disperse the radiation across the detector, their use on extended sources is restricted. Non-dispersive high resolution is obtained with microcalorimeters [19], where the X-ray photon energy is measured via the rise in temperature of a cryogenically cooled absorber at 65 mK. The only successful flight so far with this technique is Hitomi, which was short lived but did however observe and measure the velocity structure of the Perseus cluster with an unprecedented accuracy of $<10\,\mathrm{km\,s^{-1}}$. This showed, contrary to initial expectations, that cluster gas does not appear to be very turbulent [15].

Dispersive spectrometers rely on natural crystal lattice spacing and grating periods of $\sim 10^{2-3}$A and have an approximately fixed value of $\Delta\lambda$, whereas non-dispersive spectrometers rely on natural band gaps and phonon energies leading to ΔE being fixed (70–150 eV for CCDs, 5 eV for Hitomi and 1–2 eV for future microcalorimeters).

R for the current high resolution X-ray Spectrometers is shown in Fig. 1.1. A similar summary of R for different instruments can be found in [4]. Despite the original version being nearly 2 decades old, the plot is still up-to-date in early 2023! This is partly due to the sad history of non-dispersive spectrometers: AXAF-S, which would have used the first proposed such detector, was cancelled by NASA in 1993; the rocket carrying the microcalorimeter mission ASTRO-E failed to reach orbit; ASTRO-E2 suffered a Helium leak; and after operating for only a few weeks, the Hitomi spacecraft spun up and was lost.[1] Nevertheless, the Chandra and XMM gratings have been enormously productive and very important for progressing X-ray astronomy over the past 2 decades (clearly evident in other Chapters in this Volume). The brief operation of Hitomi demonstrated the tremendous potential of a 5 eV microcalorimeter. A reflight of a Hitomi class detector is expected in 2023 in the form of XRISM.

Assuming a long operational lifetime for XRISM we can expect transformational advances in X-ray astronomy arising from high resolution spectra of many classes of objects, as discussed elsewhere in this volume. The future beyond XRISM depends on our Space Agencies. Although selected by ESA in 2014 as its next Large Mission, ATHENA, with a 2.5 eV resolution X-IFU (X-ray Integral Field Unit), is undergoing a redesign as NewATHENA and a launch date in the mid 2030s. Soft X-ray Grating spectroscopy (10–60 Å, R > 3500 + EUV) ARCUS is a NASA Probe Class Mission candidate for launch in the early 2030s, as is the Line Emission Mapper with a 2 eV

[1] It almost feels as if the results to be revealed by HRXS are so precious that we have to work harder to obtain them.

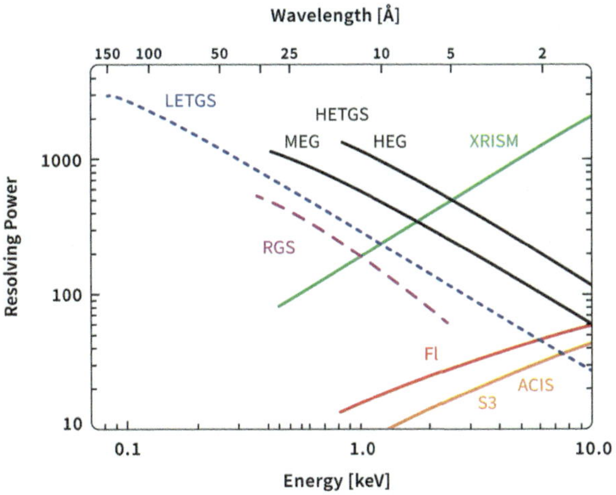

Fig. 1.1 Spectral Resolving Power of various recent instruments following [4]. Wavelength is displayed at the top and energy at the bottom. The Chandra LETGS and HETGS (consisting of the MEG and HEG) are in the upper centre. The high spatial resolution of the Chandra mirror means that their use on extended objects is restricted to where line emission dominates the spectrum (e.g. some supernova remnants). The line for the XMM-RGS lies just below the LETGS line. There are 2 RGS spectrometers each of which has lost a CCD detector chip (at different energies). The lower spectral resolution of the XMM mirrors and design of the reflection gratings allows spectra of sources extended up to about one arcmin to be obtained. ACIS refers to the Chandra CCD detectors and XRISM is expected to launch in 2023. The line is appropriate for $\Delta E = 5\,\text{eV}$. A 2.5 eV calorimeter means therefore that R lies a factor of 2 above the XRS line

microcalorimeter and 10 arcsec PSF. LYNX, with 2 m^2 of mirror effective area, 3 eV spectral resolution, <1 arcsec spatial resolution microcalorimeter and gratings with $R > 4000$, has been proposed to NASA for a later launch.

Essential to any observational spectroscopic study is an understanding of what spectra should look like and why. X-ray spectroscopy of atoms, molecules and solids (e.g. dust) is a highly complex subject involving atomic physics, quantum mechanics and relativity. Most observers rely on the results of detailed calculations gathered in databases such as AtomDB, SPEX, CHIANTI made accessible through modelling (CLOUDY [12], XSTAR [2], RELXILL [5]) and spectral-fitting packages such as XSPEC, SPEX and CIAO [1, 13, 16]. This can be backed up by laboratory work using, for example, Electron Beam Iron Traps (EBITs) which have been operating since 1986, and other devices. Advances are still being made even on well-observed X-ray lines, such as the 3C:3D ratio of FeXVII [17], which provides the brightest X-ray lines from the Sun.

An additional component to HRXS is the astronomical interpretation of the data. In some sense, this is the hardest part. You may have wonderful spectra reduced to line lists, strengths and widths but answering the question of "what does it all mean?" is often difficult and a major challenge. The observed object is not a glowing, single-temperature blob of gas but a complex multi-temperature, turbulent, plasma with

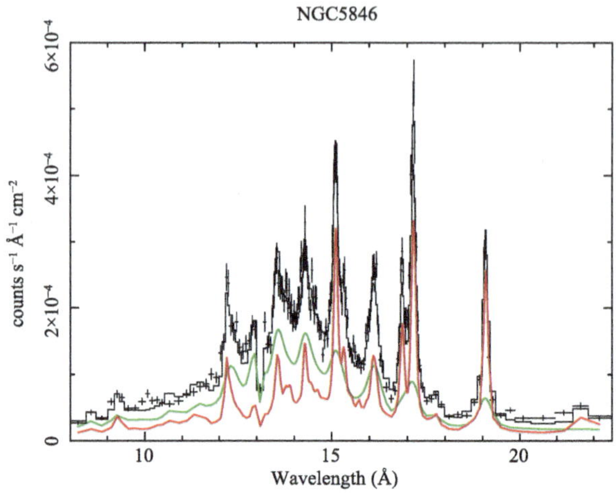

Fig. 1.2 RGS spectrum of the core of NGC5846 which is the brightest central galaxy in the nearest X-ray luminous group. The lower line shows the hidden (absorbed) cooling flow component (which is not very well hidden in this case). It represents a cooling flow of $1.3 \pm 0.3\,M_\odot\,yr^{-1}$ at the centre of this object [11]

possible embedded dust, magnetic fields, cold clouds and a range of densities, and that is just for starters. I think that we (including myself) were misled about cooling flows two decades ago by a too simplistic interpretation of high-resolution X-ray spectra. There are probably other examples awaiting an unravelling. This is just how science works.

We can all think of great targets for future HRXS observations with XRISM. My immediate preference is confirmation of hidden cooling flows at the centres of cool core clusters and elliptical galaxies [8, 10, 11, Fig. 1.2] which may be where most of the growth of massive black holes is happening in the low redshift Universe. I also look for detection of absorption features arising from lower density gas lying above the inner disc at a few gravitational radii from the black hole in some Active Galactic Nuclei [14]. We see emission lines so why not absorption lines too? They will need to absorb the brightest parts of the disc which are on the approaching side so the absorption lines should be blueshifted. Maybe they have been observed already [9] Both would change the present picture of how cool cores and AGN winds operate. X-ray spectroscopy of narrow absorption lines from the inner regions of luminous black holes at all distances can then compete with the amazing Event Horizon Telescope imaging which is limited to nearby low luminosity black holes. Exploring and testing observational data is how progress is made. If we don't look we don't find.

For Si-based detectors operating at $-100\,°\mathrm{C}$, $\Delta E \sim 47(E_\gamma/\mathrm{keV})^{0.5}$ eV. According to Eq. 2.3, the higher the incident photon energy (E_γ), the larger the energy resolution (ΔE). The energy resolution is a slow function of E_γ though. In practice, the total energy resolution should take system noise (σ) into account [26],

$$\Delta E = 2.35w\sqrt{F\ E_\gamma/w + \sigma^2}. \tag{2.4}$$

When the instrument is operated at $\sim -100\,°\mathrm{C}$, system noise contributed by leakage current is negligible. At high readout frequencies, in the absence of other degrading factors, the on-chip amplifier white noise (σ_w) dominates the system noise [26]. In this case, $\sigma_\mathrm{w} \propto \sqrt{f_\mathrm{ro}}$, where f_ro is the readout frequency [26].

The energy resolution of Chandra/ACIS spectroscopic array is ~95 eV at 1.5 keV and 150 eV at 5.9 keV. But the energy resolution of a non-diffractive spectrometer is not necessarily low. Micro-calorimeter, another type of non-diffractive spectrometer, can achieve an energy resolution of $\Delta E \lesssim 10$ eV, which is an order of magnitude better than ACIS and EPIC.

A micro-calorimeter unit consists of three basic components: an X-ray absorber, a sensitive temperature sensor, and a weak thermal link [43]. After absorbing the incident X-ray photon, the detector temperature will experience a sharp rise followed by a slow decay. The temperature sensor will convert the temperature pulse to an electronic pulse (to be read out). The weak thermal link will remove excess heat in preparation for absorbing the next incident photon. The energy resolution of the device is [16]

$$\Delta E = \eta\sqrt{\frac{kT^2C}{\alpha}}, \tag{2.5}$$

where η is a dimensionless parameter of the order of unity, k the Boltzmann constant, T the detector temperature (~50 mK, slightly above absolute zero), C the heat capacity, and α the temperature sensitivity. High energy resolution can be achieved by reducing the heat capacity and increasing the temperature sensitivity. We refer readers to Chaps. 4 (Transition Edge Sensors X-ray Spectrometers: Applications for X-ray Astronomy) and 5 (Hitomi/XRISM micro-calorimeter) of this book for more technical details.

The first astrophysical application of micro-calorimeter was conducted by McCammon et al. [43] on a sounding rocket launched in 1999. Its energy resolution ranges from 5 to 12 eV over the 0.06 to 1 keV band [43]. This micro-calorimeter was Silicon-based. The same design was adopted by Hitomi/SXS (Soft X-ray Spectrometer) [25] and XRISM/Resolve [74]. Micro-calorimeter based on transition-edge sensors (TES) has an even better performance. For instance, Athena X-ray observatory is an approved ESA large class mission [49]. Its X-ray Integral Field Unit (X-IFU) [3] aims to achieve an energy resolution of 2.5 eV over the 0.2–12 keV energy band. The Hot Universe Baryon Surveyor (HUBS) [15, 16], a mission proposed to the Chinese National Space Agency (CNSA), aims to achieve an energy resolution of 0.6 eV for its central 12×12 pixel^2 over the energy band of 0.1–2 keV.

The energy resolution (ΔE) of CCDs and micro-calorimeters is constant or only moderately dependent on energy. Accordingly, their spectra are often plotted in the energy space.

2.2.2 Past and Current Diffractive X-ray Spectrometers

The aforementioned non-diffractive spectrometers convert the incident photon energy to countable objects (e.g., electrons). Diffractive spectrometers construct interference of incident photons along different light paths based on their photon energy. This is realized by placing a diffraction element at the exit aperture of the focusing optic. Dispersed X-ray photons will then arrive at a focal plane imaging detector (usually a non-diffractive imaging spectrometer).

The first astrophysical application of a diffractive X-ray spectrometer is actually a crystal spectrometer, i.e., the Focal Plane Crystal Spectrometer (FPCS) on the Einstein Observatory [8]. FPCS has a resolving power $R = E/\Delta E = 50{-}500$ over the energy band of 0.2–3 keV [8]. Limited by the Bragg condition, crystal spectrometers cannot simultaneously provide diffract radiation for a wide range of wavelengths. To overcome this drawback, the device needs to scan through a range of Bragg angles (by rocking the FPCS crystal) [8]. Crystal spectrometers were also used on the Solar Maximum Mission [1], HINOTORI [36], P78-1 [21], SOLAR-A [17], and YOHKOH [38]. In particular, the Bragg Crystal Spectrometer aboard SOLAR-A achieved a resolving power $R = \lambda/\Delta\lambda = 3000{-}6000$ for some narrow wavelength ranges.

Unlike crystal spectrometers, grating spectrometers can obtain spectra over a wide energy interval simultaneously. The Reflection Grating Spectrometer (RGS) [20] aboard XMM-Newton, High- and Low-Energy Transmission Grating Spectrometers (HETGS [9] and LETGS [7]) aboard Chandra are the current main working horses to obtain high-resolution X-ray spectra. As can be told from the names of these instruments, classical grating spectrometers can be further divided into two types: transmission and reflection gratings.

Prior to XMM-Newton, reflection gratings were applied to the Extreme UltraViolet Explorer (EUVE) mission [4]. The three reflection gratings cover three UV wavelength ranges: 70–190 Å, 140–380 Å, and 280–760 Å, respectively. In addition, reflection gratings were used on the recent sounding rocket instrument—Marshall Grazing Incidence X-ray Spectrometer [10]. Prior to Chandra, transmission gratings were applied to Einstein [66] and EXOSAT [6].

For reflection gratings, the dispersion relation is [20]

$$m\lambda = d(\cos\beta - \cos\alpha), \tag{2.6}$$

where $m = 0, \pm1, \pm2, \ldots$ is the spectral order, λ the wavelength of the incident photon, d the grating period, β the dispersion angle, α the incident angle on the grating plane. Note that triangular grooves on the grating plane are tilted by δ (Fig. 2.1). For

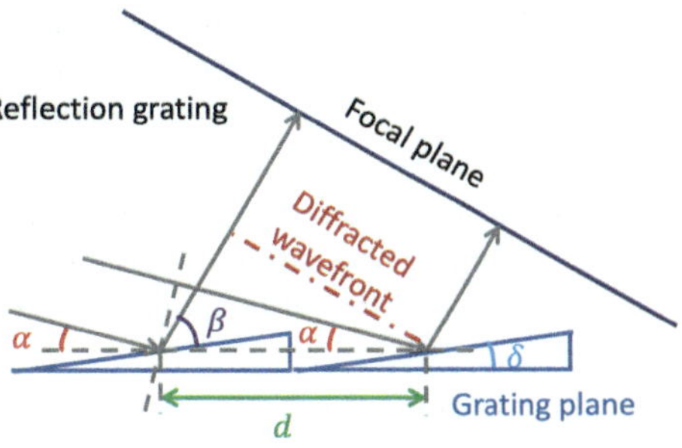

Fig. 2.1 Cartoon of the dispersion geometry for the reflection gratings. X-rays enter from the left with an incidence angle α with respect to the grating plane. The triangular grating grooves are titled by δ with respect to the grating plane

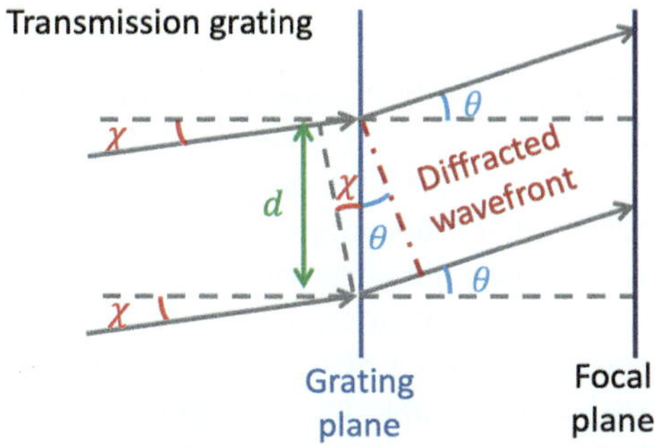

Fig. 2.2 Cartoon of the dispersion geometry for the transmission gratings. X-rays enter from the left with an incidence angle χ with respect to the normalization direction of the grating plane

the zeroth order, we have $\alpha_{m=0} = \beta_{m=0}$, and the grating acts as a mirror. For negative orders, also known as inside orders, we have $\alpha < \beta < \beta_{m=0}$. Positive orders, also known as outside orders, might not always exist, because $\cos\beta$ might be greater than unity for some combinations of α, d, and λ. In the absence of other degrading factors, the wavelength resolution of reflection gratings is [59]

$$\Delta\lambda = \frac{d}{m} \sin\alpha \Delta\alpha, \tag{2.7}$$

where $\Delta\alpha$ is the angular resolution of the telescope.

For transmission gratings, the dispersion relation is [59]

$$m\lambda = d(\sin\theta - \sin\chi), \tag{2.8}$$

where $m = 0, \pm1, \pm2, \ldots$ is the spectral order, λ the wavelength of the incident photon, d the grating period, θ the dispersion angle, χ the incident angle on the grating (Fig. 2.2). Since the incidence and dispersion angles are often small, the wavelength resolution is [59]

$$\Delta\lambda = \frac{d}{m} \Delta(\theta - \chi) \tag{2.9}$$

Both reflection and transmission gratings are constant $\Delta\lambda$ devices (Eqs. 2.7 and 2.9). Accordingly, grating spectra are often plotted in the wavelength space.

Note that for any slitless spectrometer, the incident and dispersion angles will vary across a finite-size grating placed in a focused beam. In practice, the Rowland circle was used to refocus the radiation by curving the gratings [57, 59]. Misalignment of the gratings or deviations from the flatness of the grating plane can also play a significant role in the performance of the gratings ([57], see also Sect. 2.2.3).

2.2.3 Scattering by Diffraction Gratings

Scattering by diffraction gratings is a natural consequence of imperfections in the optical arrangement, whether by roughness on the surface of a grating, or by any random variability of any of the grating properties (such as groove spacing). If this variability occurs on spatial wavelength scales that are short compared to the grating period, the main effect will be an apparent loss of light from the brightest diffraction peaks, without a detectable effect on the spatial response function (i.e., the line spread function, LSF). If, on the other hand, there is variability on spatial wavelength scales comparable to the grating period, the LSF will be affected and appear to have significant scattering 'wings'. In the following, we will develop a very simple description of scattering that will allow for an intuitive understanding and provide useful first-order quantitative scalings. We will describe the scattering properties of the XMM-Newton/RGS, but start with a simple example based on transmission gratings to introduce the analytical problem.

We will assume, as usual, that the grating period, d, is much larger than the radiation wavelength λ. We will also ignore the detailed interaction between the grating material and the radiation, meaning, we will ignore the vector character of the electromagnetic field and the boundary conditions on the solutions to Maxwell's equations. In the approximation that the diffracted radiation is detected far from the diffracting element, we will just apply Huygens' Principle in a slightly extended formulation: we will sum the spherical waves emitted along the wavefronts at the diffraction grating, possibly with an amplitude and phase that have been modulated by propagation through the material.

Let us assume a very simple model for a grating: a set of points at locations $\mathbf{r}_i$, periodically spaced, each of which radiates spherical waves in response to being illuminated by a plane wave

$$A_{\text{in}} = \exp i\mathbf{k}_{\text{in}} \cdot \mathbf{r} \tag{2.10}$$

with $\mathbf{k}_{\text{in}}$ the wave vector of the incident waves. Far from the grating, the diffracted wave propagating in direction $\mathbf{k}_{\text{out}}$ has complex amplitude (ignoring an uninteresting constant phase factor)

$$A = \sum_i \exp i(\mathbf{k}_{\text{in}} - \mathbf{k}_{\text{out}}) \cdot \mathbf{r}_i \tag{2.11}$$

and the intensity of the diffracted light will be given by $I = |A|^2$. In the following we will set $\mathbf{k}_{\text{in}} - \mathbf{k}_{\text{out}} = \mathbf{q}$. The mathematically perfect grating therefore produces a diffraction pattern

$$I_0 = AA^* = \left|\sum_i \exp i\mathbf{q}\cdot\mathbf{r}_i\right|^2 \tag{2.12}$$

Now imagine that the location of each radiating point is slightly displaced by a random displacement $\mathbf{r}_i \to \mathbf{r}_i + \delta\mathbf{r}_i$, with $\delta r_i/r_i \ll 1$ and $\langle\delta\mathbf{r}_i\rangle = 0$. The $\delta\mathbf{r}_i$ are uncorrelated, so that when averaged over the entire grating $\langle\delta\mathbf{r}_i\cdot\delta\mathbf{r}_j\rangle = 0$. The variance of the displacements is $\langle\delta\mathbf{r}_i\cdot\delta\mathbf{r}_i\rangle = \sigma^2$. Calculating the intensity of the diffracted field, and expanding to the first nonzero new term, we get

$$\begin{aligned}
I &= \langle AA^*\rangle \\
&= \left\langle \sum_i \exp i\mathbf{q}\cdot(\mathbf{r}_i+\delta\mathbf{r}_i)\cdot\sum_j \exp -i\mathbf{q}\cdot(\mathbf{r}_j+\delta\mathbf{r}_j)\right\rangle \\
&= \left\langle \sum_i\sum_j \exp i\mathbf{q}\cdot\left(\mathbf{r}_i-\mathbf{r}_j\right)\exp i\mathbf{q}\cdot\left(\delta\mathbf{r}_i-\delta\mathbf{r}_j\right)\right\rangle \\
&\approx \sum_i\sum_j \exp i\mathbf{q}\cdot(\mathbf{r}_i-\mathbf{r}_j)\cdot \\
&\quad \left\langle 1+i\mathbf{q}\cdot(\mathbf{r}_i-\mathbf{r}_j)-\frac{1}{2}q^2\left(\delta\mathbf{r}_i\cdot\delta\mathbf{r}_i+\delta\mathbf{r}_j\cdot\delta\mathbf{r}_j-2\delta\mathbf{r}_i\cdot\delta\mathbf{r}_j\right)+\cdots\right\rangle \\
&= \sum_i\sum_j \exp i\mathbf{q}\cdot(\mathbf{r}_i-\mathbf{r}_j)\cdot\left(1-q^2\sigma^2+\cdots\right) \\
&= I_0\left(1-q^2\sigma^2+\cdots\right)
\end{aligned} \tag{2.13}$$

As long as $q^2\sigma^2 \ll 1$, the only effect of the perturbations is to remove a small amount of light, $I/I_0 = q^2\sigma^2$ from the sharp diffraction pattern, and distribute it widely in between the diffraction peaks (moving the lattice points $\mathbf{r}_i$ around conserves the number of photons, so whatever is missing from the diffraction peaks must have ended up in between the peaks).

A straightforward application of this idea to small perturbations to the period of a transmission grating is given by [18, 56]. Monochromatic line radiation dispersed by the High Energy Grating that is part of Chandra/HETGS onto a CCD camera showed that there appeared to be a faint continuum between the sharp diffraction orders. Inspection of the CCD spectrum of these photons showed that in fact they were all of the same energy, indicating that they had ended up dispersed far away from the diffraction maxima. The same thing was seen in dispersed light from Chandra/LETGS.

A calculation for the diffraction pattern in the presence of small perturbations to the grating period shows that the fraction of scattered light is

$$f = q^2\sigma^2 = k^2\sigma^2 \sin^2\theta = 4\pi m^2 \left(\frac{\sigma}{d}\right)^2 \tag{2.14}$$

where we assume the radiation is incident perpendicularly to the grating, the dispersion angle θ is given by $\sin\theta = m\lambda/d$, and m is the diffraction order, d the average grating period. There is no effect in the zeroth order, and the effect grows quadratically with diffraction orders. Figure 2.3 shows the diffraction pattern of the LETG of Al Kα radiation (8.34 Å), out to $m = 9$. The bottom graph shows the spectrum in a narrow band of energies centered on 8.34 Å (so the faint continuum visible as curved bands in the third panel from the top has been filtered out). The photons in between the diffraction peaks are now almost all Al Kα, and this 'continuum' clearly rises sharply with diffraction order. This graph incidentally also shows that the Al Kα source is not strictly monochromatic; in the third order especially a second emission line is clearly visible on the high energy side of Al Kα. Most likely this is Al Kα excited in aluminum oxide present on the electron impact source!

In order to understand the spatial distribution of the scattered light, imagine that the displacements $\delta\mathbf{r}_i$ are expanded in a Fourier series or integral. Assuming at first just a single sinusoidal perturbation, $\delta\mathbf{r}_i = \mathbf{w}\sin(\mathbf{g}\cdot\mathbf{r}_i)$, with $\mathbf{g}$ the wavevector of the sinusoidal perturbation, of period $l = 2\pi/|\mathbf{g}|$. Introducing this into Eq. 2.13 and carrying through the calculation, assuming that the amplitude w of the perturbation is small (so the Bessel functions of argument $\mathbf{q}\cdot\mathbf{w}$ that will appear can be expanded to first order), it is straightforward to show that the resulting diffraction pattern, in addition to peaking at $\sin\theta = m\lambda/d$, also peaks at $\sin\theta = m\lambda/d \pm \lambda/l$ with an amplitude proportional to $|\mathbf{q}\cdot\mathbf{w}|^2$. These are the first-order terms; we ignore the higher-order terms. The wave-like perturbation will cause diffracted light to appear at an angle $\Delta\theta \approx \lambda/l$ away from the main diffraction order. Now imagine that the perturbation is made up of an ensemble of sine waves of wave vectors g, with a power spectrum $W(g)$. In the first order, each sine wave contributes diffracted light at an angle $\Delta\theta = \lambda g/2\pi$ with respect to a given diffraction order, with an amplitude determined by $W(g)$. The resulting angular distribution of scattered light therefore directly maps out the power spectral distribution of the perturbations.

This can be applied to the reflection gratings on the RGS. An abbreviated calculation is given by [33, 55, 58]. We expect to see the effect of variations in the grating properties (period, groove profile) as well as scattering by surface roughness. We based our analysis of scattering in RGS on the assumption that scattering by surface roughness is probably dominant. Scattering by variations in groove profile and period are not likely to be dominated by coherent perturbations on the scale of around a few dozen grooves or less (in which case they will cause a distribution of scattered light comparable in width to the LSF), but instead light scattered by groove profile variations probably simply merges with the surface roughness scattering, at wider angles.

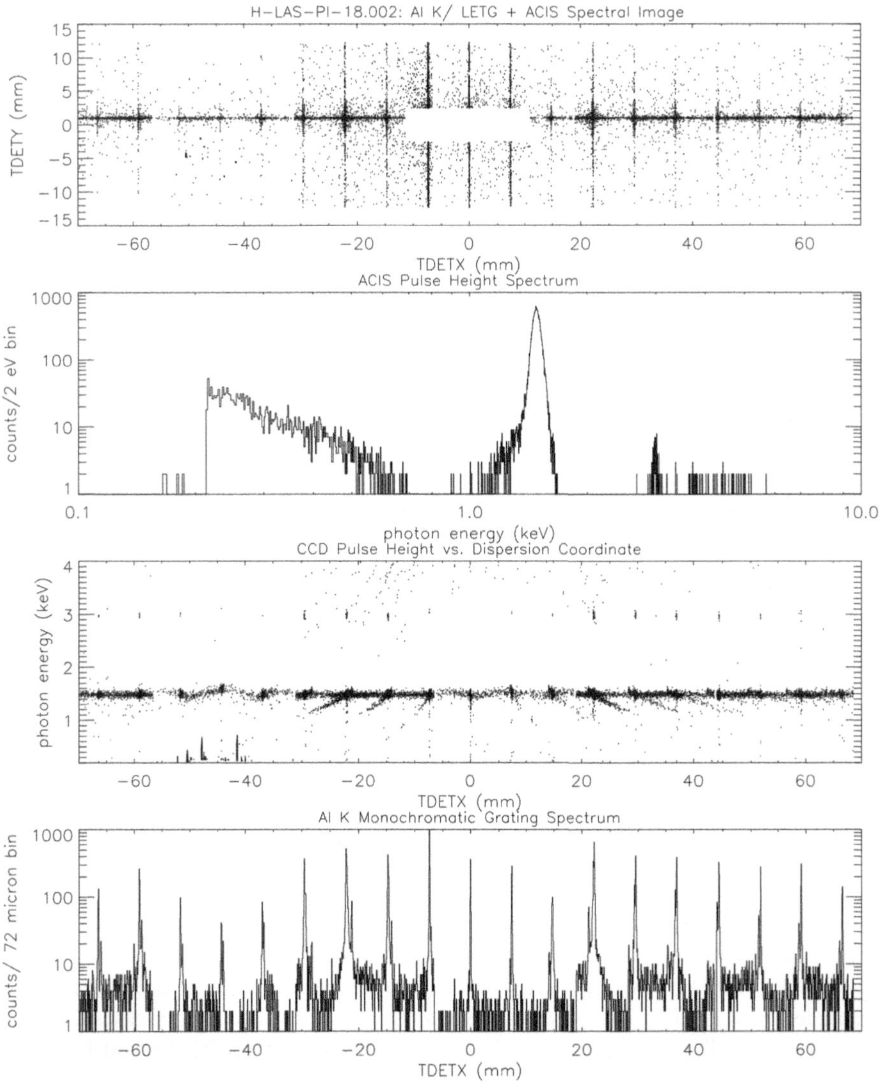

Fig. 2.3 Al Kα λ8.34 Å radiation diffracted by the *Chandra* Low Energy Transmission Grating in a ground calibration exposure. Data were recorded with a CCD camera. The top panel shows the spatial diffraction pattern. The second panel shows the CCD spectrum; the third panel shows photon energy versus dispersion coordinate (TDETX) for each photon. The bottom plot shows the diffraction pattern, filtered to exclude continuum photons, integrated over cross-dispersion coordinate TDETY. Data courtesy of Dan Dewey (MIT)

We pursued a scalar calculation for the grating diffraction pattern analogous to the one described above for the case of a transmission grating. The angle of incidence on the grating is α, the dispersion angle is β, the grating period is d, and the dispersion relation for diffraction order m is given by

$$\cos\beta_m - \cos\alpha = \frac{m\lambda}{d} \tag{2.15}$$

In the case of the gratings and the geometry on RGS, only orders $m \leq 0$ exist. The scalar perturbation theory now gives for the intensity of light dispersed at angle β out of diffraction order m:

$$\frac{1}{I_m}\frac{dI_m}{d\beta} = \frac{(\sin\alpha + \sin\beta)^4}{(\sin\alpha + \sin\beta_m)^2} \sin\beta_m\, k^3 W(p) \tag{2.16}$$

Here, I_m is the intensity in diffraction order m of light of wavelength $\lambda = 2\pi/k$, and $p = k(\cos\alpha - \cos\beta^{\pm}) \approx \pm k \sin\beta_m \Delta\beta$, and $\Delta\beta$ is the angle $\beta - \beta_m$. The total fraction of light $f = I_{\text{scattered}}/I_m$ scattered off order m is

$$f = k^2\sigma^2(\sin\alpha + \sin\beta_m)^2 \tag{2.17}$$

with σ^2 the variance of the surface roughness. The power spectrum normalization is $\int W(p)dp = \sigma^2$.

In applying this to the LSF of the RGS, we found that there appeared to be two scattering distributions, of different spatial coherence lengths l, which we termed 'large' and 'small angle' scattering. The former effectively produces a correction on the diffraction efficiencies, by scattering light far away from the diffraction peaks. The latter produces visible scattering 'wings' to the LSF. Both should of course be taken into account in modeling the response of the spectrometer to a given incident spectrum (including the large-angle scattered light, which reduces apparent contrast). Figure 2.4 illustrates the small-angle scattering for one grating. Figure 2.5 shows an 'interorder scan' for one of the gratings, showing the effect of 'large-angle scattering'. In the left panel of Fig. 2.6 we show the distribution of surface roughnesses that gives rise to the small-angle scattering, for gratings that make up RGA1 and RGA2 (in RGS1 and RGS2, respectively). The corresponding distribution of correlation lengths is shown in the right panel of Fig. 2.6. For comparison, the average grating period of the gratings is $d = 1.54895\ \mu\text{m}$.

These grating properties were incorporated into the flight instrument models, which were generated with ESA's `SciSIM` package. Several documents describing the scattering calculations and the algorithms used to model the response exist ([11, 12]; available from frits@astro.columbia.edu on request).

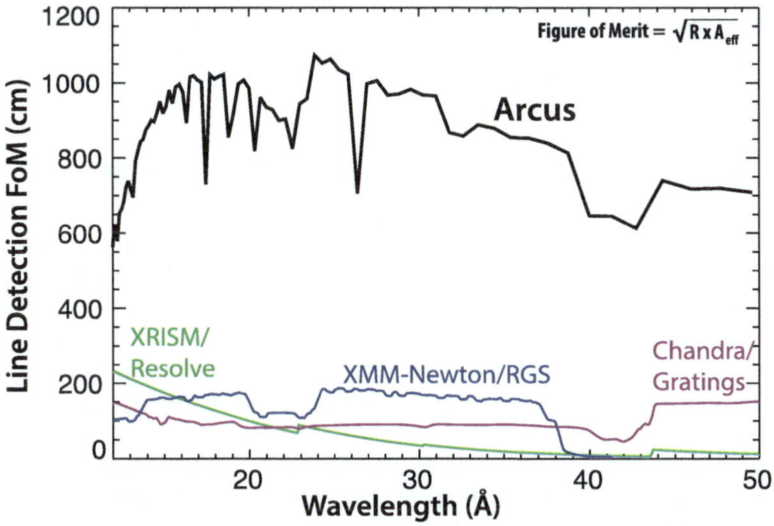

Fig. 2.9 The figure of merit (FoM) for detecting weak absorption features. The black, green, blue, and purple curves are Arcus, XRISM/Resolve, XMM-Newton/RGS, and Chandra/LEG, respectively. Figure reproduced with permission from [68], SPIE

manufacture. However, it requires precise optical alignment (cf. free-standing CAT grating bars). We refer readers to [44, 59] for technical details.

2.2.5 *X-ray Spectrometers: Diffractive or Not*

As mentioned earlier, diffractive grating spectrometers can be viewed as constant $\Delta\lambda$ devices so that their resolving power $R = \lambda/\Delta\lambda$ increases with the wavelength of the photon. Micro-calorimeters can be viewed as constant ΔE devices so that their resolving power ($R = E/\Delta E$) decreases with the wavelength of the photon. At 0.5 keV, a future micro-calorimeter with $\Delta E = 0.5$ eV (slightly better than HUBS aims to achieve for its central array) will have a resolving power of 10^3. At $\lambda = 24.797$ Å (i.e., $E = 0.5$ keV), Chandra/MEG with $\Delta\lambda = 0.023$ Å has already achieved the resolving power of $\sim 10^3$ (Table 2.1). Hence, diffractive grating spectrometers will still be the leading designs in terms of resolving power in the soft X-ray band.

The large resolving power of diffractive grating spectrometers comes with a price though. These devices are optimized for point sources. The energy resolution of the device is degraded for extended sources because it is hard to disentangle the dispersion effect from the different optical paths along which the photons of an extended source are dispersed. The effective area of diffractive grating spectrometers is much smaller than that of non-diffractive ones. Furthermore, micro-calorimeters with thousands of pixels (integral field units) are more efficient in obtaining a large number of high-resolution X-ray spectra in one single observation.

2.3 XMM-Newton Reflection Grating Spectrometer

In the following, we focus on the XMM-Newton Reflection Grating Spectrometer. XMM-Newton [29] is the second cornerstone project of the ESA's Horizon 2000 Science Programme. It has a set of three X-ray CCD cameras, comprising the European Photon Imaging Camera (EPIC). Two of them are MOS (Metal Oxide Semiconductor) [71]. The third one is a pn device [69]. All three CCD imaging spectrometers ($\sim$30 arcmin field-of-view and $\sim$6 arcsec angular resolution) are non-diffractive spectrometers. The resolving power of EPIC is $R = E/\Delta E \sim 20-50$.

Behind each MOS camera is a Reflection Grating Array (RGA, Fig. 2.10). About half of the incident soft X-ray photons are reflected onto the RGS Focal Camera (RFC). The latter consists of 9 back-illuminated MOS CCDs (similar to those of the EPIC/MOS) in a row along the dispersion direction.

Figure 2.11 shows the effective area of both RGS instruments for the first order. Dozens of narrow dips are caused by CCD gaps and hot pixels [19]. The two broad troughs $\sim$11−14 Å and $\sim$20−24 Å are due to failures of CCD #7 of RGS1 and CCD #4 of RGS2 at the beginning of the mission (and are therefore not present for observations earlier than 2000-09-02) [19]. Furthermore, over the years, a gradual decline in the effective area for the entire wavelength range can be noticed.

The second-order RGS spectra (5−19 Å or 0.65−2.5 keV) overlap with the first-order spectra (5−38 Å or 0.33−2.5 keV). The two spectra can be distinguished with the intrinsic CCD energy resolution of RFC. Although the second-order RGS spectra have a higher spectral resolution, they are less frequently used (e.g., Pinto et al. 2016 [62]). This is mainly due to the relatively small effective area of the second-order (Fig. 2.12).

2.4 RGS Data Reduction

Here, we illustrate how to access RGS data (Sect. 2.4.1) and how to reduce RGS (imaging, timing, and spectral) data for the benefit of new learners. For the latter, we include both a general recipe in Sect. 2.4.2 and guidance to handle special cases (Sect. 2.4.3).

It might take a few minutes to complete the downloading process, depending on the data file size, the network speed, and so forth. As mentioned earlier, one can also download the ODF data via the XSA website. Subsequently, we unpack the ODF data and set the SAS_ODF environment variable.

```
user$ tar -xvf $obsid_odf.tar
user$ tar -xf *.TAR
user$ SAS_ODF=${dir_main}/${obsid}/odf
user$ export SAS_ODF
```

Before leaving the $dir_main/odf directory, we have to create the ccf.cif file, which is an index file of Current Calibration Files (CCF). The SAS_CCF environment variable should refer to the ccf.cif file.

```
user$ cifbuild
user$ SAS_CCF=${dir_main}/${obsid}/odf/ccf.cif
user$ export SAS_CCF
```

Run the *odfingest* task to extend the ODF summary file, whose file name matches the pattern of *SUM.SAS, with data extracted from the instrument housekeeping data files and the calibration database. Once this task is finished, users have to update the SAS_ODF environment to the extended ODF summary file.

```
user$ odfingest
user$ SAS_ODF=$(ls ${dir_main}/${obsid}/odf/*SUM.SAS)
user$ export SAS_ODF
```

2.4.2.2 Running the RGS Data Reduction Pipeline *rgsproc*

Although the pipeline can be run directly using all default parameters, users are strongly encouraged:

- To set the source coordinates in degrees by setting the parameter withsrc to yes, as well as specifying srclabel, srcra, and srcdec parameters. This is crucial because the source coordinates have a profound influence on the accuracy of the wavelength scale as recorded in the response file generated by *rgsproc*.
- To activate the RGS effective area correction by setting the parameter witheffectiveareacorrection to yes. This correction is based on the careful analysis after

Kaastra et al. [32]. For SAS v19.0 (released at the end of 2020) and later versions, witheffectiveareacorrection is set to yes by default.

- To activate the RGS background model spectrum by setting the parameter withbackgroundmodel to yes. This is particularly useful if the source extends a large fraction of the RGS field or view or if the local background has poor statistics. For bright point-like sources, the local background is sufficient.

```
user$ cd ${dir_main}/rgs
user$ # HR 1099 sky coordinates from Simbad/NED
user$ ra_deg=54.1970
user$ dec_deg=0.5878
user$ # The source label (srclabel) is defined by the user but should avoid
PROPOSAL and ONAXIS.
user$ rgsproc withsrc=yes srclabel=USER srcra=${ra_deg}
   srcdec=${dec_deg} witheffectiveareacorrection=yes
   withbackgroundmodel=yes
```

It might take a few minutes to generate all the products. These output files are named following the Pipeline Processing Subsystem (PPS) file name convention in the current working directory ($dir_main/rgs in this example).

2.4.2.3 Extracting RGS Images

There are two types of RGS images can be extracted from the event file produced by *rgsproc*. The first type is in the M_LAMBDA versus XDSP_CORR parameter space while the second type is in the M_LAMBDA versus PI parameter space. M_LAMBDA is $m\lambda$, where $m = 1, 2$ is the spectral order and λ is the wavelength of the dispersed photon. XDSP_CORR reflects the extension of the source in the cross dispersion direction. PI is related to the energy of the dispersed photon. Both types of images are extracted in a similar way. The event file has the following name convention: P${obsid}R${rgsid}${expid}EVENLI0000.FIT, where ${rgsid} is either 1 (for RGS1) or 2 (for RGS2), ${expid} is the exposure ID of the instrument. The exposure ID starts with either S (for scheduled observations) or U (for unscheduled observations), followed by a three-digit number. For obsid=0791980501, we have expid=S004 for RGS1 and expid=S005 for RGS2. Figure 2.15 shows the RGS images for obsid=0791980501.

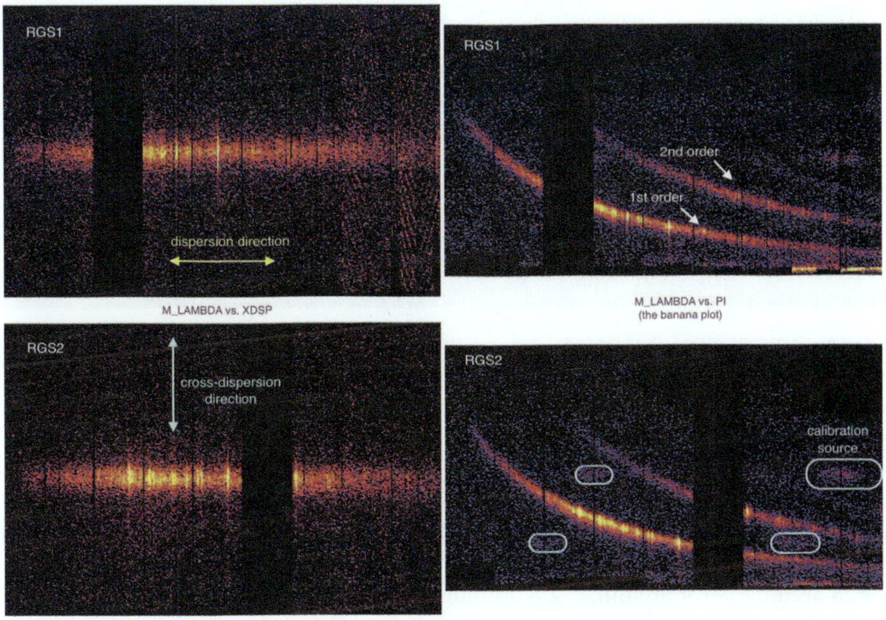

Fig. 2.15 RGS images for Obs.ID=0791980501 (HR 1099). The left two panels are displayed in the M_LAMBDA versus XDSP_CORR parameter space while the right two panels are displayed in the M_LAMBDA versus PI parameter space. The upper and lower two panels are RGS1 and RGS2, respectively. Failure of CCD #7 of RGS1 and CCD #4 of RGS2 are clearly visible

```
user$ rgsid=1
user$ expid=S004
user$ lis_evt=P${obsid}R${rgsid}${expid}EVENLI0000.FIT
user$ # The output image file name (imageset) is defined by the user.
user$ evselect table=${lis_evt}:EVENTS
    imageset=img_rgs1_xdsp_dsp.fits xcolumn='M_LAMBDA'
    ycolumn='XDSP_CORR'
user$ # Next, we extract the M_LAMBDA versus PI plot
user$ # and specify the size of the images (600 pixels × 600 pixels),
user$ # which can be defined by the user.
user$ evselect table=${lis_evt}:EVENTS
    imageset=img_rgs1_pi_dsp.fits xcolumn='M_LAMBDA'
    ycolumn='PI' imagebinning=imageSize ximagesize=900
    yimagesize=600
```

In Sect. 2.4.2.2, we used the default extraction region for the source along the cross-dispersion direction (i.e., xpsfincl = 90 for *rgsproc* by default), which is 90%

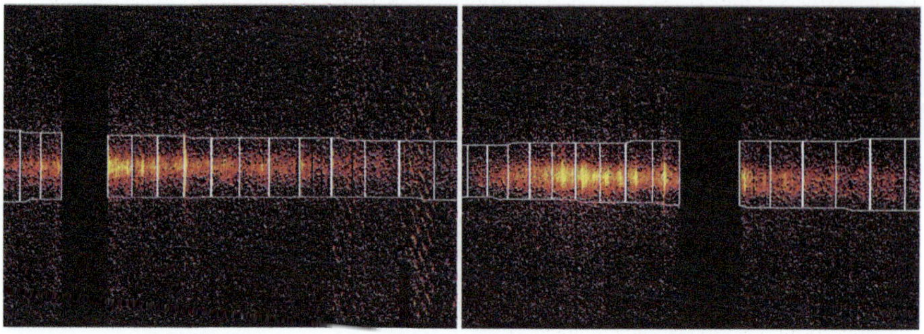

Fig. 2.16 RGS source extraction region (90% PSF along the cross-dispersion direction) for Obs.ID=0791980501 (HR 1099). The left and right panels are RGS1 and RGS2, respectively

of the telescope point spread function (PSF). This can be visualized by first generating the extract region from the source list file via the *cxctods9* task and then loading the region file along with the M_LAMBDA versus XDSP_CORR image on ds9 (Fig. 2.16). 90%, 95%, 98%, and 99% of the telescope PSF corresponds to ~0.8 arcmin, ~1.4 arcmin, ~2.8 arcmin, and ~3.4 arcmin, respectively, along the cross dispersion direction.

```
user$ rgsid=1
user$ expid=S004
user$ lis_src=P${obsid}R${rgsid}${expid}SRCLI_0000.FIT
user$ # The source ID (srcid) can be found in the source list file ${lis_src}
user$ srcid=3
user$ # The output region file name is defined by the user.
user$ cxctods9 table=${lis_src}:RGS${rgsid}_SRC${srcid}_SPATIAL
    regtype=linear -V 0 > rgs${rgsid}_src.reg
```

2.4.2.4 Extracting RGS Lightcurves

To get a barycentric corrected and background subtracted RGS lightcurve, one should use the *rgslccorr* task. In the following example, we create an RGS1 and RGS2 combined, 1st order barycentric-corrected and background-subtracted light curve with a time bin size of 100 s, for the 3rd source in the source list (Fig. 2.17):

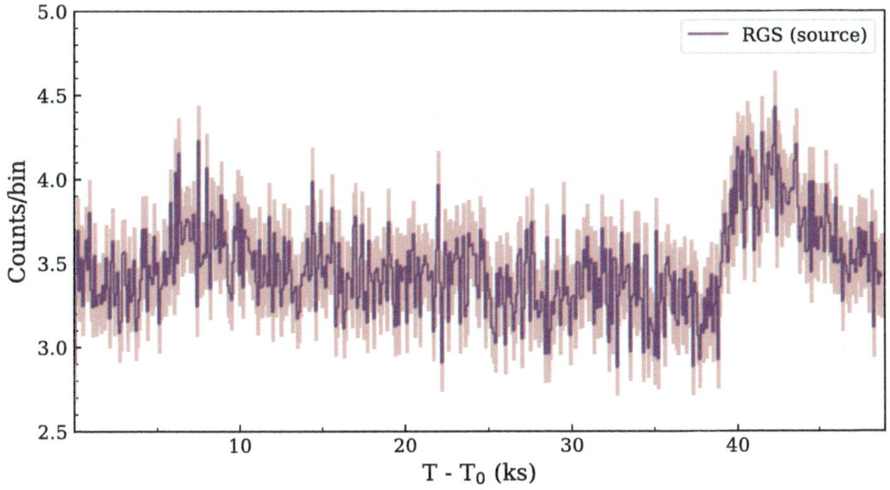

Fig. 2.17 RGS source lightcurve for Obs.ID=0791980501 (HR 1099)

```
user$ # Time bin size (100 s) is defined by the user
user$ # The output file name (ltc_rgs_src.fits) is also defined by the user
user$ rgslccorr evlist="P${obsid}R1S004EVENLI0000.FIT
   P${obsid}R2S005EVENLI0000.FIT"
   srclist="P${obsid}R1S004SRCLI_0000.FIT
   P${obsid}R2S005SRCLI_0000.FIT"
   timebinsize=100 orders='1' sourceid=3
   outputsrcfilename=ltc_rgs_src.fits
```

If users would like to study the RGS imaging and spectral data in the last ~10 ks of this observation, they can generate a Good Time Interval (GTI) file using either *gtibuild* or *tabgtigen*. The former requires a text file in the ASCII format with three columns and one to many rows. The first two columns are the start and end time (in seconds) for an interval, while the last column is either "+" (to keep in the analysis) or "–" (to discard).

```
user$ # The input file name (gti.txt) can be defined by the user
user$ cat gti.txt
699021517.798482 699030397.999542 +
user$ gtibuild file=gti.txt table=gti.fits
```

Alternatively, the *tabgtigen* task can be used as follows:

```
user$ # The input file name (gti.fits) can be defined by the user
user$ tabgtigen table=ltc_rgs_src.fits gtiset=gti.fits timecolumn=TIME
    expression='(TIME in [6.990215E8:6.990304E8])'
```

We caution that the XMM-Newton is susceptible to background flares. In Sect. 2.4.2.2, we generate imaging, timing, and spectra products using all the exposure of the observation. While RGS suffers less from the background flares than EPIC, it is still highly recommended to examine the background lightcurve. The RGS background lightcurve is extracted from CCD #9 for two reasons: (1) photons arriving at CCD #9 (with $\lambda \sim 5-7$ Å) are more sensitive to background flares; (2) CCD #9 records the least source photons due to its location close to the optical axis. Furthermore, we extract the background lightcurve in a region away from the source (along the cross-dispersion direction) as follows:

```
user$ # Time bin size (100 s) is defined by the user
user$ # The output file name (ltc_rgs1_bkg.fits) is also defined by the user
user$ evselect table=${lis_evt} timebinsize=100 rateset=ltc_rgs1_bkg.fits
    makeratecolumn=yes maketimecolumn=yes
    expression="(CCDNR==9)&&(REGION(${lis_src}:
    RGS${rgsid}_BACKGROUND,M_LAMBDA,XDSP_CORR))"
```

As shown in Fig. 2.18, this exemplary observation does not suffer from background flares. Nonetheless, the total exposure can be significantly reduced for e.g., Obs.ID= 0116710901 (HR 1099). In that case, users can create a GTI with a threshold count rate (all those below this threshold are kept as GTI) using the *tabgtigen* task. Caution that the threshold can vary significantly for different observations of the same target, let alone different targets.

```
user$ # The threshold count rate can be defined by the user.
user$ tabgtigen table=ltc_rgs_src.fits gtiset=gti.fits timecolumn=TIME
    expression="(RATE <= 4.049)"
```

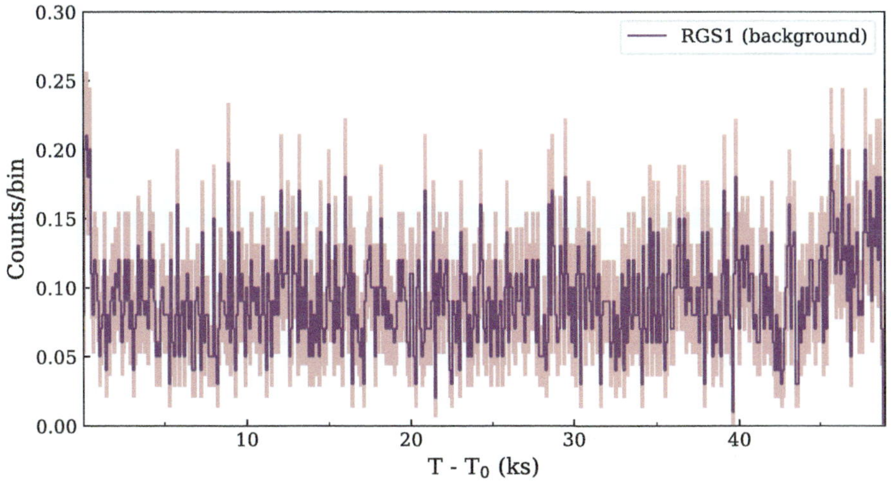

Fig. 2.18 RGS1 background lightcurve for Obs.ID=0791980501 (HR 1099)

Alternatively, users can determine the GTI using the so-called sigma clipping method via the *deflare* task provided by CIAO (Chandra Interactive Analysis of Observations).[12] Figure 2.19 shows an example of such an application (see also [42]). In order to apply the outcome of *deflare* to the RGS data reduction with SAS, one needs to delete the first extension of the *deflare* output file. This can be realized using the *fdelhdu*[13] or *ftdelhdu*[14] task available from the Heasoft package.

```
user$ # Assuming gti.fits is the output file of deflare
user$ fdelhdu gti.fits+1 N Y
user$ # Use ftdelhdu as an alternative
user$ ftdelhdu infile="gti.fits[FILTER]" outfile=gti.fits clobber=1
```

To apply the GTI filtering, we need to run *rgsproc* again. The first two steps (events and angles) of *rgsproc* might be skipped by specifying the entry and final stages.

```
user$ rgsproc auxgtitables=gti.fits entrystage=3:filter finalstage=5:fluxing
    witheffectiveareacorrection=yes withbackgroundmodel=yes
```

[12] https://cxc.cfa.harvard.edu/ciao/ahelp/deflare.html.

[13] https://heasarc.gsfc.nasa.gov/lheasoft/ftools/fhelp/fdelhdu.html.

[14] https://heasarc.gsfc.nasa.gov/lheasoft/ftools/headas/ftdelhdu.html.

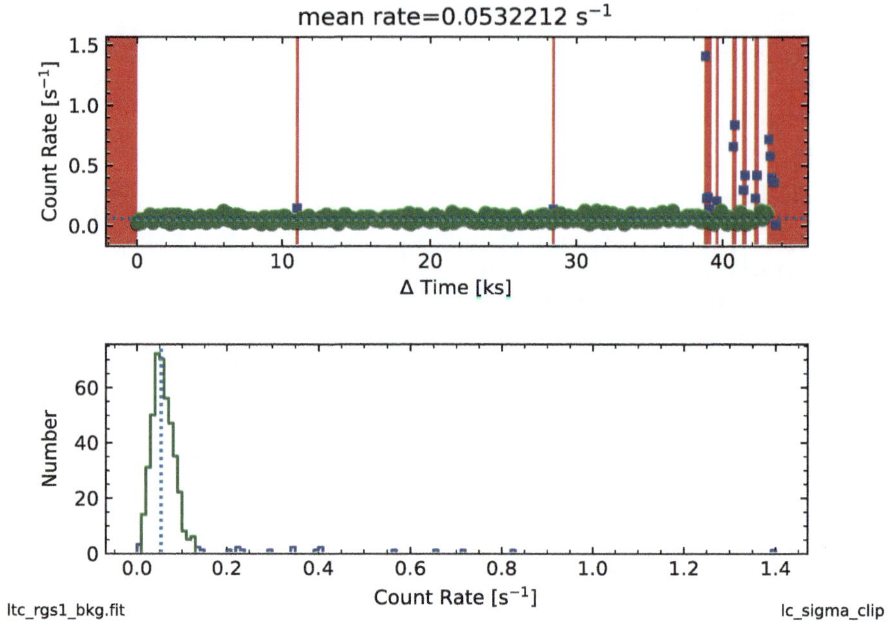

Fig. 2.19 RGS1 background lightcurve for Obs.ID=0134540101 (HR 1099). Time intervals impacted by flares are identified using the sigma clipping method via the *deflare* task provided by CIAO. The threshold used in this example is 3σ. The shaded area in red also includes time intervals with zero count rates. Good time intervals are those highlighted in green

2.4.2.5 Extracting RGS Spectra

The pipeline products of *rgsproc* contain RGS spectra files for both instruments (rgsid = 1 or 2) and both spectral orders. Note that for RGS, the response matrix file (rmf) and ancillary response file (arf) are combined into one response file.

- P${obsid}R${rgsid}${expid}BGSPEC100${srcid}.FIT: 1st-order local background spectrum file
- P${obsid}R${rgsid}${expid}BGSPEC200${srcid}.FIT: 2nd-order local background spectrum file
- P${obsid}R${rgsid}${expid}MBSPEC1000.FIT: 1st-order model background spectrum
- P${obsid}R${rgsid}${expid}MBSPEC2000.FIT: 2nd-order model background spectrum
- P${obsid}R${rgsid}${expid}RSPMAT100${srcid}.FIT: 1st-order response file
- P${obsid}R${rgsid}${expid}RSPMAT200${srcid}.FIT: 2nd-order response file
- P${obsid}R${rgsid}${expid}SRSPEC100${srcid}.FIT: 1st-order source spectral file
- P${obsid}R${rgsid}${expid}SRSPEC200${srcid}.FIT: 2nd-order source spectral file

To combine RGS1 and RGS2 spectra of the same spectral order, one can take advantage of the *rgscombine* task. This applies to either one observation or multiple observations. The example below combines the first-order RGS1 and RGS2 spectra (with the model background) for Obs.ID=0791980501:

```
user$ # First, create some lists of files to be combined (src.lis, rsp.lis, bkg.lis)
user$ cat src.lis
P0791980501R1S004SRSPEC1003.FIT P0791980501R2S005SRSPEC1003.
FIT
user$ cat rsp.lis
P0791980501R1S004RSPMAT1003.FIT P0791980501R2S005RSPMAT1003.
FIT
user$ cat bkg.lis
P0791980501R1S004MBSPEC1000.FIT P0791980501R2S005MBSPEC1000.
FIT
user$ # Output files (filepha, filermf, and filebkg) are defined by the user
user$ rgscombine pha="$(cat src.lis)" rmf="$(cat rsp.lis)"
    bkg="$(cat bkg.lis)" filepha=rgs_o1_src.fits
    filermf=rgs_o1.rsp ilebkg=rgs_o1_bkg.fits
```

Figure 2.20 compares the 1st- and 2nd-order combined and folded spectra of Obs.ID=0791980501 (HR 1099). Due to the smaller effective area (Fig. 2.12), the 2nd-order spectrum is lower than the 1st-order in this plot.

RGS fluxed spectra should also be available among the *rgsproc* pipeline products. They can also be generated with the *rgsfluxer* task:

```
user$ # Output file (rgs_o1_flux.fits) is defined by the user
user$ cat src.lis
P0791980501R1S004SRSPEC1003.FIT P0791980501R2S005SRSPEC1003.
FIT
user$ cat rsp.lis
P0791980501R1S004RSPMAT1003.FIT P0791980501R2S005RSPMAT1003.
FIT
user$ # User can include the background files via the keyword agrument "bkg"
user$ # We skip the background setting here for simplicity.
user$ rgsfluxer pha="$(cat src.lis)" rmf="$(cat rsp.lis)" file=rgs_o1_flux.fits
```

In Fig. 2.21, we compare the combined (RGS1 and RGS2 coadded for the same spectral order) and fluxed RGS spectra for Obs.ID=0791980501 (HR 1099). If the 1st- and 2nd-order fluxed spectra differ by more than 10%, this indicates that the observation targeting a bright source is piled up. Pileup occurs when two or more

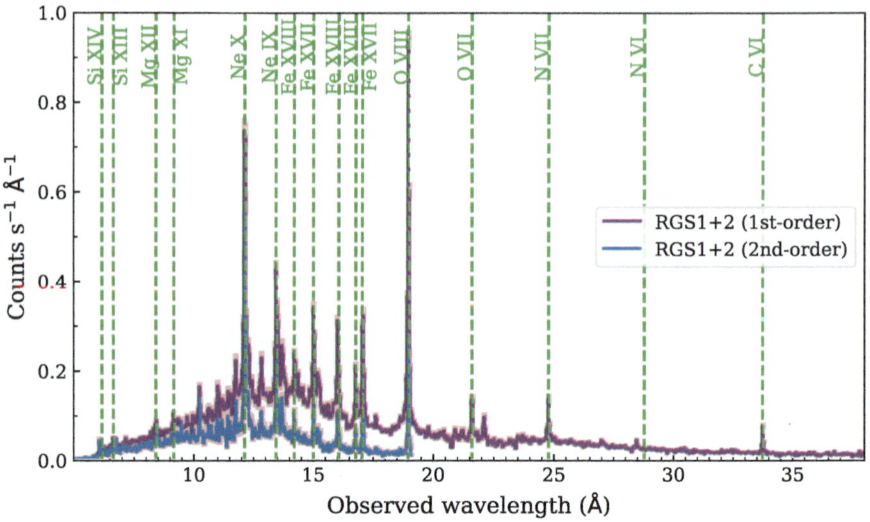

Fig. 2.20 RGS combined and folded spectrum for Obs.ID=0791980501 (HR 1099). The 1st- and 2nd-order spectra are shown in purple and blue, respectively. Uncertainties are shown in pink. Some key diagnostic lines are labeled in green

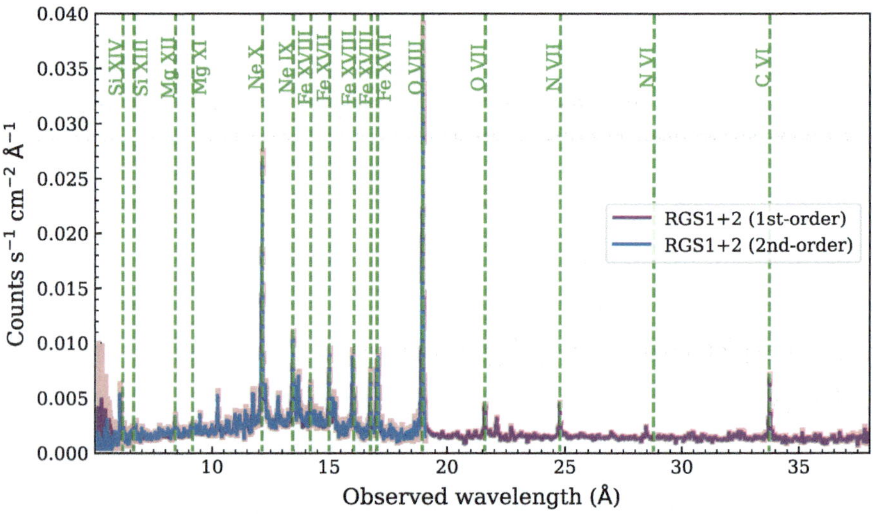

Fig. 2.21 Combined and fluxed (i.e., unfolded) RGS spectra for Obs.ID=0791980501 (HR 1099). The 1st- and 2nd-order spectra are shown in purple and blue, respectively. Uncertainties are shown in pink. Some key diagnostic lines are labeled in green. The 1st- and 2nd- order spectra should agree in flux unless the target is piled up

events arrive at the same (or neighboring) pixel during the same readout frame. Unfortunately, there is not much to do to alleviate the issue in existing observations [51]. To mitigate the pile-up effect, proposers can request the RGS small window configuration. In this mode, only a quarter of the RGS FOV along the cross-dispersion direction will be used. Accordingly, the readout time is reduced by a factor of 4.

2.4.3 Special Guide for RGS Data Reduction

In the following, we provide some guidance to handle special cases: multiple X-ray bright sources in the RGS field of view (Sect. 2.4.3.1) and line broadening for spatially extended sources (Sect. 2.4.3.2).

2.4.3.1 Multiple X-ray Bright Sources in the RGS Field of View

In some cases, there might be more than one X-ray bright source in the RGS field of view. If these sources are well-separated along the cross-dispersion direction, they might be dealt with. For instance, Obs.ID=0601781401, targeting Mrk 817, has an X-ray bright star (RX J1436.6+5843) nearby. This star falls in the field of view of RGS (Fig. 2.22). To exclude this star from the background region, users need to first identify it based on its coordinates in the EPIC source list generated by the *edetect_chain* task.[15] Subsequently, users need to run the *rgsproc* pipeline with the following parameters:

```
user$ # Assuming the EPIC source list is called emmlist.fits
user$ # and the source (RX J1436.6+5843 here)
user$ # to be excluded from the background region
user$ # has the index of 3 in the source list.
user$ rgsproc orders='1 2' withepicset=yes epicset=emllist.fits
   exclsrcsexpr='INDEX==3'
```

Caution that the RGS field of view along the dispersion direction extends beyond the field of view of EPIC. In Obs.ID=0158160201 (targeting GRB 031203), the standard RGS local background spectrum (as shown in XSA) will be contaminated by the X-ray bright star zeta Puppis (Fig. 2.23).

[15] https://xmm-tools.cosmos.esa.int/external/sas/current/doc/edetect_chain/edetect_chain.html.

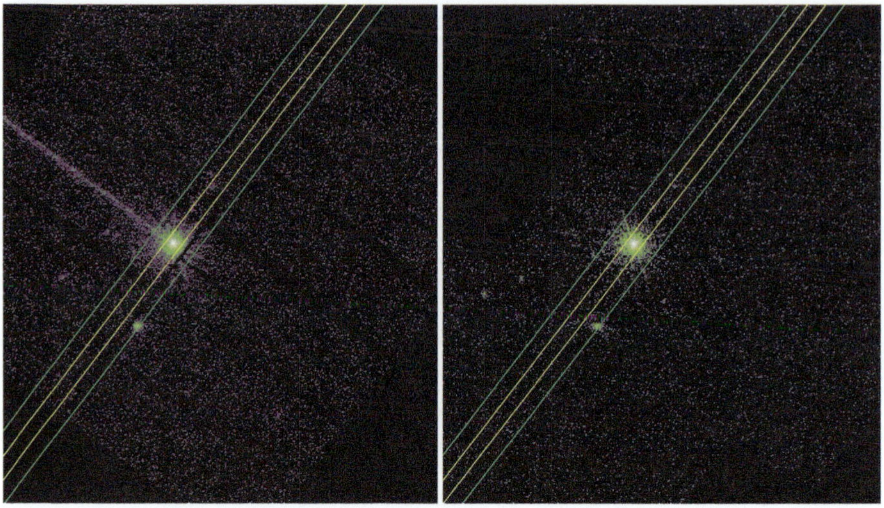

Fig. 2.22 EPIC images of Obs.ID=0601781401 (Mrk 817). EPIC pn and MOS1 images are on the left and right panels, respectively. The RGS source extraction region (~0.8 arcmin along the cross-dispersion direction) is shown as tilted boxes in yellow. The X-ray bright star RX J1436.6+5843 also falls in the field of view of RGS (tilted green boxes). The roll angle of RGS (~143° in Fig. 2.22) is given by the fits header keyword PA_PNT of the RGS source list file (P${obsid}R${rgsid}${expid}SRCLI_0000.FIT)

Fig. 2.23 EPIC images of Obs.ID=0158160201 (GRB 031203). The RGS field of view (tilted green box) extends beyond the field of view of EPIC/pn along the dispersion direction. The yellow circle indicates the position of zeta Puppis, which is an X-ray bright star. RGS spectra of zeta Puppis can still be obtained here by setting the target coordinates properly

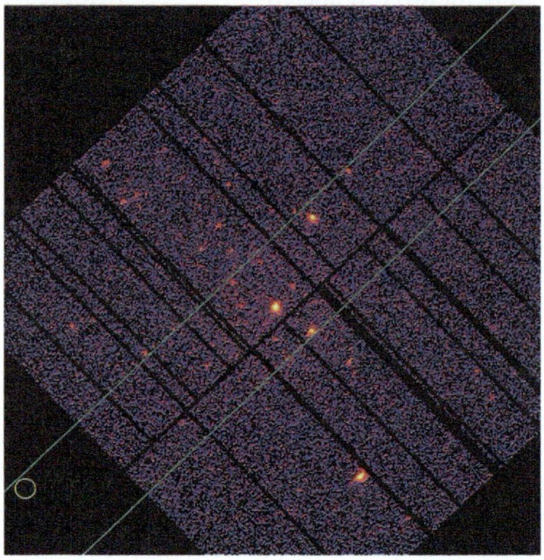

Fig. 2.24 EPIC/MOS1 images of Obs.ID=0037950101 (NGC 5044). The RGS source extraction, ~3.4 arcmin along the cross dispersion direction, is shown as the tilted yellow box

2.4.3.2 Line Broadening for Extended Sources

Emission lines of HR 1099 in Fig. 2.20 or 2.21 are narrow. The line broadening is limited by the spectral resolution of the instrument, which is ~0.06–0.07 Å for RGS in the first order.

For extended sources like NGC 5044 (see also [40]), spatial broadening dominates (Eq. 2.19). Whether the target is an extended source can be verified from the RGS or EPIC images (Figs. 2.15 and 2.24), where the source region (if symmetric) extends well beyond the 90% PSF (~0.8 arcmin). In this case, emission lines appear to be broader in Fig. 2.25. To be more specific, emission lines are broadened by [70]

$$\frac{\Delta\lambda}{\text{Å}} = \frac{0.138}{m} \frac{\Delta\theta}{\text{arcmin}}, \tag{2.19}$$

where m is the spectral order, $\Delta\theta$ the spatial extent (in arcmin) of the source [70].

To account for such kind of spatial broadening, users can adapt the RGS response file according to the spatial extent of a moderately extended source (~1 arcmin) via the *ftrgsrmfsmooth*.[16] This tool is developed by Andy Rasmussen of the Columbia University XMM-Newton RGS instrument team. Alternatively, users can take advantage of the *rgsvprof*[17] tool. This tool is part of the SPEX code [30].

[16] https://heasarc.gsfc.nasa.gov/lheasoft/ftools/fhelp/ftrgsrmfsmooth.html.

[17] https://spex-xray.github.io/spex-help/tools/rgsvprof.html.

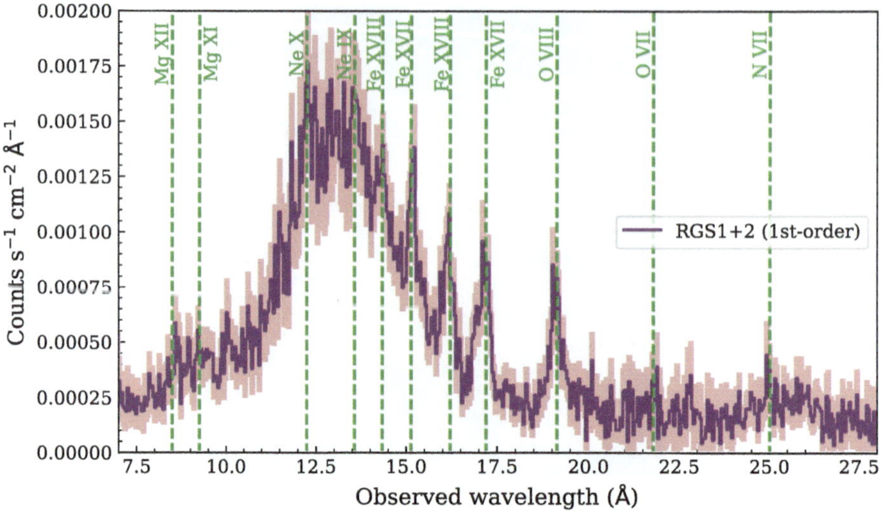

Fig. 2.25 First-order RGS spectrum (RGS1 and RGS2 combined) of Obs.ID=0037950101 (NGC 5044). The source extraction region is 99% along the cross-dispersion direction. Spectral lines are broadened due to spatial broadening

2.5 Summary

Since the launch of XMM-Newton in 1999, RGS has delivered thousands of high-quality high-resolution (soft) X-ray spectra [41]. Considering both the effective area and spectral resolution (Figs. 2.8 and 2.9), RGS will keep playing an important role even in the era of XRISM.

Acknowledgements We would like to thank Randall K. Smith for providing figures related to Arcus, Rosario Gonzalez-Riestra for valuable inputs, Guan-Fu Liu and Chunyang Jiang for feedback after careful reading. This work is based on observations obtained with XMM-Newton, an ESA science mission with instruments and contributions directly funded by ESA Member States and the USA (NASA). SRON is supported financially by NWO, the Netherlands Organization for Scientific Research.

References

1. L.W. Acton, J.L. Culhane, A.H. Gabriel et al., Solar Phys. **65**, 53 (1980). https://doi.org/10.1007/BF00151384
2. C. Argiroffi, F. Reale, J.J. Drake et al., Nat. Astron. **3**, 742 (2019). https://doi.org/10.1038/s41550-019-0781-4
3. D. Barret, T. Lam Trong, J.-W. den Herder et al., Proc. SPIE **10699**, 106991G (2018). https://doi.org/10.1117/12.2312409
4. S. Bowyer, R.F. Malina, Adv. Space Res. **11**, 205 (1991). https://doi.org/10.1016/0273-1177(91)90077-W

Chapter 3
Chandra X-ray Observatory: The High Energy Transmission Grating Spectrometer

Mark T. Reynolds

3.1 Introduction

The *Chandra* X-ray Observatory is the 3rd mission in NASAs great observatories program, dedicated to the study of the X-ray sky [99, 105, 112].[1] *Chandra* was launched on 23rd July 1999 aboard the space shuttle Columbia (STS 93) and placed into Earth orbit from where it was subsequently maneuvered into its final science orbit. The *Chandra* orbit is highly elliptical and is coincidentally approaching its mission maximum of 0.9 in 2024. In principle, this enables long observations as $\sim$70% of the 63.5 h orbit is spent outside the Earth radiation belt system.

Chandra is a high spatial/spectral resolution X-ray observatory, sensitive to X-ray photons in the 0.3–10.0 keV energy range. The X-ray mirrors (HRMA) provide spatial resolution better than 0.5 arc-seconds (on-axis) and insertable spectroscopic gratings enable high resolution spectroscopy with r $\gtrsim$ 500 across the bandpass. Focused X-ray photons are registered by two independent detectors, which may be moved to the telescope focal point. In Fig. 3.1, we display a schematic drawing of *Chandra* illustrating the spacecraft geometry and primary components.

[1] AXAF (*Chandra*) was the number one priority in astronomy and astrophysics in the 1980s National Academies decadal survey.

M. T. Reynolds (✉)
Department of Astronomy, University of Michigan, 1085 S. University Ave., Ann Arbor, MI 48103, USA
e-mail: markrey@umich.edu

Department of Astronomy, The Ohio State University, 140 W. 18th Ave., Columbus, OH 43210, USA

C. Bambi and J. Jiang (eds.), *High-Resolution X-ray Spectroscopy*, Springer Series in Astrophysics and Cosmology, https://doi.org/10.1007/978-981-99-4409-5_3

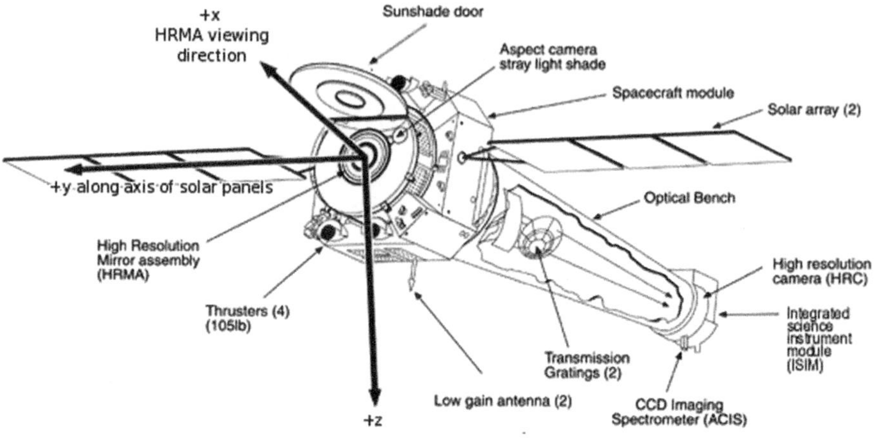

Fig. 3.1 Schematic of the *Chandra* observatory defining the spacecraft coordinate system. The optical axis is the x-axis, the y-axis points along the solar panels, and the z-axis completes the right-handed system and points toward the cold side of the spacecraft. Adapted from [99]

3.1.1 The High Resolution Mirror Assembly (HRMA)

The Chandra mirror system—the HRMA[2]—utilizes a Wolter-I geometry as previously employed by *Einstein* [38] and *Rosat* [6, 104]. The HRMA consists of 4 nested Wolter-I grazing incidence telescopes with the inner shell diameter of ~0.65 m and the outer shell diameter of ~1.2 m. The mirror system provides a focus at a distance of 10 m. The mirrors are coated with an Iridium layer over ~300 Å in depth. The total mirror mass is almost 1.5 metric tonnes [108].

The HRMA provides an unobscured aperture of 1145 cm^2 that is reduced by ~10% by the telescope support structure. Further energy dependent reductions in effective area are a result of the dependence of mirror reflectivity on photon energy and the X-ray detectors. The point spread function (PSF) produced by the HRMA is position and energy dependent. The enclosed energy fraction (eef) within a 1 arcsec on-axis region varies from ~0.75 at 1.5 keV to ~0.64 at 6.3 keV. The PSF has a strong position dependence, a fundamental property of the HRMA Wolter-I mirror geometry. This varies from $\lesssim$0.5 arcsec full width half maximum (FWHM) for an on-axis observation to $\gtrsim$4 arcsec for an observation 10 arcmin off-axis (for an assumed energy of 1.5 keV, 50% eef or $\gtrsim$6 arcsec at 6.4 keV). Ongoing calibration observations have demonstrated stability of the PSF provided by the HRMA with respect to that measured after *Chandra* launch [15, 76].

These effects are typically not important for a grating spectroscopy observation at the ACIS nominal aim point. However, observations of sources at an offset zeroth order position may require significant additional modeling to verify the nature of the observed astrophysical signal processed through the *Chandra* optics.

[2] https://cxc.harvard.edu/cal/Hrma/UsersGuide.html.

There is evidence for low level layer of contamination of the mirror assembly by a hydrocarbon molecule in the HRMA mirrors in orbit (~22 Å). This is accounted for in the standard *Chandra* calibration products [76].

3.2 The Advanced CCD Imaging Spectrometer—ACIS

The *ACIS* detectors comprise ten 1 k × 1 k pixel CCDs (0.492"/pixel) organized in two arrays (i) the imaging array—*ACIS-I*—and the spectroscopy array—*ACIS-S*. *ACIS-I* is organized in a 2 × 2 geometry and provides a 17′ × 17′ field of view (FoV) [1]. The detectors in this array are all front illuminated (FI) and are slightly tilted such that the array is optimized for imaging photons from the HRMA. The *ACIS-I* aimpoint is centered on the I3 detector. The *ACIS-S* detectors are organized in a linear 1 × 6 geometry and are optimized to sample the *HETG* Rowland circle for high resolution spectroscopy. Two of the detectors are back illuminated (BI) while the other 4 are FI. The *ACIS-S* aimpoint is centered on the BI S3 detector. The detector chip-gaps are well defined on each array and their position and its potential impact on a scientific investigation should be noted.

Imaging spectroscopy is possible on both detector arrays with the *ACIS-I* array excelling in wider field observations. As the S3 detector is BI, its soft response excels and this detector is often utilized for imaging spectroscopy as a result providing an 8′ × 8′ field of view. The sharp PSF provided by the HRMA in combination with the spacecraft housing and elliptical orbit lead to superior point source sensitivity of $\sim 4 \times 10^{-15}$ erg s^{-1} cm^{-2} in 10 ks. The *ACIS* background is low in imaging mode and negligible in high resolution spectroscopy observations. Astrophysical backgrounds are present [40].

The *ACIS* detectors are highly flexible devices and offer many operating modes. In standard full-frame mode, the detectors are read-out after a nominal frame time of 3.2 s. Shorter frame times are possible with a choice of sub-array mode. For the *ACIS-S3* chip, sub windows are typically defined around the aimpoint and may go as low as 128 rows for an effective exposure time of 0.4 s. Frame times lower than this are achieved in 2 manners (i) during frame read-out (~40 μs duration) data is still recorded and will result in scientifically useful information in the so-called "read-out" streak. These are typically present only for the brightest sources where pile-up is an issue, and (ii) cc-mode—here the *ACIS* detector are collapsed to a single dimension, and the resulting loss of spatial information comes at the benefit of 3 ms time resolution. This is an important detector mode for observations of bright astrophysical sources [97].

In standard operation, observations with *Chandra* occur in a dithered mode, i.e., the spacecraft will follow a Lissajous pattern during an observation. This is done to provide some coverage in the detector gaps and to smooth out pixel to pixel variations in the detector response. These patterns are 32 × 32 pixels (Y, Z) in full frame observations and can change to 32 × 16 in sub-array observations. The effect of dither is removed during ground processing. Of note, these motions can inject

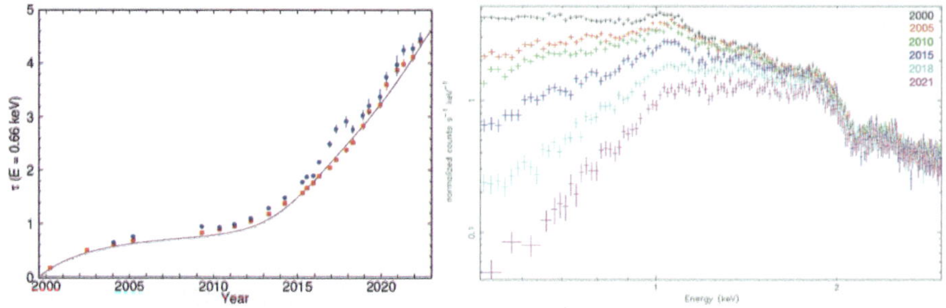

Fig. 3.2 **Left:** Optical depth of the *ACIS* contaminant measured at 0.66 keV from observations of the galaxy cluster Abell 1795. Red squares denote observations with the target on *ACIS-S* and blue disks denote observations with the target on *ACIS-I*. The solid curve shows the optical depth of the contaminant as modeled in the most recent contamination model (circa 2023). **Right:** *ACIS-S* spectrum of the galaxy cluster Abell 1795 at six different epochs ranging from 2000 to 2021. The decreasing count rates in the spectrum of a stable calibration source demonstrate the loss of effective area due to the build-up of the contaminant on the *ACIS* OBF. Adapted from [1]

timing signatures into astrophysical data and signals near the *ACIS* dither periods of 1.4 ks and 2 ks should be treated with caution.

The harsh space radiation environment has led to changes in the conditions of the spacecraft insulting layers and exterior facing surfaces. This has resulted in rising temperatures at the telescope focal plane. In an effort to maintain temperature stability, observations with *Chandra* now face enhanced constraints w.r.t. spacecraft pitch/roll angles and hence continuous observation length. In addition, observations are recommended to require only 4 CCDs detectors to be turned on in the absence of strong scientific justification at the current time, due to ongoing evolution of the thermal constraints/profile of the observatory.

The soft X-ray spectral response of the ACIS detectors has been discovered to be slowly decreasing as a function of time [75]. This is thought to be due to the build-up of out-gassed molecular compounds on the optical blocking filter (OBF, see Fig. 3.2). Calibration studies indicate the contaminant to primarily consist of Carbon. Updated response files and effective area curves are released annually to account for this effect.[3]

3.3 The High Energy Transmission Grating—HETG

The *HETG*[4] is a transmission grating spectrometer which provides for spectral resolution $E/\Delta E$ up to 1000 in the 0.4–10.0 keV bandpass [16]. The spectrometer consists

---

[3] https://cxc.cfa.harvard.edu/ciao/why/acisqecontamN0015.html.

[4] https://cxc.harvard.edu/cal/Hetg/.

Table 3.1 The high level characteristics of the *HETGS*

	Energy range keV [Å]	Effective area cm^2 @1.5 keV [cm^2 @6.5 keV]	Resolving power E/ΔE [$\lambda/\Delta\lambda$]
HEG	0.8–10.0 [15.0–1.2]	200 [28]*	1070–65 (1000 @1 keV)
MEG	0.4–5.0 [31.0–2.5]		970–80 (660 @0.826 keV)

*The effective area is for the combination of MEG+HEG, see Fig. 3.4

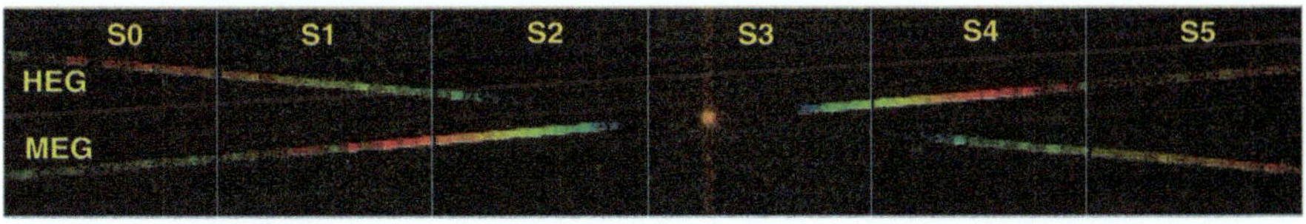

Fig. 3.3 *HETG* observation of Capella, read-out on the *ACIS-S* CCD array. The dispersed X-rays from the HEG and MEG form a crossed pattern that can easily be distinguished spatially. Energy bands are color coded. The highest energies (shortest wavelengths) are nearest the zero order point image. Since the HEG has twice the dispersion of the MEG, equivalent energies are twice as far from zero order. The plot shows raw data, not yet corrected for aspect, dither, or removal of the *ACIS* readout streak from the zero order image. Adapted from [99]

of 2 gratings arranged in a concentric circular configuration, these are (i) Medium Energy Grating (MEG), and (ii) the High Energy Grating (HEG), see Table 3.1.

The gratings are mounted on a single structure which can be inserted into the light path. The gratings are arranged such that the MEG disperses X-rays from the outer 2 HRMA mirror shells while the HEG disperses X-rays from the inner 2 HRMA shells. The grating ruling angles are offset by $\sim 10°$ such that the X-rays are dispersed in differing directions onto the *ACIS-S* array enabling simultaneous measurement of the spectra from both gratings (see Fig. 3.3). The energy resolution of the *ACIS-S* CCD detector enables the identification and separation of the signal from the various spectral orders.[5] The grating Rowland geometry works to maintain the telescope focus in the dispersion direction ensuring high quality data [9, 16].

The *HETG* gratings are made of polymide backed gold bars. The dimensions of the grating elements are chosen to maximize the grating scientific signal. The element width is chosen to maximize the signal in the first order spectrum in contrast to the zeroth order image and by choosing the width to be one-half of the grating period, the odd orders are enhanced at the expense of even orders. The MEG elements are of a width that maximizes this effect. The elements in the HEG are larger than those in the MEG and this results in larger 2nd order efficiency and reduced 3rd order efficiency. The vertical extent of the grating elements determines the energy efficiency peak, which lies in the region above 1.2 keV for the *HETGS*. The MEG

[5] The *HETG* can also be dispersed onto the HRC in principle. This option is becoming relevant with the ongoing degradation of the *ACIS* low energy effective area due to the optical blocking filter contamination build-up (Fig. 3.2). There are significant issues with this operating mode due to the lack of energy resolution in the *HRC* [46].

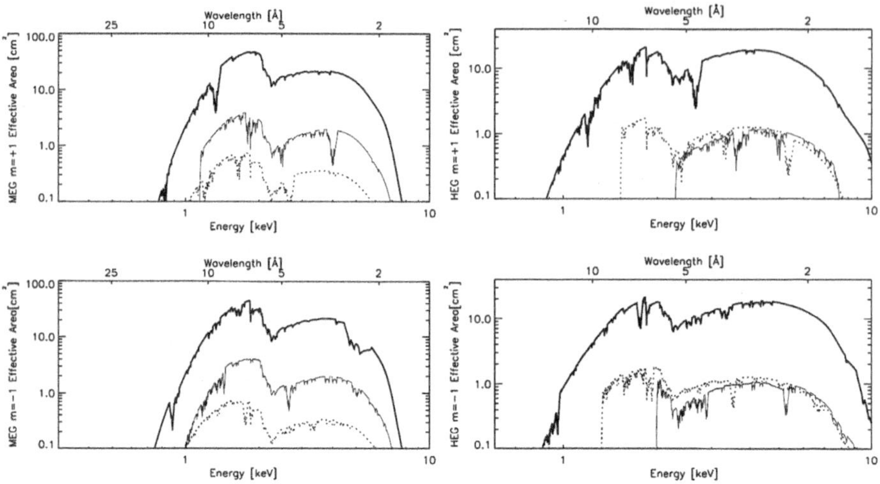

Fig. 3.4 The HETGS effective area curves (circa 2023). The thick solid lines are first order; the thin solid line is third order; and the dotted line is second order. **Left:** The MEG effective area, integrated over the PSF is shown with energy and wavelength scales. The m = +1, +2, +3 orders (falling on ACIS chips S5, S4, S3; left to right) are displayed in the top panel and the m = −1, −2, −3 orders (falling on ACIS chips S0, S1, S2; left to right) are in the bottom panel. **Right:** The HETGS MEG effective area, integrated over the PSF is shown with energy and wavelength scales. The m = +1, +2, +3 orders (falling on ACIS chips S5, S4, S3; left to right) are displayed in the top panel and the m = −1, −2, −3 orders (falling on ACIS chips S0, S1, S2; left to right) are in the bottom panel. Adapted from [48]

provides spectra sensitive to X-ray photons with energies 0.4 keV $\lesssim$ E $\lesssim$ 5.0 keV, while the HEG provides high resolution spectra between energies 0.8 keV $\lesssim$ E $\lesssim$ 10.0 keV. The background illuminated detectors S1 and S3 provide the improved lower energy sensitivity of these detectors for MEG observations [99, 112].

The *HETGS* line response function is successfully modeled as a Gaussian with FWHM equal to the instrument resolution ($E/\Delta E$ or $\lambda/\Delta\lambda$). The resolving power of the *HETGS* ($E/\Delta E$) is approximately a linear function of wavelength. High resolution spectra should be considered in wavelength units (as opposed to energy) as this is natural unit in which to analyze the spectra provided by these detectors.[6]

The *ACIS* detectors have 2 primary modes for use when detecting photons dispersed by the *HETGS*: (i) timed exposure mode and (ii) continuous clocking mode. Timed exposure mode (TE) is utilized in the majority of observations with the *ACIS* detectors, including high resolution observations with the *HETGS*. Here the detectors or sub-sections thereof accumulate events for a set exposure time before the active area of the array is read-out. In continuous clocking mode (CC), spatial resolution is sacrificed and the array is collapsed in a single direction enabling the detector to be read out with ~3 ms time resolution. Hence, spectroscopic observations of bright targets are possible.

[6] $E \times \lambda = hc = 12.39852$ keV Å.

Initial calibration of the *HETGS* occurred during a series of ground testing campaigns [111]. Post launch, observations of astrophysical sources in the HEG and MEG overlap region demonstrate agreement at order 1% level and the systematic wavelength error has been measured to be $\sim$100 km s^{-1}. The *HETG* calibration is continuously refined via in orbit calibration observations [74, 76, 77]. Current *HETGS* calibrations updates and ongoing areas of concern can be found at the *HETGS* calibration webpage.[7] These include updates to the Si-K edge and the confirmation that the *HETGS* effective area is not impacted by the potential for a warm *ACIS* focal plane [78].

The IACHEC collaboration[8] has carried out observations and analysis to cross calibrate *Chandra* with *XMM-Newton/NuSTAR/Swift* and *Suzaku*. It has been demonstrated that the observatory responses are consistent at the 10% level below 3 keV, with a larger inconsistency observed at energies above 4 keV [73].

The *HETG* is subject to the same background effects as imaging spectroscopy observations with the *ACIS* detectors, i.e., instrumental/observatory backgrounds, local orbit backgrounds and the astrophysical/cosmic background. The excellent spatial resolution of the HRMA allows the use of narrow spectral extraction regions, which when combined with the energy resolution of the *ACIS* detectors serve to provide a very low background. Astrophysical observations show the background to be consistent with that observed in the stowed configuration demonstrating the dominant background component to be of intrinsic instrumental origin. The *HETG* background has been measured to be $\leq$1.0 counts/Ms/FWHM/4 arcsec (10% uncertainty) [39, 56]. In continuous clocking mode (CC), as the array is collapsed to a single dimension, the background is also multiplied by this $\sim 10^3$ factor. In practice, CC-mode targets are typically bright enough such that background is not a prominent concern.

Observations of brighter sources require careful consideration of the count rate arriving at the detector. Pile-up [19, 20] is absent in the dispersed spectra of sources of typical brightness; where an equivalent imaging observation would exhibit pile-up. In contrast, the zeroth order events will be piled up. At larger fluxes, such as that observed from Galactic sources, pile-up will impact the dispersed spectra. A number of mitigation strategies are available. For example, a sub-array may be utilized to observe source $\lesssim$350 mCrab. For brighter sources, observation in continuous clocking mode may be used. A number of complications arise in this observing mode. In particular, bright Galactic sources often have extended emission in the form of a scattering halo. As spatial information is no longer present in cc-mode, this emission will also be dispersed by the *HETGS* in a complicated manner, which may require significant additional modeling. A detailed consideration of the observation of bright sources with the *HETGS* is presented in [97].

A useful feature of the *ACIS* detector is that the detector is still recording events while being read-out. For bright sources, this can result in useful high time resolution data recorded at $\sim$40 μs resolution. Thus a readout streak is typically present in

---

[7] https://space.mit.edu/CXC/calib/hetg_user.html.

[8] https://iachec.org.

observations of brighter sources. As the readout time is short, such events do not suffer from pile-up (in contrast to the zeroth order image) and analysis of this data enable the measurement of the non pile-up biased intrinsic continuum flux observed from the scientific target.

When planning or analyzing observations with the *HETGS* a number of additional details should be carefully considered.

Extended Sources: The standard *HETGS* analysis and calibration pipeline assumes the observation of a point source at/near the focal point of the *ACIS-S3* array. As such, care should be taken to ensure only a single source in the FoV. This may require the usage of a specific roll angle constraint. Observations of extended sources are feasible, but require care during subsequent analysis [23, 30].

Chip Gaps: The *HETGS* disperses light across the 6 detector *ACIS-S* array. There are well defined gaps between each detector. Care should be taken to ensure spectral features of potential interest are not placed in these gaps, e.g., by a Y/Z offset of the aimpoint. Similarly, potential spectral features of interest in *HETGS* spectra should be compared to the known detector chip gaps in order to ensure they are of astrophysical origin.

Spacecraft Dither: As with imaging observations, the observatory executes a dither for observations with the *HETGS*. The spacecraft Lissajous pattern exhibits an ~8" amplitude pattern. This results in periodic signatures at ~1 ks and ~0.7 ks in the Y and Z directions respectively (Fig. 3.1) and these should be noted when time resolved spectroscopic analyses are undertaken.

Analysis of high resolution spectroscopy generally proceeds towards increasing levels of physical insight. Naive analysis, will identify and analyze spectral features in isolation. However, the true power of high resolution spectroscopy is revealed when the data is compared to detailed models of astro-physical plasmas. Initially, this will comprise emission line models of collisionally ionised plasmas such as `apec`[9] and progress to more detailed models including prescriptions for absorption and photo-ionisation such as `warmabs` [59]. Detailed consideration and full scientific exploitation of the spectra requires the calculation of custom models for the response of plasma to the observed input X-ray continuum spectrum, such as may be achieved with `cloudy` [28], `xstar` [59] and `spex` [101].

There remain fundamental uncertainties in our understanding of the physics at the atomic level in astrophysical plasmas, e.g., [42, 49, 51]. Ongoing theoretical and laboratory studies are working to address this and should be monitored, e.g., [68].

3.3.1 HETG Analysis

Analysis of spectroscopic data is at the heart of the process that takes us from a *HETG* high resolution observation to a furthering of astrophysical knowledge. High

[9] http://atomdb.org.

resolution spectra can contain an incredible depth of information on the temperature, density, kinematics, geometry and content of astrophysical plasmas and colder atomic/molecular material. As such, a detailed description of the analysis process is beyond the scope of this document. Below, we mention a number of key steps, that will result in the initial inspection of a *Chandra/HETGS* spectrum. The steps thereafter are at the heart of the scientific process and are left to the readers ingenuity.

(i) Use `dmkeypar` (or equivalent) to check the observation parameters e.g,
`dmkeypar acisf11111N001_evt2.fits.gz DETNAM echo+`
to inspect the number of active *ACIS* detectors.
The `heasoft fv` package provides a quick method to inspect an astronomical fits-format file. Alternatively, various `astropy` tasks may be utilized. One should also inspect the observation specific V&V report, where any issues identified by the initial observatory level processing will be highlighted.

(ii) Use `chandra_repro` to ensure that the latest calibrations and standard reduction procedures are applied to the dataset, e.g.,
`chandra_repro indir=11111 outdir=11111/repro cleanup =no verbose=1 clobber=yes`
The first order spectra are now ready to analyze, with the spectra and associated response files residing in `11111/repro/tg/`. One is advised to inspect the individual plus and minus orders and to compare with the combined spectrum. Responses for the higher order spectra are not generated (see below).

(iii) It is important that the correct bad pixel map is properly identified for `ciao` in the case of any subsequent analysis. Use `ardlib` to ensure that the correct bad pixel map is being used, e.g.,
```
cd 11111/repro/
punlearn ardlib
acis_set_ardlib acisf11111_repro_bpix1.fits
```

(iv) Typically, it is useful to combine the positive and negative orders to increase the S/N. Use `combine_grating_spectra` to separate the grating orders from the pha2 file for analysis, e.g.,
`combine_grating_spectra infile=acisf11111_repro_pha2.fits outroot=acisf11111 add_plusminus=yes garm =HEG order=1 arf="tg/*heg*.arf" rmf="tg/*heg*.rmf" clobber =yes`
This command should be repeated for the MEG and the 2nd and 3rd order spectra as required.

(v) The spectra can be rebinned. This may be necessary in order to ensure sufficient counts in each bin for the adopted test statistic when fitting the spectra. The spectra produced by the pipeline are oversampled by default (see Table 3.2). The spectra may be rebinned using standard tools, e.g., `dmgroup`, `grppha`, `ftgrouppha` or this step would be unnecessary if using `sherpa`.

Additional points of consideration:
After initial inspection and evaluation of the data, a number of further processing issues should be considered.

Table 3.2 Spectra resolution (FWHM) and default grids for *HETGS* spectra

	Resolution Å	Default binsize Å	Oversampling factor
HEG	0.012	0.0025	4.8
MEG	0.023	0.005	4.6

(a) Verify that the zeroth order position used in the extraction of the dispersed spectra is correct. This is particularly important for bright source observations where the zeroth order image may be distorted due to pile-up.
(b) Inspect the background extraction region. It may be possible to increase the S/N in your science spectrum over a specific wavelength range if the background region can be modified due to the scientific characteristics of the observed science target.

Further Resources:
The *Chandra* office has prepared an introduction to X-ray astronomy document. This is an excellent place to start or refresh your X-ray knowledge: "The X-ray Data Primer"

– https://cxc.cfa.harvard.edu/cdo/xray_primer.pdf

Analysis of *Chandra* data will require the use of `ciao` [32]. Detailed guides exist on the *Chandra* webpage demonstrating how `ciao` can be utilized in the analysis of *HETG* observations:

– https://cxc.cfa.harvard.edu/ciao/.

Check the watch-out page for current software or data processing issues that may impact your analysis. This page is regularly updated:

– https://cxc.cfa.harvard.edu/ciao/watchout.html.

The *HETG* user information page contains the latest information regarding the evolving performance of the *HETG* and ongoing calibration updates:

– https://space.mit.edu/ASC/calib/hetg_user.html.

Analysis of X-ray spectroscopic data in `xspec` [5],[10] `sherpa`[11] [31] or `isis`[12] [54] are similar; however, there are syntax and process differences. Be certain that the environment definitions required by your chosen analysis package do not clash with `ciao` or other software in your system, e.g., `xspec`. This point has particular importance with the widespread usage of `python` in the scientific community.

[10] https://heasarc.gsfc.nasa.gov/docs/software/lheasoft/.
[11] https://cxc.cfa.harvard.edu/sherpa/.
[12] https://space.mit.edu/ASC/ISIS/.

3.3.2 Science Highlights

Herein, we briefly discuss a number of *Chandra/HETGS* observations, which illustrate the capabilities of this instrument and the diverse fields of astronomy in which the high spectral resolution X-rays provided by it have made an impact.[13]

• *Stellar Physics—Capella*
The Capella system[14] is a binary consisting of a pair of G1 and G8 giants stars in a ~104 day orbit. It is known as the brightest star in the Auriga constellation and the 6th brightest in the night sky (V ~ 0.1, d ~ 13pc). It was one of the first sources of X-rays characterized in the initial period of exploratory observations in the 1960s [29]. As a bright X-ray source, Capella presents an excellent laboratory in which to study the X-ray emission physics of stellar coronae [43], as such it has been studied extensively by subsequent X-ray missions [12, 22, 26, 53, 109].

Capella was one of the first sources observed with the *HETGS* and the resulting spectra demonstrated the resolving power of the instrument, revealing a rich line dominated spectrum (see Fig. 3.5). In particular, observation of the He-like O VII lines facilitated direct constraints on the plasma density and temperature ($N_e \sim 10^{10}\,\mathrm{cm}^{-3}$, $T_e \sim 2 \times 10^6$ K) [15]. Subsequent analysis of the Fe L complex suggested this plasma component to have a temperature of $T_e \sim 6 \times 10^6$ K [8].

The combination of data from the *XMM-Newton/RGS* (see Chap. 2) and the *HETGS* have produced stringent constraints on the differential emission measure (DEM) and abundances. This has revealed the presence of a solar like first ionisation potential (FIP) effect to be present in the Capella corona, e.g., Fig. 3.6 [7, 41].

• *Star Formation—TW Hya*
TW Hya is a nearby (~57 pc) t-tauri star (M ~ 0.8 $M_\odot$, $\tau \sim 8$ Myr). It lies in a nearly face-on orientation ($i \sim 7^o$). TW Hya is the archetype of its class and observations with *ALMA* have revealed disk structure suggestive of ongoing planet formation (down to earth-sun scales), in addition to the presence of complex organic molecules [2, 27, 110].

TW Hya is one of the brightest t-tauri stars at X-ray energies and as such has been subject to detailed study. In the magnetospheric accretion model for t-tauri stars, a shock is expected to form on the stellar surface when matter lifted from the disk free-falls onto the star following magnetic field lines. For the typical free-fall velocities, soft X-rays form in the shock, some of which may escape and correspond to the cooler emission observed in X-rays. Observations by *XMM-Newton* and *Chandra* have revealed a complex accretion flow configuration [44, 45, 63, 102]. In particular deep *HETGS* observations detected emission associated with the stellar corona, the accretion shock and an extended component attributed to post-shock plasma [13, 14].

---

[13] This discussion contains a biased list of some of the many measurements made by the *HETGS* over the past two decades.

[14] As a bright star, Capella is familiar to many cultures globally, e.g, https://en.wikipedia.org/wiki/Capella#Etymology_and_culture.

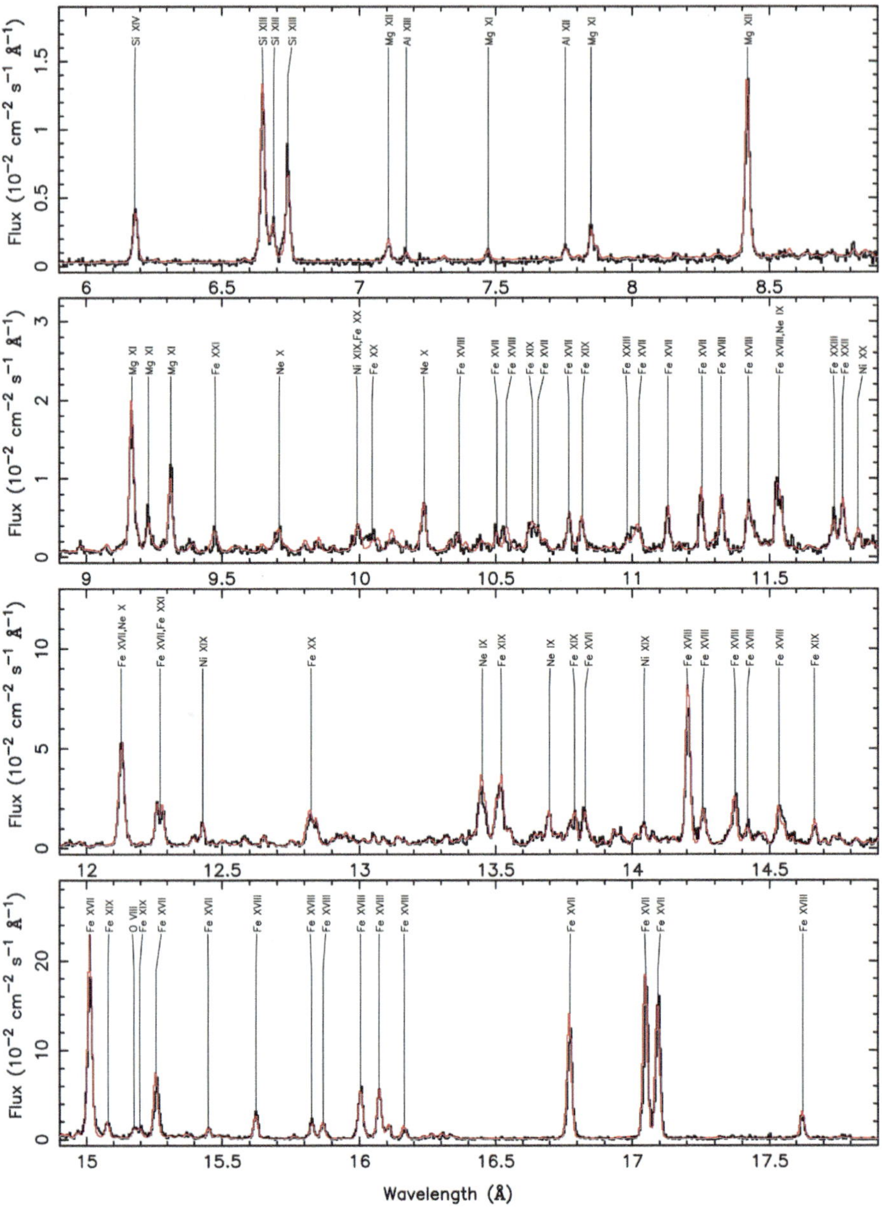

Fig. 3.5 *Chandra/HETGS* X-ray spectrum of Capella. The MEG spectrum measured from the Capella corona in the 6–18 Å region is plotted. The black line is the sum of ±1 orders of the MEG spectra; the red line is the model using the DEM and abundances reconstructed from the *HETGS* data. Adapted from [41]

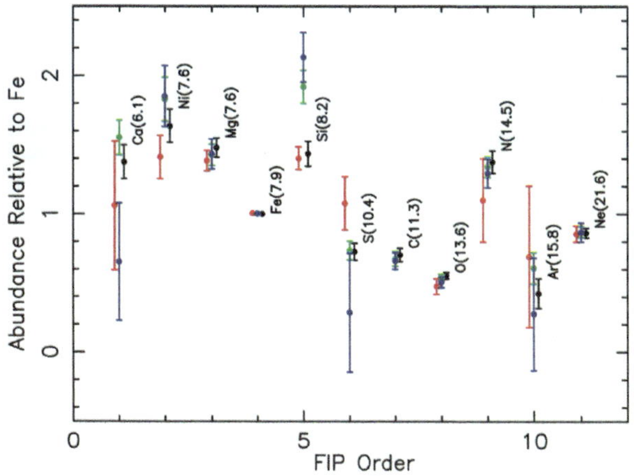

Fig. 3.6 Derived abundances relative to that of Fe for the Capella corona, demonstrate the presence of a solar like FIP effect. The black symbols are results of the joint fit of *Chandra/HETGS* and *XMM-Newton/RGS* data, the green and blue are for the *RGS* alone, and the red symbols are results of *HETGS* data alone. The numbers in the parentheses are the first ionisation potentials of the elements in eV. Adapted from [41]

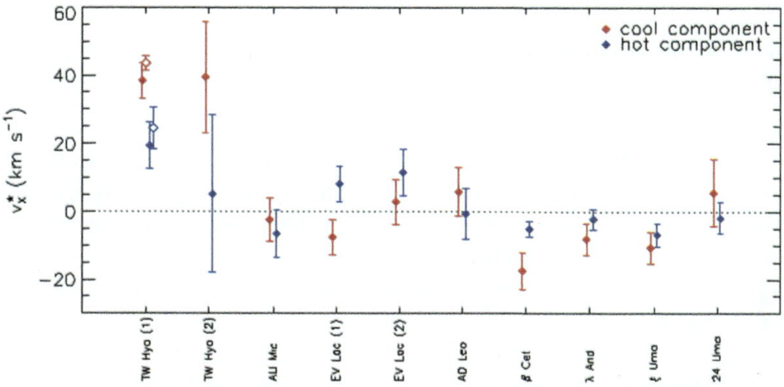

Fig. 3.7 Cool and hot X-ray emitting plasma velocities v_X^* in the stellar reference frame for TW Hya and a comparison sample of stars (with 1σ errors). The filled/open symbols denote differing analysis methods. A redshift of ~40 km s^{-1} is present in the TW Hya data. Adapted from [4]

While the *HETGS* excels at constraining the physical conditions of X-ray plasmas, it is also adept at detailed studies of spectral line kinematics. This has been demonstrated by the measurement of X-ray emission line velocities with respect to the stellar photosphere in TW Hya. [4] measured a ~40 km s^{-1} relative velocity for the 2–4 $\times 10^6$ K plasma in the accretion shock (see Fig. 3.7). As the expected free-fall velocity in the accretion flow is order 100 km s^{-1}, these X-ray spectral constraints suggest much of the flow impacts the stellar surface at low latitudes. As can be observed in Fig. 3.7, this result relied on the broader archive of high quality

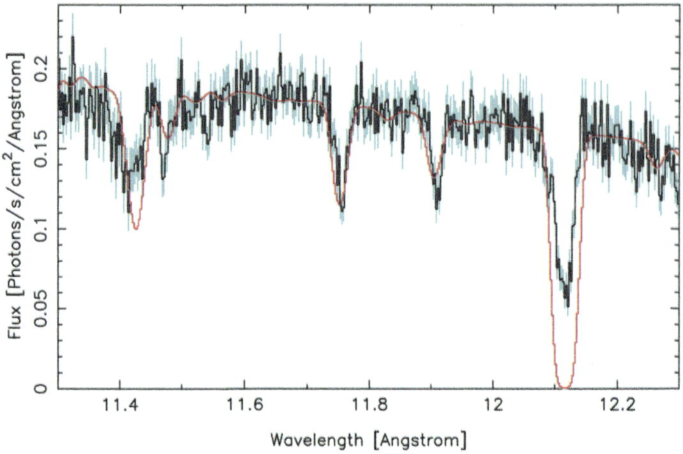

Fig. 3.8 *HETGS* detection of the density-sensitive Fe XXII lines at 11.77 and 11.92 Å in the spectrum of the Galactic black hole GRO J1655-40. The panel shows the best-fit XSTAR model for this region. These models strongly suggest that the density is log(n) $\sim$ 13.7–14.0 and that magnetic driving is an important wind launching mechanism. Adapted from [81]

spectral observations of stellar X-ray sources that have been made during the *HETGS* lifetime to date.

• *Black Hole Accretion Flows—GRO J1655-40*
GRO J1644-40 is a Galactic stellar mass black hole discovered during an outburst in 1994 [47]. The black hole, with mass 7.0±0.2 $M_\odot$, has a F-type mass donor in a 2.6 day orbit and lies at a distance of $\sim$3.2 kpc [92]. This system is of particular interest as apparently super-luminal radio jets were detected during this 1994 outburst, thus placing GRO J1655-40 in the micro-quasar class [52, 103]. Observations with *ASCA* and *RXTE* had hinted at the existence of X-ray absorbing gas above the accretion disk during the 1994 and a subsequent smaller outburst in 1996 [69, 107].

A second large outburst was detected from this source in 2005 [100], the first since the launch of *Chandra*. GRO J1655-40 was observed by the *HETGS* for 65 ks on April 1st. The resulting spectrum revealed the spectacular detection of a rich absorption line spectrum across the *HETGS* bandpass, containing 90 lines detected at the 5σ level [80], e.g., Fig. 3.13. Analysis of the data demonstrated that the absorbing plasma must lie close to the black hole and that magnetic forces play an important role in launching the observed dense wind [80, 81]. Of critical importance, the *HETGS* facilitated the detection of a density sensitive state of Fe XXII (Fig. 3.8), which allowed a direct constraint on the density of the accretion disk wind only a few hundred gravitational radii from the black hole, $n_e \gtrsim 10^{13.6}$ cm^{-3} [81].

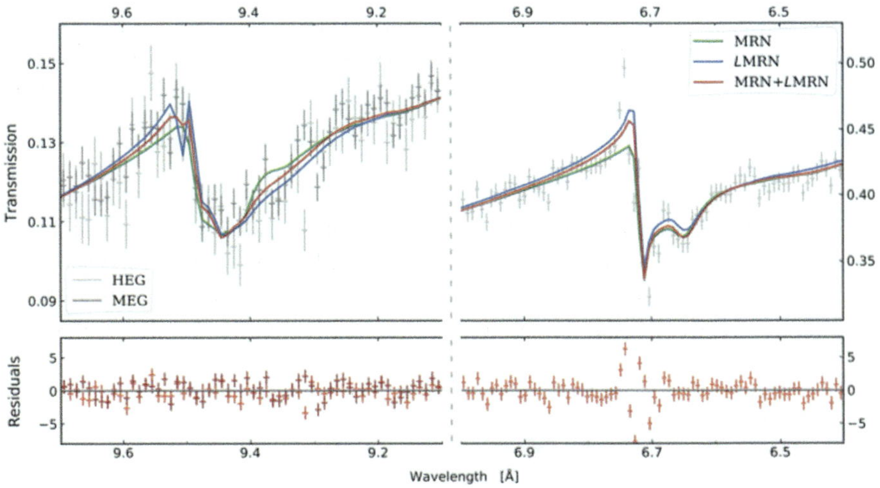

Fig. 3.9 Magnesium and silicon K-edges as measured in the spectrum of the Galactic neutron star LMXB GX 3+1. The HEG and MEG data are shown in light and dark grey, respectively. The two edges are fit using models with different grain size distributions. Bottom panel: the residuals defined as (observed-model)/error of the best fit obtained using Mathis–Rumpl–Nordsieck (MRN) and larger-MRN type grain models. The HEG and MEG data are shown in light and dark red, respectively. Adapted from [96]

- *The Inter-Stellar Medium—Dust and Gas*

The inter-stellar medium (ISM) plays a key role in the process of star formation. High resolution X-ray spectroscopy enables the detailed study of the properties of the ISM and Galactic X-ray binaries provide luminous backlights enabling such studies with the *HETGS*. The dust content of the ISM can be studied as the size, composition and structure of the dust grains are encoded in their resulting absorption features [24]. Observations of the Galactic microquasar GRS1915+105 presented the first tentative evidence for the presence of dust absorption [71]. Observations in subsequent years have confirmed this approach [17, 94, 95, 98, 114].

In Fig. 3.9, we plot model fits consisting of varied grain composition and size distribution to *HETGS* observations of the X-ray binary GX 3+1, where evidence for dust grain absorption is detected in the Si and Mg absorption edges. Here the dust is determined to be consistent with a model containing amorphous olivine, crystalline fayalite and amorphous quartz mixture (mixed as ~71/16/13%) [96]. Additional structure remains with respect to the model for the Si K line. The nature of this structure remains to be determined by future observations.

The gas phase content of the ISM may also be studied in this manner [58, 113]. Constraints on the Si, O, Ne and Fe content of the ISM are presented in analysis of *HETGS* spectra of a sample of X-ray binaries in [35, 36].

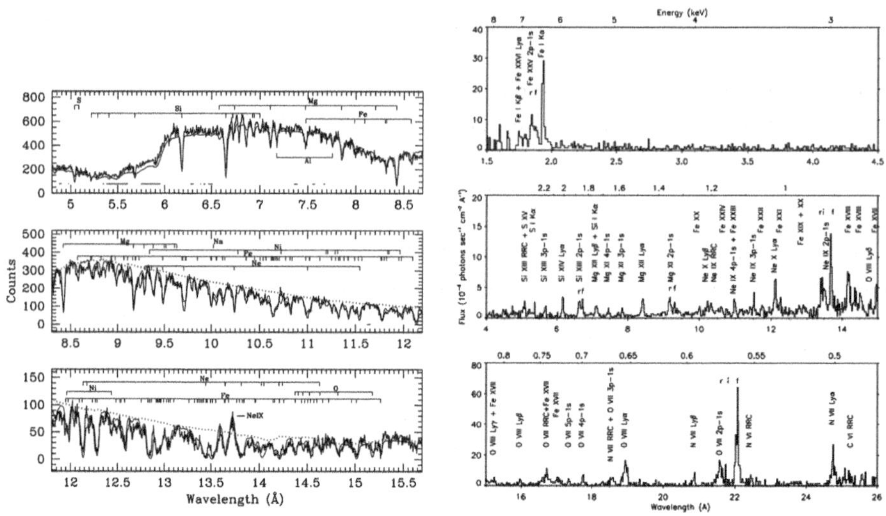

Fig. 3.10 **Left:** Two-phase absorber model plotted against the first-order MEG spectrum of NGC 3783. Absorption lines predicted are marked in the top. Single labels stand for emission lines. The spectrum is presented in the rest-frame system of the absorbing gas. Adapted from [67]. **Right:** *Chandra/HETGS* spectra of the central 1.5" radius nuclear region in the Sy 2 galaxy NGC 1068. The top panel is from the HEG; bottom two are from MEG. The strong forbidden emission lines (f) and narrow recombination continuum features (RRC) indicate recombination following photoionisation by the hidden nucleus. Fe I, Si I, and S I Kα emission lines indicate fluorescence from a low-ionisation region, which may be identified with the obscuring molecular torus. Adapted from [91]

• *Active Galactic Nuclei—Supermassive Black Hole Fueling*

Supermassive black holes appear to be intrinsically involved in the process of galaxy formation [25]. Studies of black holes in the local Universe allow us to probe how black holes grow and probe the nature of their coupling to the broader evolution of their host galaxy. The *HETGS* has made many contributions to this field via detailed studies of the plasma conditions in the nuclear regions of galaxies as exemplified by the studies of the Seyfert 1 (Sy 1) galaxy NGC 3783, the Sy 1.5 NGC4151 and the Sy 2 NGC 1068 [3].

NGC 3783 is a Sy 1 galaxy containing a black hole of mass $M_{BH} \sim 2.8 \times 10^7\ M_\odot$ [10]. As a bright X-ray source, NGC 3783 has been extensively studied, revealing the presence of a dynamic absorbing system associated with the accretion flow onto the black hole [37, 106]. Deep observations with the *HETG* have provided constraints on the nature of this warm absorber [61]. The high resolution spectrum reveals a rich set of absorption lines ($\gtrsim$100, e.g., see Fig. 3.10, left). Multiple components are required to model the absorbing gas. This gas is found to be outflowing at a velocity of $v \sim 750$ km s^{-1} and the most ionised component is constrained to lie at sub-parsec distances from the central SMBH [62, 67, 89].

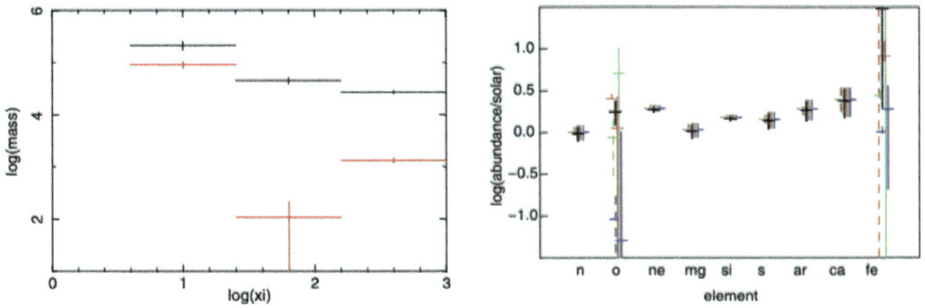

Fig. 3.11 Constraints on the plasma properties in the nuclear region of NGC 1068. **Left:** Distribution of mass among the emission components used to model the spectrum. Black points correspond to the component with column 3×10^{23} cm^{-2}, and red corresponds to the component with column 3×10^{22} cm^{-2}. **Right:** Element abundances from the best-fit models to the *HETGS* spectrum of NGC 1068. Colors red, green, and blue correspond to $\log(\xi) = 1, 1.8, 2.6$ respectively; solid corresponds to column 3×10^{23} cm^{-2}, and dashed corresponds to column 3×10^{22} cm^{-2}. Black points denote the average over the best-fit model. Adapted from [60]

NGC 4151 is a Sy 1.5 galaxy containing an approximately $1.7 \times 10^7\ M_\odot$ black hole [11]. *HETG* observations have revealed the X-ray gas in NGC4151 to be spatially extended and dominated by photo-ionisation from the hard X-ray continuum originating near the black hole. The X-ray emission line gas has been found to have a similar mass and outflow rate to gas observed in optical and UV, but with evidence indicating that the X-ray outflow extends to larger distance from the black hole [18, 64, 90]. In addition to placing constraints on the physical conditions of the ionised gas, long term observations are now revealing how this gas responds to changes in the accretion luminosity from the black hole [65].

NGC 1068 is a Sy 2 galaxy, containing a black hole of mass $M_{BH} \sim 10^7\ M_\odot$ [33, 34, 84] and is known to be a source of neutrinos [57]. Initial *Chandra* observations showed the X-ray plasma to be concentrated in the nuclear region and in a photo-ionised condition (see Fig. 3.10, right). The observed spectra are consistent with models of gas being illuminated by a hidden central source as expected for a Sy 2 AGN [91]. Deeper observations have facilitated detailed measurements of the nuclear gas properties in this Sy 2 galaxy, e.g., see Fig. 3.11 for constraints on the mass of the X-ray emitting gas and its composition [60].

• *The Highest Resolution Frontier—HETGS 3rd Order Spectra*
The 2nd and 3rd order *HETGS* spectra have an effective area of only ~7% of the first order spectrum [16]; however, owing to the larger resolving power, compelling scientific investigations are possible for the brightest X-ray sources.

GRS 1915+105 is the canonical Galactic microquasar [83] and has been the subject of numerous *HETGS* investigations [70, 86–88, 115]. In a long observation during a soft spectral state in 2015, the source flux was such that significant S/N was present in the 3rd order spectrum. Four separate photo-ionised model components (with absorption and re-emission) are required to successfully re-produce the observed spectrum, see Fig. 3.12 (left), and these components demand high densities

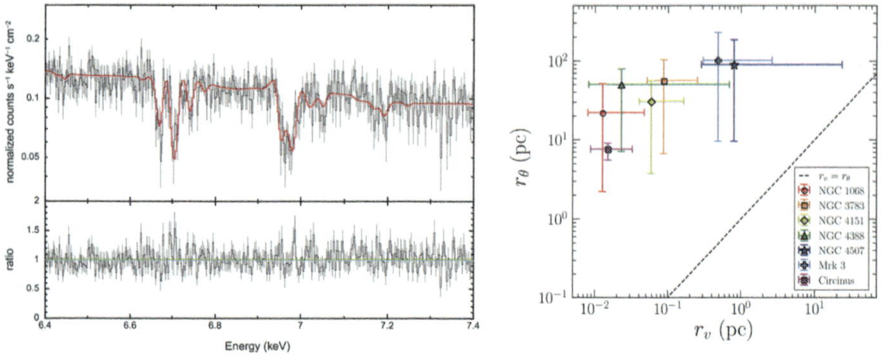

Fig. 3.12 **Left:** *HETGS* third-order spectrum of the microquasar GRS 1915+105. Four photo-ionisation zones with paired absorption and re-emission are required to fit the data. The He-like Fe XXV line is resolved into intercombination and resonance lines (rest-frame energy: 6.700 keV). Instances of H-like Fe XXVI absorption lines close to the rest-frame value of 6.970 keV and blueshifted up to 7.05 and 7.2 keV are apparent. The Fe XXVI line shape is a doublet owing to the expected spin-orbit splitting in the H-like atom. Adapted from [82]. **Right:** Comparison of the distance of the emitting gas from the black hole in a sample of bright AGN. The black dashed line shows equality between the two estimates. The distances measured by the spatial extent, r_θ, are consistently larger than the distances measured by the velocity width, r_v, under the assumption of gravitational motions for all seven objects in the sample. Adapted from [79]

($n \sim 10^{13-16}$ cm^{-3}) and a location in the inner accretion flow ($r \sim 10^{2-4}$ GM/c^2) [82].

The narrow Fe K emission line is a ubiquitous feature in the X-ray spectra of AGN and can be used to constrain the hot plasma and the distant cold gas [66, 85, 93]. In studies of nearby luminous AGN, the difference between the line widths in the first and third order spectra from the *HETGS* has been used to constrain the spatial extent of Fe K emission (see Fig. 3.12, right). The line is found to be spatially extended on scales of 5–100 pc. This supports the interpretation of the narrow Fe K line as originating from reflection off distant material. However, the 3rd order data provide velocity constraints that suggest distances an order of magnitude smaller. This study confirms that a significant fraction of the Fe K emission is generated in regions much closer to the black hole [72, 79].

The above studies present examples of the potential constraints that can be placed on the dynamics, kinematics and geometry of astrophysical plasma flows with high resolution spectroscopy. These are mere teasers of the science that was initially highlighted by *Hitomi* [50] and will become common in coming years with the advent of high resolution calorimeters and the next generation of grating spectrometers.

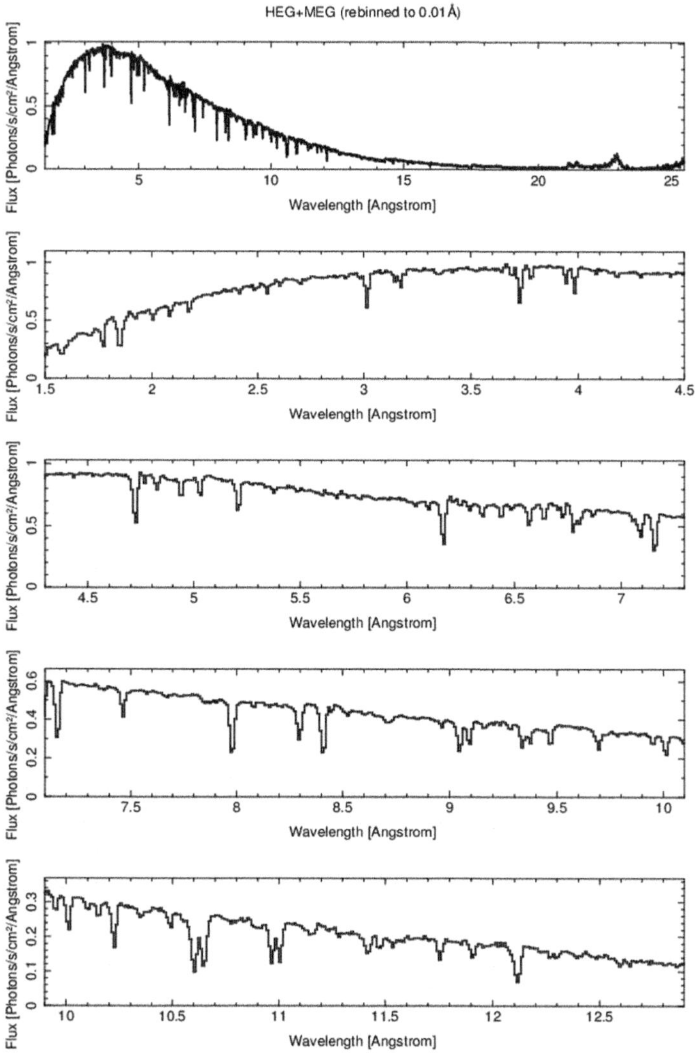

Fig. 3.13 Archival *Chandra*/HETGS observation of the Galactic stellar mass black hole GRO J1655-40 (obsid: 6329, [80, 81]) as obtained in science ready format from TGCat. This plot is in the diagnostic format provided by TGCat and displays the broadband HEG/MEG spectrum (top), with the panels below focusing on successively longer wavelengths

3.4 TGCat

The Transmission Grating Catalog and Archive (TGCat [55]) provides a searchable interface of archived *HETGS* observations.[15] The resulting spectra have been processed, extracted and are downloadable. The data in this format are science ready and present an efficient means to begin an analysis of archival observations. As of the beginning of 2023, almost 1800 spectra are available and further spectra are added to the archive on a regular basis.

In Fig. 3.13, we plot the *HETGS* spectrum of GRO J1655-40 obtained during the soft state in 2004 and which reveals a dense highly ionised outflow [80, 81]. Science ready products such as these are available at TGCat. The included spectra and responses can be directly loaded into `xspec/isis` for scientific analysis.

3.5 Simulations with MARX

Marx[16] simulates the on-orbit performance of *Chandra* utilizing ray tracing in combination with detailed models of the mirror assembly, gratings and detector systems. The MARX end to end simulator can prove useful in the case where a non-standard observation is planned or has been carried out, e.g., a spatially extended source, an unusual placement of the zeroth order image, a cc-mode observation or in the case of multiple point sources. In such cases, the dispersal of X-rays by the gratings will need to be modeled to properly understand the resulting dispersed spectra. MARX is able to produce standard fits format files, which can be analyzed in the normal manner with `ciao/isis/xspec` [21].

Acknowledgements The *Chandra* X-ray Observatory is the result of the work of generations of scientists and engineers, who have provided the scientific community with an excellent tool to study the wonders of the X-ray sky. Particular recognition should go to the instrument teams of ACIS PI Gordon Garmire and HETGS PI Claude Canizares. Finally, our gratitude extends to the *Chandra* directors Harvey Tananbaum, Belinda Wilkes and Pat Slane who have steered the great ship adeptly over the past 24 years.

References

1. The ACIS detector. https://cxc.harvard.edu/proposer/POG/html/chap6.html
2. S.M. Andrews, D.J. Wilner, Z. Zhu, T. Birnstiel, J.M. Carpenter, L.M. Pérez, X.-N. Bai et al., ApJL **820**, L40 (2016). https://doi.org/10.3847/2041-8205/820/2/L40
3. R. Antonucci, ARA&A **31**, 473 (1993). https://doi.org/10.1146/annurev.aa.31.090193.002353

[15] http://tgcat.mit.edu.

[16] https://space.mit.edu/cxc/marx/.

4. C. Argiroffi, J.J. Drake, R. Bonito, S. Orlando, G. Peres, M. Miceli, A&A **607**, A14 (2017). https://doi.org/10.1051/0004-6361/201731342
5. K.A. Arnaud, ASPC **101**, 17 (1996)
6. B. Aschenbach, RvMA **4**, 173 (1991). https://doi.org/10.1007/978-3-642-76750-0_13
7. M. Audard, M. Güdel, A. Sres, A.J.J. Raassen, R. Mewe, A&A **398**, 1137 (2003). https://doi.org/10.1051/0004-6361:20021737
8. E. Behar, J. Cottam, S.M. Kahn, ApJ **548**, 966 (2001). https://doi.org/10.1086/318999
9. K.P. Beuermann, H. Braeuninger, J. Truemper, ApOpt **17**, 2304 (1978). https://doi.org/10.1364/AO.17.002304
10. M.C. Bentz, P.R. Williams, R. Street, C.A. Onken, M. Valluri, T. Treu, ApJ **920**, 112 (2021). https://doi.org/10.3847/1538-4357/ac19af
11. M.C. Bentz, P.R. Williams, T. Treu, ApJ **934**, 168 (2022). https://doi.org/10.3847/1538-4357/ac7c0a
12. N.S. Brickhouse, A.K. Dupree, R.J. Edgar, D.A. Liedahl, S.A. Drake, N.E. White, K.P. Singh, ApJ **530**, 387 (2000). https://doi.org/10.1086/308350
13. N.S. Brickhouse, S.R. Cranmer, A.K. Dupree, G.J.M. Luna, S. Wolk, ApJ **710**, 1835 (2010). https://doi.org/10.1088/0004-637X/710/2/1835
14. N.S. Brickhouse, S.R. Cranmer, A.K. Dupree, H.M. Günther, G.J.M. Luna, S.J. Wolk, ApJL **760**, L21 (2012). https://doi.org/10.1088/2041-8205/760/2/L21
15. C.R. Canizares, D.P. Huenemoerder, D.S. Davis, D. Dewey, K.A. Flanagan, J. Houck, T.H. Markert et al., ApJL **539**, L41 (2000). https://doi.org/10.1086/312823
16. C.R. Canizares, J.E. Davis, D. Dewey, K.A. Flanagan, E.B. Galton, D.P. Huenemoerder, K. Ishibashi et al., PASP **117**, 1144 (2005). https://doi.org/10.1086/432898
17. E. Costantini, C. Pinto, J.S. Kaastra, J.J.M. in't Zand, M.J. Freyberg, L. Kuiper, M. Méndez et al., A&A **539**, A32 (2012). https://doi.org/10.1051/0004-6361/201117818
18. J.D. Couto, S.B. Kraemer, T.J. Turner, D.M. Crenshaw, ApJ **833**, 191 (2016). https://doi.org/10.3847/1538-4357/833/2/191
19. J.E. Davis, ApJ **562**, 575 (2001). https://doi.org/10.1086/323488
20. J.E. Davis, SPIE **4851**, 101 (2003). https://doi.org/10.1117/12.461476
21. J.E. Davis, M.W. Bautz, D. Dewey, R.K. Heilmann, J.C. Houck, D.P. Huenemoerder, H.L. Marshall et al., SPIE **8443**, 84431A (2012). https://doi.org/10.1117/12.926937
22. R.C. Dempsey, J.L. Linsky, J.H.M.M. Schmitt, T.A. Fleming, ApJ **413**, 333 (1993). https://doi.org/10.1086/173001
23. D. Dewey, hrxs.conf, 14 (2002)
24. B.T. Draine, ApJ **598**, 1026 (2003). https://doi.org/10.1086/379123
25. A.C. Fabian, ARA&A **50**, 455 (2012). https://doi.org/10.1146/annurev-astro-081811-125521
26. F. Favata, R. Mewe, N.S. Brickhouse, R. Pallavicini, G. Micela, A.K. Dupree, A&A **324**, L37 (1997). https://doi.org/10.48550/arXiv.astro-ph/9707061
27. C. Favre, D. Fedele, D. Semenov, S. Parfenov, C. Codella, C. Ceccarelli, E.A. Bergin et al., ApJL **862**, L2 (2018). https://doi.org/10.3847/2041-8213/aad046
28. G.J. Ferland, K.T. Korista, D.A. Verner, J.W. Ferguson, J.B. Kingdon, E.M. Verner, PASP **110**, 761 (1998). https://doi.org/10.1086/316190
29. P.C. Fisher, A.J. Meyerott, ApJ **139**, 123 (1964). https://doi.org/10.1086/147742
30. K.A. Flanagan, C.R. Canizares, D. Dewey, J.C. Houck, A.C. Fredericks, M.L. Schattenburg, T.H. Markert et al., ApJ **605**, 230 (2004). https://doi.org/10.1086/382145
31. P. Freeman, S. Doe, A. Siemiginowska, SPIE **4477**, 76 (2001). https://doi.org/10.1117/12.447161
32. A. Fruscione, J.C. McDowell, G.E. Allen, N.S. Brickhouse, D.J. Burke, J.E. Davis, N. Durham et al., SPIE **6270**, 62701V (2006). https://doi.org/10.1117/12.671760
33. C.M.V. Impellizzeri, J.F. Gallimore, S.A. Baum, M. Elitzur, R. Davies, D. Lutz, R. Maiolino et al., ApJL **884**, L28 (2019). https://doi.org/10.3847/2041-8213/ab3c64
34. L.J. Greenhill, C.R. Gwinn, R. Antonucci, R. Barvainis, ApJL **472**, L21 (1996). https://doi.org/10.1086/310346

35. E. Gatuzz, J.A. García, T.R. Kallman, C. Mendoza, A&A **588**, A111 (2016). https://doi.org/10.1051/0004-6361/201527752
36. E. Gatuzz, T.W. Gorczyca, M.F. Hasoglu, N.S. Schulz, L. Corrales, C. Mendoza, MNRAS **498**, L20 (2020). https://doi.org/10.1093/mnrasl/slaa119
37. I.M. George, T.J. Turner, R. Mushotzky, K. Nandra, H. Netzer, ApJ **503**, 174 (1998). https://doi.org/10.1086/305988
38. R. Giacconi, G. Branduardi, U. Briel, A. Epstein, D. Fabricant, E. Feigelson, W. Forman et al., ApJ **230**, 540 (1979). https://doi.org/10.1086/157110
39. C.E. Grant, M.W. Bautz, S.N. Virani ASPC **262**, 401 (2002). https://doi.org/10.48550/arXiv.astro-ph/0202086
40. C.E. Grant, M.W. Bautz, P.G. Ford, P.P. Plucinsky, SPIE **9144**, 91443Q (2014). https://doi.org/10.1117/12.2055652
41. M.F. Gu, R. Gupta, J.R. Peterson, M. Sako, S.M. Kahn, ApJ **649**, 979 (2006). https://doi.org/10.1086/506602
42. L. Gu, C. Shah, J. Mao, A.J.J. Raassen, J. de Plaa, C. Pinto, H. Akamatsu et al., A&A **664**, A62 (2022). https://doi.org/10.1051/0004-6361/202039943
43. M. Güdel, A&ARv **12**, 71 (2004). https://doi.org/10.1007/s00159-004-0023-2
44. M. Güdel, Y. Nazé, SSRv **157**, 211 (2010). https://doi.org/10.1007/s11214-010-9713-4
45. H.M. Günther, J.H.M.M. Schmitt, J. Robrade, C. Liefke, A&A **466**, 1111 (2007). https://doi.org/10.1051/0004-6361:20065669
46. H.M. Günther, ChNew, 25 (2018)
47. B.A. Harmon, C.A. Wilson, W.S. Paciesas, S.N. Zhang, G.J. Fishman, IAUC 6084 (1994)
48. The HETG. https://cxc.harvard.edu/proposer/POG/html/chap8.html
49. K. Heuer, A.R. Foster, R. Smith, ApJ **908**, 3 (2021). https://doi.org/10.3847/1538-4357/abcaff
50. Hitomi Collaboration, F. Aharonian, H. Akamatsu, F. Akimoto, S.W. Allen, N. Anabuki, L. Angelini et al., Nature **535**, 117 (2016). https://doi.org/10.1038/nature18627
51. Hitomi Collaboration, F. Aharonian, H. Akamatsu, F. Akimoto, S.W. Allen, L. Angelini, M. Audard et al. PASJ **70**, 12 (2018). https://doi.org/10.1093/pasj/psx156
52. R.M. Hjellming, M.P. Rupen, Nature **375**, 464 (1995). https://doi.org/10.1038/375464a0
53. S.S. Holt, N.E. White, R.H. Becker, E.A. Boldt, R.F. Mushotzky, P.J. Serlemitsos, B.W. Smith, ApJL **234**, L65 (1979). https://doi.org/10.1086/183110
54. J.C. Houck, L.A. Denicola, ASPC **216**, 591 (2000)
55. Huenemoerder, D.P., A. Mitschang, D. Dewey et al., AJ **141**, 129 (2011). https://doi.org/10.1088/0004-6256/141/4/129
56. D.P. Huenemoerder, Chandra X-ray Center memo dated 7/7/2017 (2017). https://space.mit.edu/cxc/docs/hetg_bg/hetg_background.pdf
57. IceCube Collaboration, R. Abbasi, M. Ackermann, J. Adams, J.A. Aguilar, M. Ahlers, M. Ahrens et al., Science **378**, 538 (2022). https://doi.org/10.1126/science.abg3395
58. A.M. Juett, N.S. Schulz, D. Chakrabarty, ApJ **612**, 308 (2004). https://doi.org/10.1086/422511
59. T. Kallman, M. Bautista, ApJS **133**, 221 (2001). https://doi.org/10.1086/319184
60. T. Kallman, D.A. Evans, H. Marshall, C. Canizares, A. Longinotti, M. Nowak, N. Schulz, ApJ **780**, 121 (2014). https://doi.org/10.1088/0004-637X/780/2/121
61. S. Kaspi, W.N. Brandt, H. Netzer, I.M. George, G. Chartas, E. Behar, R.M. Sambruna, G.P. Garmire, J.A. Nousek, ApJ **554**, 216 (2001). https://doi.org/10.1086/321333
62. S. Kaspi, W.N. Brandt, I.M. George, H. Netzer, D.M. Crenshaw, J.R. Gabel, F.W. Hamann et al., ApJ **574**, 643 (2002). https://doi.org/10.1086/341113
63. J.H. Kastner, D.P. Huenemoerder, N.S. Schulz, C.R. Canizares, D.A. Weintraub, ApJ **567**, 434 (2002). https://doi.org/10.1086/338419
64. S.B. Kraemer, I.M. George, D.M. Crenshaw, J.R. Gabel, T.J. Turner, T.R. Gull, J.B. Hutchings et al., ApJ **633**, 693 (2005). https://doi.org/10.1086/466522
65. S.B. Kraemer, T.J. Turner, J.D. Couto, D.M. Crenshaw, H.R. Schmitt, M. Revalski, T.C. Fischer, MNRAS **493**, 3893 (2020). https://doi.org/10.1093/mnras/staa428
66. J.H. Krolik, T.R. Kallman, ApJL **320**, L5 (1987). https://doi.org/10.1086/184966

67. Y. Krongold, F. Nicastro, N.S. Brickhouse, M. Elvis, D.A. Liedahl, S. Mathur, ApJ **597**, 832 (2003). https://doi.org/10.1086/378639
68. S. Kühn, C. Cheung, N.S. Oreshkina, R. Steinbrügge, M. Togawa, S. Bernitt, L. Berger et al., PhRvL **129**, 245001 (2022). https://doi.org/10.1103/PhysRevLett.129.245001
69. E. Kuulkers, R. Wijnands, T. Belloni, M. Méndez, M. van der Klis, J. van Paradijs, ApJ **494**, 753 (1998). https://doi.org/10.1086/305248
70. J.C. Lee, C.S. Reynolds, R. Remillard, N.S. Schulz, E.G. Blackman, A.C. Fabian, ApJ **567**, 1102 (2002). https://doi.org/10.1086/338588
71. J.C. Lee, B. Ravel, ApJ **622**, 970 (2005). https://doi.org/10.1086/428118
72. J. Liu, MNRAS **463**, L108 (2016). https://doi.org/10.1093/mnrasl/slw164
73. K.K. Madsen, A.P. Beardmore, K. Forster, M. Guainazzi, H.L. Marshall, E.D. Miller, K.L. Page et al., AJ **153**, 2 (2017). https://doi.org/10.3847/1538-3881/153/1/2
74. H.L. Marshall, D. Dewey, K. Ishibashi, SPIE **5165**, 457 (2004). https://doi.org/10.1117/12.508320
75. H.L. Marshall, A. Tennant, C.E. Grant, A.P. Hitchcock, S.L. O'Dell, P.P. Plucinsky, SPIE **5165**, 497 (2004). https://doi.org/10.1117/12.508310
76. H.L. Marshall, Improving the relative accuracy of the HETGS effective area (2005). https://space.mit.edu/ASC/calib/heg_meg/meg_heg_report.pdf
77. H.L. Marshall, SPIE **8443**, 844348 (2012). https://doi.org/10.1117/12.927209
78. H.L. Marshall, Chandra X-ray center memo on effects of warm ACIS on the HETGS effective area (2020). https://space.mit.edu/CXC/calib/rmf_effects.pdf
79. M. Masterson, C.S. Reynolds, ApJ **936**, 66 (2022). https://doi.org/10.3847/1538-4357/ac83ae
80. J.M. Miller, J. Raymond, A. Fabian, D. Steeghs, J. Homan, C. Reynolds, M. van der Klis et al., Nature **441**, 953 (2006). https://doi.org/10.1038/nature04912
81. J.M. Miller, J. Raymond, C.S. Reynolds, A.C. Fabian, T.R. Kallman, J. Homan, ApJ **680**, 1359 (2008). https://doi.org/10.1086/588521
82. J.M. Miller, J. Raymond, A.C. Fabian, E. Gallo, J. Kaastra, T. Kallman, A.L. King et al., ApJL **821**, L9 (2016). https://doi.org/10.3847/2041-8205/821/1/L9
83. I.F. Mirabel, L.F. Rodríguez, Nature **371**, 46 (1994). https://doi.org/10.1038/371046a0
84. Y. Morishima, H. Sudou, A. Yamauchi, Y. Taniguchi, N. Nakai, PASJ **75**, 71 (2023). https://doi.org/10.1093/pasj/psac092
85. K. Nandra, MNRAS **368**, L62 (2006). https://doi.org/10.1111/j.1745-3933.2006.00158.x
86. J. Neilsen, J.C. Lee, Nature **458**, 481 (2009). https://doi.org/10.1038/nature07680
87. J. Neilsen, R.A. Remillard, J.C. Lee, ApJ **737**, 69 (2011). https://doi.org/10.1088/0004-637X/737/2/69
88. J. Neilsen, A.J. Petschek, J.C. Lee, MNRAS **421**, 502 (2012). https://doi.org/10.1111/j.1365-2966.2011.20329.x
89. H. Netzer, S. Kaspi, E. Behar, W.N. Brandt, D. Chelouche, I.M. George, D.M. Crenshaw et al., ApJ **599**, 933 (2003). https://doi.org/10.1086/379508
90. P.M. Ogle, H.L. Marshall, J.C. Lee, C.R. Canizares, ApJL **545**, L81 (2000). https://doi.org/10.1086/317890
91. P.M. Ogle, T. Brookings, C.R. Canizares, J.C. Lee, H.L. Marshall, A&A **402**, 849 (2003). https://doi.org/10.1051/0004-6361:20021647
92. J.A. Orosz, C.D. Bailyn, ApJ **477**, 876 (1997). https://doi.org/10.1086/303741
93. K.A. Pounds, K. Nandra, G.C. Stewart, I.M. George, A.C. Fabian, Nature **344**, 132 (1990). https://doi.org/10.1038/344132a0
94. C. Pinto, J.S. Kaastra, E. Costantini, C. de Vries, A&A **551**, A25 (2013). https://doi.org/10.1051/0004-6361/201220481
95. D. Rogantini, E. Costantini, S.T. Zeegers, C.P. de Vries, W. Bras, F. de Groot, H. Mutschke et al., A&A **609**, A22 (2018). https://doi.org/10.1051/0004-6361/201731664
96. D. Rogantini, E. Costantini, S.T. Zeegers, C.P. de Vries, M. Mehdipour, F. de Groot, H. Mutschke et al., A&A **630**, A143 (2019). https://doi.org/10.1051/0004-6361/201935883
97. M.S. Schulz, and the Chandra ACIS calibration team, Calibration properties of CHANDRA HETG spectra observed in CC-mode (2014). https://space.mit.edu/CXC/calib/ccmode_final_doc.pdf

98. N.S. Schulz, L. Corrales, C.R. Canizares, ApJ **827**, 49 (2016). https://doi.org/10.3847/0004-637X/827/1/49
99. D.A. Schwartz, RScI **85**, 061101 (2014). https://doi.org/10.1063/1.4881695
100. N. Shaposhnikov, J. Swank, C.R. Shrader, M. Rupen, V. Beckmann, C.B. Markwardt, D.A. Smith, ApJ **655**, 434 (2007). https://doi.org/10.1086/509755
101. J.S. Kaastra, R. Mewe, H. Nieuwenhuijzen, uxsa.conf, 411 (1996)
102. B. Stelzer, J.H.M.M. Schmitt, A&A **418**, 687 (2004). https://doi.org/10.1051/0004-6361:20040041
103. S.J. Tingay, D.L. Jauncey, R.A. Preston, J.E. Reynolds, D.L. Meier, D.W. Murphy, A.K. Tzioumis et al., Nature **374**, 141 (1995). https://doi.org/10.1038/374141a0
104. J. Trümper, AdSpR **3**, 483 (1984). https://doi.org/10.1016/0273-1177(84)90134-0
105. W. Tucker, K. Tucker, rumc.book (2001)
106. T.J. Turner, K. Nandra, I.M. George, A.C. Fabian, K.A. Pounds, ApJ **419**, 127 (1993). https://doi.org/10.1086/173466
107. Y. Ueda, H. Inoue, Y. Tanaka, K. Ebisawa, F. Nagase, T. Kotani, N. Gehrels, ApJ **492**, 782 (1998). https://doi.org/10.1086/305063
108. L.P. van Speybroeck, J. Diab, R.J. Edgar, T.J. Gaetz, P. Zhao, P.B. Reid, SPIE **3113**, 89 (1997). https://doi.org/10.1117/12.278890
109. P.W. Vedder, C.R. Canizares, ApJ **270**, 666 (1983). https://doi.org/10.1086/161157
110. C. Walsh, R.A. Loomis, K.I. Öberg, M. Kama, M.L.R. van 't Hoff, T.J. Millar, Y. Aikawa, et al., ApJL **823**, L10 (2016). https://doi.org/10.3847/2041-8205/823/1/L10
111. M.C. Weisskopf, S.L. O'Dell, SPIE **3113**, 2 (1997). https://doi.org/10.1117/12.278838
112. M.C. Weisskopf, H.D. Tananbaum, L.P. Van Speybroeck, S.L. O'Dell, SPIE **4012**, 2 (2000). https://doi.org/10.1117/12.391545
113. Y. Yao, N.S. Schulz, M.F. Gu, M.A. Nowak, C.R. Canizares, ApJ **696**, 1418 (2009). https://doi.org/10.1088/0004-637X/696/2/1418
114. S.T. Zeegers, E. Costantini, C.P. de Vries, A.G.G.M. Tielens, H. Chihara, F. de Groot, H. Mutschke et al., A&A **599**, A117 (2017). https://doi.org/10.1051/0004-6361/201628507
115. A. Zoghbi, J.M. Miller, A.L. King, M.C. Miller, D. Proga, T. Kallman, A.C. Fabian et al., ApJ **833**, 165 (2016). https://doi.org/10.3847/1538-4357/833/2/165

Chapter 4
Transition Edge Sensors X-ray Spectrometers: Applications for X-ray Astronomy

Hiroki Akamatsu and Jan van der Kuur

4.1 Introduction: Cryogenic X-ray Microcalorimeters

High-resolution X-ray spectroscopy is one of the best tools to understand the physical condition of cosmic plasma such as temperature, ionization state, ion-electron non-equilibrium state and the impacts of extreme environments [for reviews, see Chaps. 7–13 in this book and 9, 18, 24, 34, 58]. Till 2016, High-resolution X-ray spectroscopy was only possible with X-ray grating spectrometers onboard on Chandra (Low Energy Transmission Grating: LETG and High-Energy Transmission Grating: HETG [10]) and XMM-Newton (Reflection Grating Spectrometer: RGS [16]) satellites.

Current X-ray observatories use charge-coupled devices (CCD) for energy dispersive imaging. In these detectors for each absorbed X-ray photon an integer number of charge carriers is created proportional to its energy. Poisson statistics on the number of created charge carriers in the form of electron-hole pairs, fundamentally limits the energy resolving power of this type of detector. The absolute value of the energy resolution scales with the electron-hole pair creation energy, which typically is of the order of a few electron volt, and driven by material properties.

Grating-based spectrometers are used to further increase the energy resolving power beyond this intrinsic limit of CCD pixels, by translating wave length, or equivalently, photon energy information into position information. As a result, one of two geometrical dimensions needed for 2D imaging capability, is traded to obtain

H. Akamatsu (✉) · J. van der Kuur
SRON Netherlands Institute for Space Research, Leiden, The Netherlands
e-mail: h.akamatsu@sron.nl; hirokia@post.kek.jp

J. van der Kuur
e-mail: J.van.der.Kuur@sron.nl

H. Akamatsu
International Center for Quantum-field Measurement Systems for Studies of the Universe and Particles (QUP) at High Energy Accelerator Research Organization (KEK), Tsukuba, Japan

C. Bambi and J. Jiang (eds.), *High-Resolution X-ray Spectroscopy*, Springer Series in Astrophysics and Cosmology, https://doi.org/10.1007/978-981-99-4409-5_4

spectral information with a higher resolution, while reducing the imaging capability of the CCD from 2D imaging to 1D line scanning.

Cryogenic transition edge sensor (TES) based thermal detectors (e.g., TES X-ray microcalorimeters) provide intrinsically a high energy resolving power as a result of their low operating temperature (typically 0.1 K). Because the pixels can be made sufficiently energy dispersive, TES based detector arrays provide a combination of high spectral resolution and imaging capability. Simultaneously, the low operating temperature in combination with the detection principle puts constraints on the size and scalability of the number of pixels. As a result, only narrow field imaging is achievable under the current state of the technology.

The idea of thermal detectors as X-ray spectrometer was proposed 1980's by Moseley, Mather, and McCammon [45]. The main concept of the thermal equilibrium detector has been reviewed in McCammon [43].

As the earth atmosphere is not transparent for X-ray photons from the sky, X-ray spectrometers must be operated outside of the earth atmosphere, i.e. in space. Operation of TES-based detector arrays in space imposes different, and in general more stringent, engineering boundary conditions on the instrument design in comparison with ground-based applications. The most noticeable differences are the lack of accessibility after launch, which drives the need for high reliability designs, components, and redundancy, the radiation environment, and the acoustic and vibrational loads during launch. Furthermore, the maximum weight and power consumption of an instrument are strictly limited, which drives the need to minimize the electrical and thermal power dissipation per pixel.

In this chapter we will describe the state of the technology, the new science windows opened by this detector technology, and an overview of the applicable design considerations imposed by space-based operation. The reviews of semiconductor based X-ray microcalorimeter are available at Chap. 5 "Hitomi/XRISM Micro-Calorimeter" in this book, McCammon 2005 [42] and Porter et al. [49, 50] (Fig. 4.1).

4.2 Transition Edge Sensors

`Transition Edge Sensors (TESs):` TESs are sensitive thermistor by using a shape resistance change across the phase transition of the superconducting material. The critical temperature T_c of superconductor is one of the key parameters of TESs as well as the normal resistance R. Depending on the applications, the T_c is defined by the material used for TESs or a bilayer of superconducting and normal materials to use lateral proximity effect [20, 40, and, references there in]. For X-ray astronomy, typically 100 mK is selected to balance the available cooling power on the orbit and required spectral resolution. For the bolometer applications (e.g., mainly for Cosmic Microwave Background: CMB science), around 300–400 mK is also used because such temperature can be realized by ^{3}He sorption pump coolers rather than dilution coolers or adiabatic demagnetisation refrigerators (Fig. 4.2).

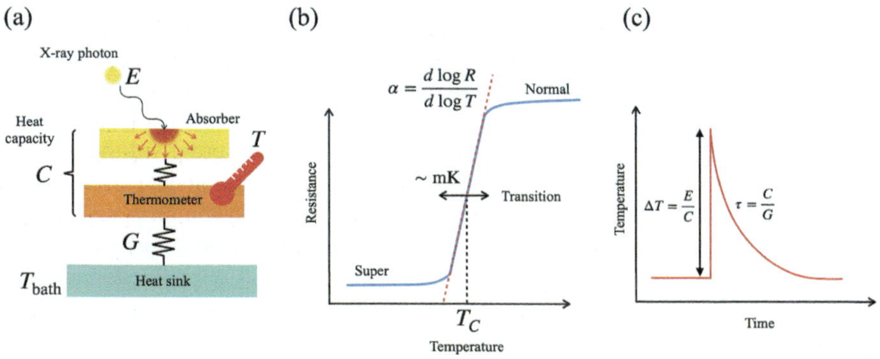

Fig. 4.1 **a** Schematic of transition-edge sensor. An incident X-ray of energy E heats a capacitor consisting of an absorber and a thermometer above the temperature of a heat reservoir. **b** A superconducting resistive transition edge. **c** A pulse in temperature excited by an incident energy deposition. Adapted from Y. Nakashima's Thesis University of Tokyo

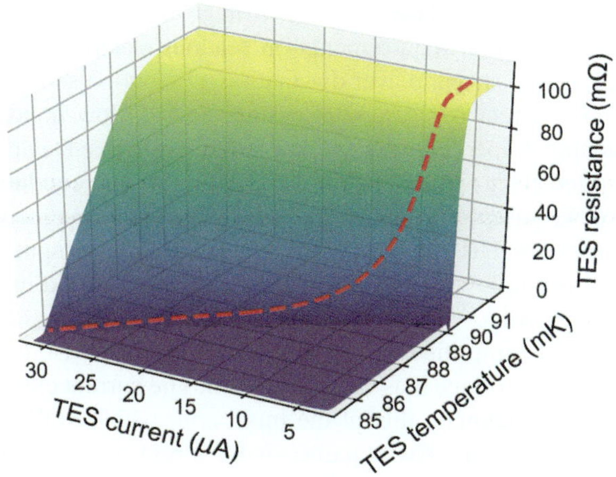

Fig. 4.2 Calculated resistive transition curve $R(T, I)$ for a TES microcalorimeter as described by Kozorezov et al. [38]. The color represents the TES resistance. The dashed red line is the equilibrium curve across the transition following the detector power law. Adapted from [22]

By tuning shape, thickness and material of the absorber, TESs are able to detect a wide range of electromagnetic waves from sub-mm, infrared, optical, UV, X-ray and Gamma-ray wavelengths. For astronomy, TESs are commonly used in Cosmic Microwave Background science [1, 27, 28, 30, 47, 53]. For X-ray astronomy, TES calorimeters are still future instrument in upcoming projects such as Athena/X-IFU [7, 8], LynX [6], LEM [39], HUBS [13] and SuperDIOS [52] although there are several rocket experiments [2]. Limited electrical power availability and mass allocation on the space orbit/satellite environment, the space instrumentation requires more engineering than ground-based instrumentation [57]. An example of allocated power consumption of the Athena X-IFU instrument [7] is about 1.3 kW, which is only a fraction of the typical power consumption of a laboratory-grade pulse tube cooler (6–10 kW). The instrument system needs to be run with such limited electrical power. Details will be discussed in Sect. 4.4.

`Electro-thermal feedback:` Although TESs are powerful thermistors, for a long time it was considered difficult to use them in radiation detectors because of their narrow operational temperature range. In 1995, K. Irwin proposed the use of negative electrical thermal feedback (ETF) [31] in TES based detectors by using voltage bias. In this mode of operation the ETF fixes the TES operating temperature at a value, which is selected by its bias voltage. When the temperature of TESs rises, e.g. due to incident X-rays, the resistance of the TESs rises rapidly. Since TESs is under constant voltage bias, TESs's Joule heat is reduced. This generates the feedback, which is in the direction of preventing a temperature change. A pseudo constant voltage is achieved by inserting a shunt resistor in parallel to the TES with a value that is sufficiently smaller than the resistance of the TESs.

`Readout:` TESs have typically low-resistance spreading from several mΩ to $\sim\Omega$. As mentioned above, TESs require voltage bias for the stable operation. These situations set several requirements on readout electronics: (1) can work at low-temperature, (2) no strong power dissipation, (3) low-noise (lower than TESs's thermal noise) (4) current measurement system and (5) low-input impedance for impedance matching. Superconducting Quantum Interference Device (SQUID) [12] is one of most suited electronics. A SQUID consists of a superconducting ring which is interrupted by two Josephson Junctions, and is sensitive to magnetic flux penetrating the ring. Because SQUID shows a periodic response to the input flux, a feedback via feedback coil is employed to increase response linearity and its dynamic range (Flux Locked Loop: FLL).

Figure 4.3 shows a simplified circuit of TESs calorimeter. Employing a parallel shunt resistance, quasi-voltage bias can be realized. The current change of TESs will generate a change of magnetic flux at the input coil, which SQUID will measure. Even in this simplified circuit, 4 wire pairs (total 8 wires) are required for the readout of the TES X-ray calorimeters.

`Performance:` Here we introduce a simple scaling relationship to get a rough parameter trade-off. With simple assumptions, the thermal energy of the entire element is CT. Since the average energy per phonon is $k_B T$, the number of phonons N is

$$N = \frac{CT}{k_B T}, \tag{4.1}$$

where k_B is the Boltzmann constant. From this, the statistical fluctuation of the phonon number can be

$$\sqrt{N} = \sqrt{\frac{CT}{k_B T}}. \tag{4.2}$$

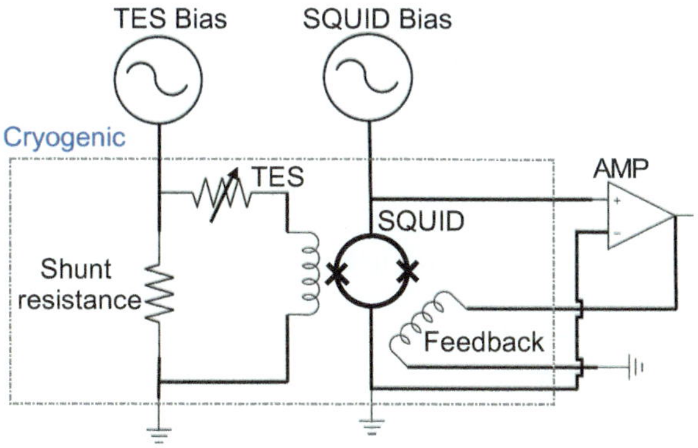

Fig. 4.3 Schematic of the readout of a TES via a single-stage dc-SQUID. Cryogenic components are inside the dashed box. This is the simplest reasonable implementation of (non-multiplexed) readout of a TES, and requires three cryogenic wire pairs per TES (for TES bias, SQUID bias/output, and SQUID feedback). However, typically, a two-stage-SQUID architecture is needed to achieve the required gain without dissipation of too much Joule power on the coldest cryogenic stage; in this case, at least five cryogenic wire pairs per TES are needed. Adapted from [5]

This means the energy fluctuation of the thermal detector due to this phonon number fluctuation is

$$\Delta E = \sqrt{\frac{CT}{k_B T}} k_B T = \sqrt{k_B T^2 C} \tag{4.3}$$

The effect of the increase in fluctuation due to incident X-rays is small, and the statistical fluctuation of the phonon number of the thermal device is dominant. Therefore, in the first-order approximation, the energy resolution does not depend on the incident X-ray energy and depends only on the statistical fluctuation of the phonon number of the device. The spectral resolution is typical defined as a full-width-half-maximum ΔE_{FWHM}:

$$\Delta E_{\mathrm{FWHM}} = 2.35\sqrt{k_B T^2 C} \tag{4.4}$$

Since the thermal equilibrium detector is characterising the temperature difference due to photon hits, sensitive thermistor can improve the performance.

$$\Delta E_{\mathrm{FWHM}} = 2.35\sqrt{\frac{k_B T^2 C}{\alpha}}, \tag{4.5}$$

where α is a temperature sensitivity of the thermistor. Similar to temperature sensitivity α, the thermistor can be current dependent, which is described by β. These two parameters are defined as follows

$$\alpha \equiv \frac{\mathrm{d}\ln R}{\mathrm{d}\ln T} \equiv \frac{T}{R}\frac{\mathrm{d}R}{\mathrm{d}T}, \qquad \beta \equiv \frac{\mathrm{d}\ln R}{\mathrm{d}\ln I} \equiv \frac{I}{R}\frac{\mathrm{d}R}{\mathrm{d}I} \tag{4.6}$$

Adapting typical values, TESs X-ray microcalorimeters could archive great spectral resolution as good as 1 eV

$$\Delta E_{\mathrm{FWHM\ @\ 6\ keV}} \sim 1.3 \left(\frac{|\alpha|}{100}\right)^{0.5} \left(\frac{C}{1\ \mathrm{pJ/K}}\right)^{0.5} \left(\frac{T_C}{100\ \mathrm{mK}}\right) \ [\mathrm{eV}]. \tag{4.7}$$

Although it is a rough scaling relation, it illustrates the potential of TESs calorimeters compared with conventional space X-ray spectrometers (e.g., X-ray CCD camera: $\Delta E_{\mathrm{FWHM}} \sim 160$ eV @ 6 keV). Actual situation will be more complicated due to the impact of the non-linear response of TES calorimeters, readout electronics and the optimisations for the signal multiplexing (see Akamatsu et al. [5] for a review of readout technologies of TES calorimeters).

`Pulse processing`: In TES calorimeters, there are several noise contributions such as thermal fluctuations, resistive Johnson noise and white noise from the readout electronics (see Figs. 2 and 3 in [43]). With these noise contributions, a pulse processing is a key to achieve high-spectral resolution. For the TES calorimeters, optimal filtering is typically applied to extract pulse energy information [3, 19, 55]. In this processing, not only the pulse-peak value but also the complete information of the pulse will be used to estimate the input energy. The optimal filtering process assumes (1) all the pulses have the same shape, (2) the noise component does not change with different energies and, (3) The time domain experimental data $D(t)$ can be expressed as

$$D(t) = H \times S(t) + N(t), \tag{4.8}$$

where H, $S(t)$ and $N(t)$ represent the optimal pulse hight, the pulse shape model, and the noise component i the time domain expression, respectively. Thus the test function to be minimized in the frequency domain expression will be introduced as follows:

$$\chi^2 \equiv \sum \frac{|D(f) - H \times S(f)|^2}{|N(f)|^2}, \tag{4.9}$$

which is in general same approach as the least-squares minimization method.

The optimal pulse estimate can be defined as the deviation of χ^2 to be zero, which is

$$H = \frac{\sum D(f) \frac{S^*(f)}{|N(f)|^2}}{\sum |\frac{S(f)}{N(f)}|^2}, \tag{4.10}$$

where $S^*(f)$ represents the complex conjugate of $S(f)$. Equation 4.10 can be transformed into time-domain expression as

$$H = \sum D(t) \times T(t), \tag{4.11}$$

which is the optimal estimation of the pulse height and a template for the optimal filtering defined as $T(t) \equiv F^{-1}[\frac{S*}{|N(f)|^2}]$. Note that the pre-factor $\sum |\frac{S(f)}{N(f)}|^2$ is dropped

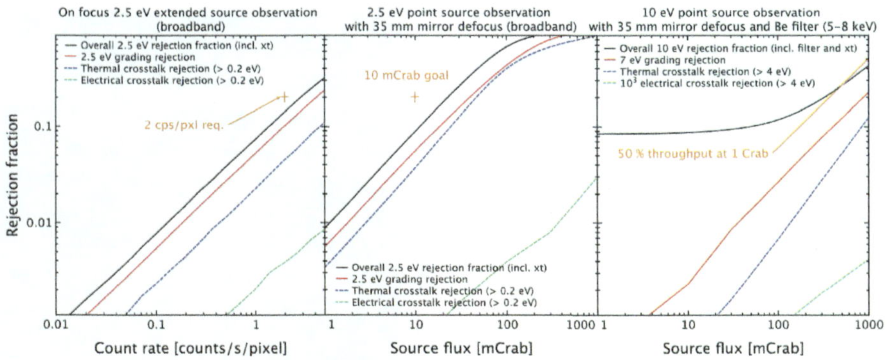

Fig. 4.4 The count-rate capability of the X-IFU instrument from [48]

for the simplicity in the above calculation. F^{-1} represents the inverse-Fourier transformation. For the TES calorimeters, an averaged pulse and averaged noise spectrum density are typically used for $S(t)$ and $N(f)$, respectively. The optimal filtering process assumes (1) The pulse shape does not change with different input energies, (2) the noise does not change with different energies and, (3) the noise is stationary during the pulse.

`Imaging capability:` Another advantage of TESs calorimeters is the imaging capability by making an array of TESs thanks to its non-dispersive property. This is different from other high-resolution spectrometer such as grating X-ray spectrometer, which can be used for point sources and limited diffuse sources with performance degradations. Furthermore, TESs calorimeters can archive high-quantum efficiency (photon collecting power) by tuning material and thickness of the absorber.

`Count-rate capability:` Thermal equilibrium detectors have rather long time constants to return to their original state. To estimate proper input energy, certain amount of quiet time is required before and after the pulse. This means that thermal equilibrium detectors will be suffered by pileups rather quickly. The resulting pileup in the data will slightly degrade the spectral performance. High-count rate also causes the high electrical and thermal cross-talks, which leads worse spectral performance [48] (Fig. 4.4).

In the case of the observation of point sources, the maximum count rate per pixel can be reduced by defocussing the telescope. In this way, the photons are distributed over multiple pixels, so that the count rate per pixel reduces. Figure 4.5 left shows an example of defocusing functionality which will be implemented in Athena X-IFU instrument [35]. By adapting the defocusing functionality, X-IFU instrument is able to handle rather high flux objects (see Fig. 4.5 right)

`Energy dynamic range:` The energy dynamic range of the spectrometer is also an important aspect. For TES calorimeters, the energy dynamic range is defined by the saturation energy (E_{Max}), which can drive TES to the normal state of the thermistor (e.g., no sensitivity). Assuming that the electrothermal feedback is strong, the saturation energy (E_{Max}) can be approximated by

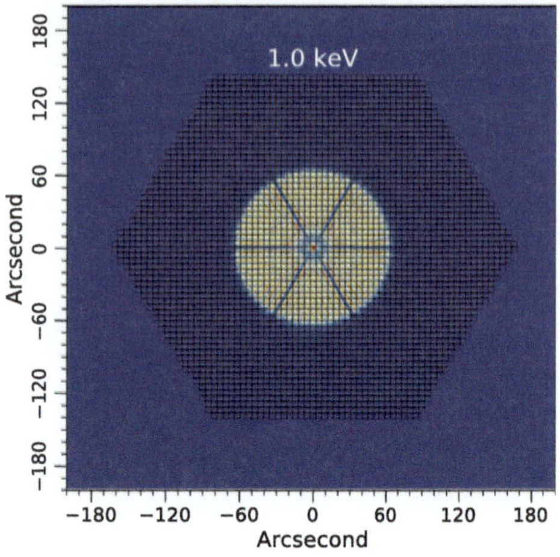

Fig. 4.5 Full X-IFU field of view image with the defocusing functionality at 1 keV. Some detailed features and the spatial variations in the image are due to the size of TES microcalorimeters. From [35]

$$E_{\rm Max} \sim \frac{CT}{\alpha} \sim 6.3 \left(\frac{|\alpha|}{100}\right)^{-1} \left(\frac{\rm C}{\rm 1\ pJ/K}\right) \left(\frac{\rm T}{\rm 100\ mK}\right)\ [\rm keV] \tag{4.12}$$

Above $E_{\rm Max}$, TES calorimeters start reacting with non-linear response due to the nature of the superconducting transition. Additionally, the impact of the electrical circuit will also contribute to the non-linear response (See Fig. 4 in [36] or Fig. 4 in [22]). This does not mean that TES calorimeters are not able to provide photon energy information above $E_{\rm Max}$, it only implies that energy resolution degrades, and pulse processing becomes more complicated.

As shown in Eqs. 4.7 and 4.12, there is a trade-off between the spectral resolution and energy dynamic range (energy range as an instrument). This means that optimal temperature sensitivity α and heat capacity C need to select to satisfy the required spectral resolution and energy bandwidth.

4.3 State-of-Art Performance of TES X-ray Calorimeters

In this section, we introduce state-of-art X-ray spectral resolutions from various groups. CAVITE: Although all groups show the great spectral resolution well below 3 eV at 6 keV, it is not easy to compare each other because the spectral resolution depends on several parameters as described in Sect. 4.2. It is noted that the spectral resolution will only reflect one aspect of the device performance.

In 2010's, there were several progress in the understanding of the physics related to the phase transition of TESs (see Ullom and Bennett [56], Gottardi and Nagayoshi [21], Gottardi and Smith [23] for reviews). As a result, the spectral resolution of TESs calorimeter approaches now to its theoretical limit (e.g. Eq. 4.7) including the

Table 4.1 Summary of the best performance of different groups

Group	NASA/GSFC [54]	SRON [46]	NIST	ISAS/JAXA [59]
TES bilayer material	Mo/Au	Ti/Au	Mo/Au	Ti/Au
Normal registance R_N (mΩ)	8–40	80–400	XX	80–120
Critical temperature T_c (mK)	107	84	XX	110
Heat capacity C at 90 mK (pJ/K)	0.5	0.8	XX	0.4
$\Delta E_{\mathrm{FWHM@6\,keV}}$ (eV)	1.6±0.1	1.5±0.1	2.4	2.8±0.3
Best performance ref	[44]	[15]		[4]

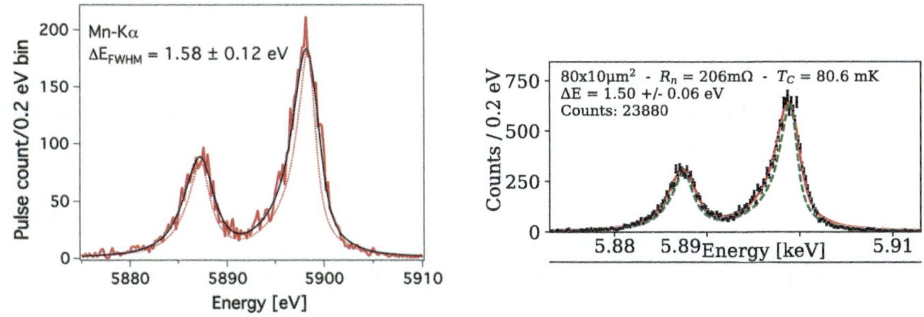

Fig. 4.6 State-of-art spectral resolutions at 6 keV from GSFC/NASA group (left [44]) and SRON (right [15]). The histograms show the Mn-K_α line complex fitted with the Mn-Kα line model [29] convolved with the detector resolution. Note that there is a report on better spectral performance at NASA/GSFC in [23]. The original article is not published yet at the time of writing (as of 2023/04/30). Therefore, we kept the results in the published literatures

impact of the readout scheme. The summary of state-of-art performance of single pixel TESs calorimeters can be found in Table 4.1 and Fig. 4.6.

4.4 Technological Frontiers for Cryogenic Space-Based TES Spectrometers

Application of cryogenic detectors in space-based telescopes opens exciting new science windows. Simultaneously, spaced based operation puts technological boundary conditions on the telescope and on its cryogenic instruments in the focal plane. In this section we discuss the origin and impact of the most significant factors influencing the design and the achievable size cryogenic focal plane instrument.

4.4.1 Cryogenic Detector Operation in Space

Operation of cryogenic detectors in space requires the use of a cryo cooling chain which is suitable, and qualified for space-based operation. Such cooling chains have been developed and are operational, but provide limited cooling power in comparison with what is available in ground-based systems. As a result, the scalability (pixel count) of a cryogenic focal plane is much stronger constrained in a space-based system, in comparison with ground-based systems. The driving reason for this is that the available electrical power in a satellite, which is provided by the available solar panel area, is constrained by mass budgets in combination with boundary conditions on the mechanical stiffness of the construction as a result of the mechanical loads induced by the launch of the spacecraft.

The constraints on the cooling power drive the need to minimise the heat load of the detector system. Heat loads are caused by (electrical) dissipation, thermal radiation, and thermal conduction. This has implications for the design, which will be discussed next. Somewhat arbitrarily, we have chosen to separate the discussion in static heat loads which occur independent of operation of the instrument, and dynamic heat loads which occur only when the detectors are being operated.

4.4.1.1 Passive Heat Loads

When two bodies are at different temperatures, there will be a net non-zero radiative heat load between them following Planck's law, as long as the bodies are in direct sight of each other. To minimise the radiative heat loads in a cryogenic instrument, the exposed area at each temperature level provided by the cooling system, and their absorptivity, must be minimised. This implies a compact design consisting of highly reflective components. The compactness of the design must be properly balanced against conductive heat loads through struts and wiring, as the latter benefit from increasing aspect ratio of the elements, leading to a minimum size, specifically when taking into account minimum stiffness constraints which originate from minimal eigenfrequency constraints following from mechanical launch loads.

The mechanical launch loads, together with the suspended mass drive the required stiffness of the mechanical suspension. The stiffness, in turn, drives the cross section and length of the struts and thereby the conductive heat load. This motivates the need to keep the suspended mass at each temperature level as low as possible, so that a lower stiffness is needed, and therefore a lower thermal conduction is obtained. This applies in general, but specifically at lowest temperatures where cooling power is energetically the most expensive because the thermodynamically limited Carnot efficiency of coolers. The optimal material for the suspension possesses a low specific thermal conductivity, combined with a high strength and stiffness.

Thermal conduction through electrical wiring, which is needed to operate the cryogenic detectors and to transport their science signals, constrains the amount of metallic conductors which can be used to carry the electrical signals between the cryogenic detector stage and electronics which typically needs to be operated at room temperature. The constraint is caused by the fact that the thermal conductance and electrical conductance scale with each other, as described by the Wiedemann–Franz law. As a result, to maximise the amount of pixels in the focal plane of a cooling power limited configuration, the amount of wires per pixel should be minimised. This implies that increasing the multiplexing factor, i.e. the number of independent signals which can be transported through a single wire or amplifier, is attractive, as long as it leads to a lower net heat load at the cryogenic stages.

The electrical signals also need to be shielded against both conductive and radiative electromagnetic interferences, which typically requires the use of low-ohmic metallic shields around the wiring bundles acting as a Faraday cage, with inevitable additional conductive heat loads as a result. The minimal thickness of the metal tube required for shielding, depends on the frequencies at which shielding is needed. This follows from the nature of the skin depth [26], which scales inversely with the square root frequency. This implies that thermally effective shielding can only be obtained when the detector signals, which are typically are located in the audio bandwidth, are shifted to higher frequencies using modulation techniques with a sufficiently high carrier frequency ($>\sim$100 kHz).

Mechanical (micro) vibrations originating from moving parts in the spacecraft, such as compressors of the cooling chain, do excite movements in the mass-spring systems formed by the cryogenic suspensions. Native damping (friction) leads to dissipation in those springs, with a parasitic heat load on the cryogenic stages as a result. Especially beating between similar mechanical excitation frequencies in the system can lead to significant variations in the thermal loads on the system. Therefore careful mechanical engineering is needed to minimise these effects below the acceptable threshold. In general this is done by creating a proper cascade of mechanical impedance mismatches, and by keeping suspended masses as low as possible. Note that a minimal suspended cryogenic mass not only helps to lower the static heat load of the suspension, but also to manage the heat load from micro vibrations.

The last source we need to discuss in the context of static heat loads is cosmic radiation. This radiation, consisting of both particles and photons, deposits energy as a result of interaction of the radiation with the matter of the spacecraft. Again its impact is minimised by keeping the suspended mass as small as possible. Furthermore, choosing materials with a low atomic mass helps to further lower the interaction with cosmic radiation.

4.4.1.2 Active Heat Loads

Active power dissipation at cryogenic temperatures, resulting from currents running through resistive materials, is another contributor to heat loads on the cryogenic detector system. We will discuss the main dissipative sources next.

The operating principle of TES-based detectors involves active power dissipation at a level which is for each pixel equal to the maximum expected thermal load of the sky signal on that pixel. By measuring the dissipative current which is needed to keep the detector pixels at a fixed temperature, an inverse thermal copy of the sky signal is obtained, and the heat load on the cooling system is to the first order kept constant, independent of thermal load induced by the sky signal, under the assumption that the TESs are operated in the (usual) limit of strong electrothermal feedback.

The latter property of TES-based detectors is attractive for astronomy applications as it helps to keep the instrument thermally stable, independent of the thermal signal load from the sky. It also helps to minimise thermal cross talk between adjacent pixels, as the net heat load on the thermal support structure is to the first order independent of the signal load on the pixels.

Another consequence of detector pixel operation in the limit of strong electrothermal feedback is that the detectors do not provide power gain. This results in an output noise temperature of a detector pixel which is approximately equal to its operating temperature, i.e. typically around 100 mK. As room temperature electronics with such low input noise temperatures does not exist, and in order to rise the signal levels above interference levels induced by other equipment in the satellite, an amplification step is needed at cryogenic temperatures. This amplification step implies power dissipation, which in turn loads the cryogenic cooling chain. The amount of dissipation is proportional to the sky signal power and the amount of required power gain. The operating temperature of the amplifier should be as high as possible, within the applicable noise power constraints.

4.4.2 *Detector Readout in the Perspective of Operation in Space*

As discussed in the previous section, power dissipation in the detector readout is inevitable to transport signals to room temperature through wires, and to amplify the detector signals above the noise level of the room temperature electronics, and above both radiative and conductive electromagnetic interference (EMI) levels. Multiplexing techniques are applied to use cables and amplifiers for the readout of multiple pixels, by encoding the signals of the pixels in a single readout channel such that they become (mathematically) independent. With a suitable encoding scheme, the signals can be added without loss of information of the individual pixels.

For the implementation of encoding of TES-based mircro-calorimeter signals, several different schemes have been proposed and implemented [11, 32, 33, 37]. Under all of the schemes, the detector pixel signals are multiplied with carriers which form the orthogonal basis of a subspace of the available multiplex space. The dimensionality of the multiplex space should be al least equal to the sum of the dimensionalities of the subspaces, of which each contains the information of a single pixel. In practice this implies that, to multiplex M pixels, the multiplex bandwidth should be at least M times the bandwidth of a single pixel. The different schemes differ significantly in terms of the technical implementation, most notably in the

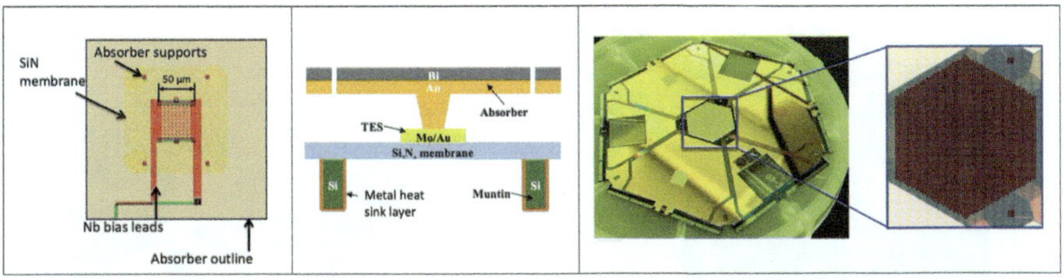

Fig. 4.7 (Left) Top-view of TES microcalorimeter pixel of the Athena X-IFU instrument. (Center) The cross-section view of the TES pixel. Bismuth and gold absorber is connected to Mo/Au TES located on the membrane. Bismuth was used to increase quantum efficiency and keeping heat capacity low. (Right) Photograph of X-IFU prototype TES calorimeter array with more than 3000 pixels. Adapted from [7]

circuit element which is used for the multiplication step, and the orthogonal set of base functions (carriers) which is being used. As mentioned earlier, all of the schemes require extra bandwidth and dynamic range to accommodate the multiplexed signals, as the information content of the pixels is being added. Although the schemes differ in multiplex factor and implementation details, they turn out to be very similar in terms readout dissipation per pixel.

Cosmic radiation in the form of energetic particles also has interaction with X-ray micro-calorimeters gives a background at energies in overlap with the science signal from the sky. To lower this background signal, an anti-coincidence detection scheme is typically applied to be able to make a separation between signals created by the targeted photons, and signals created by energetic background particles. With the scheme the different nature of the interaction between matter and photons or particles is being exploited. The X-ray photons which form the signal from the sky are either completely absorbed, or pass through matter without any interaction with the detector, whereas cosmic particles deposit energy distributed over the full trajectory of its path through the matter of the detector. When a second TES-based detector is placed in close proximity behind the X-ray detector with the same surface area, cosmic particles create a coincident signal in both detectors, whereas X-ray photons do not. By combining the information from both detectors, the particle induced background can be separated from photonic signals (Figs. 4.7 and 4.8).

4.4.3 TES-Based X-ray Space Missions

TES-based micro-calorimeters have for the first time flown in space on the Micro-X sounding rocket [2]. Its focal plane consisted of 128 pixels, with an intrinsic energy resolution of $\Delta E \sim 4.5$eV in the energy rang of interest, i.e. 0.1–2.5 keV. Time domain multiplexed readout, originally developed for ground based applications, was used as multiplex scheme with a multiplex factor of 16. Because the attitude control system of the rocket failed during the observation time of $\sim$5 min, pointing

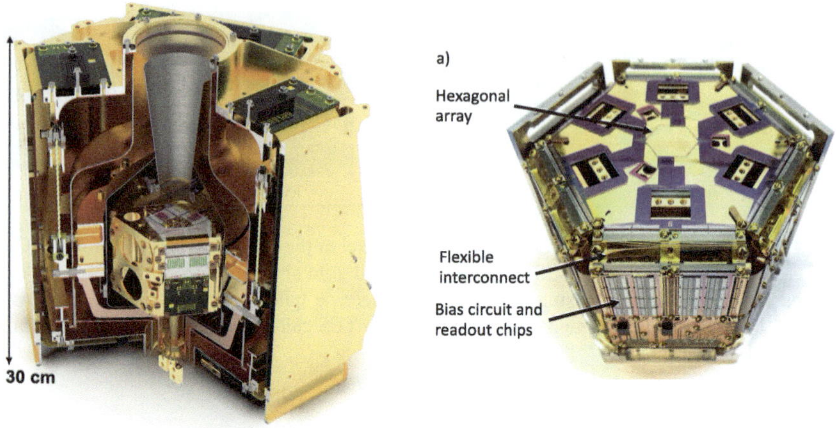

Fig. 4.8 (Left) A demonstration model of the Athena X-IFU instrument was developed by SRON for the temperature, electrical, magnetic field behaviour and the impact of the environmental effect on the performance. (Right) A demonstration model which was developed by NASA/GSFC and NIST for the performance demonstration. Adapted from [5, 22]

of the telescope to the intended target at the sky was not possible. Despite this, it was shown that the system survived the mechanical launch loads well, including its its cooling chain. It also demonstrated that maintaining the detector performance as observed in a laboratory systems in a space-based experiment is not a trivial task. The system was designed to be limited by readout noise from the time domain multiplexer to $\Delta E \sim 7$ eV, and this performance was observed in the laboratory environment for more than 90% of the pixels. However, during flight it was observed that only a selection of the best performing ~5% of the pixels showed $\Delta E \sim 13$ eV, while the rest of the pixels performed (much) worse. The degradation during flight was attributed to a combination of unforeseen electromagnetic interference in the electronics and possibly also from other sources in the sounding rocket, and inadequate magnetic field attenuation by the shield which was not foreseen to become exposed to a slowly varying field as a result of the slow tumble of the payload in the earth magnetic field.

The payload was recovered intact, and it was decided to make another flight incorporating the experiences of the first flight in the design of the payload. In the mean time, the second flight has been performed, but to our knowledge no results have been reported yet.

Several large scale future space missions for X-ray astronomy based on TES technology have been proposed [6–8, 13, 39, 51] over the last years. Although the readout schemes are based on different technologies, the total number of electrically readout pixels is very similar when normalised on the available cooling powers of the different missions. This leads to the observation that for future missions more

V. Capobianco, N. Cardiel, F. Castellani, O. Cheatom, J. Chervenak, F. Chiarello, L. Clerc, N. Clerc, B. Cobo, O. Coeur-Joly, A. Coleiro, S. Colonges, L. Corcione, M. Coriat, A. Coynel, F. Cuttaia, A. D'Ai, F. D'anca, M. Dadina, C. Daniel, L. Dauner, N. DeNigris, J. Dercksen, M. DiPirro, E. Doumayrou, L. Dubbeldam, M. Dupieux, S. Dupourqué, J.L. Durand, D. Eckert, V. Eiriz, E. Ercolani, C. Etcheverry, F. Finkbeiner, M. Fiocchi, H. Fossecave, P. Franssen, M. Frericks, S. Gabici, F. Gant, J.R. Gao, F. Gastaldello, L. Genolet, S. Ghizzardi, M.A.A. Gil, E. Giovannini, O. Godet, J. Gomez-Elvira, R. Gonzalez, M. Gonzalez, L. Gottardi, D. Granat, M. Gros, N. Guignard, P. Hieltjes, A.J. Hurtado, K. Irwin, C. Jacquey, A. Janiuk, J. Jaubert, M. Jiménez, A. Jolly, T. Jourdan, S. Julien, B. Kedziora, A. Korb, I. Kreykenbohm, O. König, M. Langer, P. Laudet, P. Laurent, M. Laurenza, J. Lesrel, S. Ligori, M. Lorenz, A. Luminari, B. Maffei, O. Maisonnave, L. Marelli, D. Massonet, I. Maussang, A.G. Melchor, I. Le Mer, F.J.S. Millan, J.P. Millerioux, T. Mineo, G. Minervini, A. Molin, D. Monestes, N. Montinaro, B. Mot, D. Murat, K. Nagayoshi, Y. Nazé, L. Noguès, D. Pailot, F. Panessa, L. Parodi, P. Petit, E. Piconcelli, C. Pinto, J.M.E. Plaza, B. Plaza, D. Poyatos, T. Prouvé, A. Ptak, S. Puccetti, E. Puccio, P. Ramon, M. Reina, G. Rioland, L. Rodriguez, A. Roig, B. Rollet, M. Roncarelli, G. Roudil, T. Rudnicki, J. Sanisidro, L. Sciortino, V. Silva, M. Sordet, J. Soto-Aguilar, P. Spizzi, C. Surace, M.F. Sánchez, E. Taralli, G. Terrasa, R. Terrier, M. Todaro, P. Ubertini, M. Uslenghi, J.G.B. de Vaate, D. Vaccaro, S. Varisco, P. Varnière, L. Vibert, M. Vidriales, F. Villa, B.M. Vodopivec, A. Volpe, C. de Vries, N. Wakeham, G. Walmsley, M. Wise, M. de Wit, G. Woźniak, The Athena X-ray Integral Field Unit: a consolidated design for the system requirement review of the preliminary definition phase. Exp. Astron. **55**(2), 373–426 (2023). https://doi.org/10.1007/s10686-022-09880-7. arXiv:2208.14562

8. D. Barret, et al., The ATHENA X-ray integral field unit (X-IFU), in *SPIE, Society of Photo-Optical Instrumentation Engineers (SPIE) Conference Series*, vol. 10699 (2018), p. 106991G. https://doi.org/10.1117/12.2312409
9. H. Böhringer, N. Werner, X-ray spectroscopy of galaxy clusters: studying astrophysical processes in the largest celestial laboratories. Astron. Astrophys. Rev. **18**(1–2), 127–196 (2010). https://doi.org/10.1007/s00159-009-0023-3
10. C.R. Canizares, J.E. Davis, D. Dewey, K.A. Flanagan, E.B. Galton, D.P. Huenemoerder, K. Ishibashi, T.H. Markert, H.L. Marshall, M. McGuirk, M.L. Schattenburg, N.S. Schulz, H.I. Smith, M. Wise, The Chandra high-energy transmission grating: design, fabrication, ground calibration, and 5 years in flight. Publ. Astron. Soc. Pacif. **117**(836), 1144–1171 (2005). https://doi.org/10.1086/432898. arXiv:astro-ph/0507035
11. J.A. Chervenak, et al., Superconducting multiplexer for arrays of transition edge sensors. Appl. Phys. Lett. **74**(26), 4043 (1999). https://doi.org/10.1063/1.123255
12. B. Chesca, R. Kleiner, D. Koelle, *SQUID Theory*, Chap. 2 (Wiley, 2004), pp. 29–92. https://doi.org/10.1002/3527603646.ch2
13. W. Cui, et al., HUBS: a dedicated hot circumgalactic medium explorer, in *Society of Photo-Optical Instrumentation Engineers (SPIE) Conference Series*, vol. 11444 (2020), p. 114442S. https://doi.org/10.1117/12.2560871
14. T. de Haan et al., Recent advances in frequency-multiplexed TES readout: vastly reduced parasitics and an increase in multiplexing factor with sub-Kelvin SQUIDs. J. Low Temper. Phys. **199**(3–4), 754–761 (2020). https://doi.org/10.1007/s10909-020-02403-8
15. M. de Wit, L. Gottardi, K. Nagayoshi, H. Akamatsu, M.P. Bruijn, M.L. Ridder, E. Taralli, D. Vaccaro, J.R. Gao, J.W.A. den Herder, Performance of the SRON Ti/Au transition edge sensor x-ray calorimeters, in *Society of Photo-Optical Instrumentation Engineers (SPIE) Conference Series*, vol. 12181 (2022), p. 121812B. https://doi.org/10.1117/12.2629368. arXiv:2208.12556
16. J.W. den Herder, A.C. Brinkman, S.M. Kahn, G. Branduardi-Raymont, K. Thomsen, H. Aarts, M. Audard, J.V. Bixler, A.J. den Boggende, J. Cottam, T. Decker, L. Dubbeldam, C. Erd, H. Goulooze, M. Güdel, P. Guttridge, C.J. Hailey, K.A. Janabi, J.S. Kaastra, P.A.J. de Korte, B.J. van Leeuwen, C. Mauche, A.J. McCalden, R. Mewe, A. Naber, F.B. Paerels, J.R. Peterson, A.P. Rasmussen, K. Rees, I. Sakelliou, M. Sako, J. Spodek, M. Stern, T. Tamura, J. Tandy, C.P. de Vries, S. Welch, A. Zehnder, The Reflection Grating Spectrometer on board XMM-Newton. Astron. & Astrophys. **365**, L7–L17 (2001). https://doi.org/10.1051/0004-6361:20000058

17. M. Dobbs, E. Bissonnette, H. Spieler, Digital frequency domain multiplexer for millimeter-wavelength telescopes. IEEE Trans. Nuclear Sci. **55**(1), 21–26 (2008). https://doi.org/10.1109/TNS.2007.911601
18. Y. Ezoe, T. Ohashi, K. Mitsuda, High-resolution X-ray spectroscopy of astrophysical plasmas with X-ray microcalorimeters. Rev. Modern Plasma Phys. **5**(1), 4 (2021). https://doi.org/10.1007/s41614-021-00052-2
19. J.W. Fowler, B.K. Alpert, W.B. Doriese, Y.I. Joe, G.C. O'Neil, J.N. Ullom, D.S. Swetz, The practice of pulse processing. J. Low Temper. Phys. **184**(1–2), 374–381 (2016). https://doi.org/10.1007/s10909-015-1380-0. arXiv:1511.03950
20. A.A. Golubov, E.P. Houwman, J.G. Gijsbertsen, V.M. Krasnov, J. Flokstra, H. Rogalla, M.Y. Kupriyanov, Proximity effect in superconductor-insulator-superconductor josephson tunnel junctions: theory and experiment. Phys. Rev. B **51**, 1073–1089 (1995). https://doi.org/10.1103/PhysRevB.51.1073
21. L. Gottardi, K. Nagayashi, A review of x-ray microcalorimeters based on superconducting transition edge sensors for astrophysics and particle physics. Appl. Sci. **11**(9) (2021). https://doi.org/10.3390/app11093793. https://www.mdpi.com/2076-3417/11/9/3793
22. L. Gottardi, S. Smith, Transition-Edge Sensors for cryogenic X-ray imaging spectrometers (2022). https://doi.org/10.48550/arXiv.2210.06617. arXiv:2210.06617
23. L. Gottardi, S. Smith, Transition-edge sensors for cryogenic X-ray imaging spectrometers, in *Handbook of X-ray and Gamma-ray Astrophysics. Edited by Cosimo Bambi and Andrea Santangelo* (2022), p. 103. https://doi.org/10.1007/978-981-16-4544-0_22-1
24. L. Gu, I. Zhuravleva, E. Churazov, F. Paerels, J. Kaastra, H. Yamaguchi, X-Ray spectroscopy of galaxy clusters: beyond the CIE modeling. Space Sci. Rev. **214**(7), 108 (2018). https://doi.org/10.1007/s11214-018-0544-z. arXiv:1811.02101
25. M. Hazumi, et al., LiteBIRD satellite: JAXA's new strategic L-class mission for all-sky surveys of cosmic microwave background polarization, in *Society of Photo-Optical Instrumentation Engineers (SPIE) Conference Series, Society of Photo-Optical Instrumentation Engineers (SPIE) Conference Series*, vol. 11443 (2020), p. 114432F. https://doi.org/10.1117/12.2563050
26. O. Heaviside, On the forces, stresses, and fiuxes of energy in the electromagnetic field. Philos. Trans. R. Soc. Lond. Ser. A **183**, 423–480 (1892). https://doi.org/10.1098/rsta.1892.0011
27. S.W. Henderson, Advanced ACTPol cryogenic detector arrays and readout. J. Low Temper. Phys. **184**(3–4), 772–779 (2016)
28. J. Henning, SPT-3G: the third generation camera and survey for the South Pole Telescope, in *American Astronomical Society Meeting Abstracts*, vol. 225 (2015), p. 220.02
29. G. Hölzer, et al., $K\alpha_{1,2}$ and $K\beta_{1,3}$ x-ray emission lines of the 3d transition metals. Phys. Rev. A **56**(6), 4554–4568 (1997). https://doi.org/10.1103/PhysRevA.56.4554
30. Y. Inoue, POLARBEAR-2: an instrument for CMB polarization measurements, in *Millimeter, Submillimeter, and Far-Infrared Detectors and Instrumentation for Astronomy VIII, Society of Photo-Optical Instrumentation Engineers (SPIE) Conference Series*, ed. by W.S. Holland, J. Zmuidzinas, vol. 9914 (2016), p. 99141I. https://doi.org/10.1117/12.2231961. arXiv:1608.03025
31. K.D. Irwin, An application of electrothermal feedback for high resolution cryogenic particle detection. Appl. Phys. Lett. **66**(15), 1998–2000 (1995). https://doi.org/10.1063/1.113674
32. K.D. Irwin, K.W. Lehnert, Microwave SQUID multiplexer. Appl. Phys. Lett. **85**(11), 2107 (2004). https://doi.org/10.1063/1.1791733
33. K.D. Irwin, et al., Time-division SQUID multiplexers, in *Low Temperature Detectors, American Institute of Physics Conference Series*, ed. by F.S. Porter, D.. McCammon, M. Galeazzi, C.K. Stahle, vol. 605 (2002), pp. 301–304. https://doi.org/10.1063/1.1457650
34. J.S. Kaastra, High-resolution X-ray spectroscopy: the coming-of-age. Astronomische Nachrichten **338**(146), 146–152 (2017). https://doi.org/10.1002/asna.201713322. arXiv:1611.05924
35. E.S. Kammoun, D. Barret, P. Peille, R. Willingale, T. Dauser, J. Wilms, M. Guainazzi, J.M. Miller, The defocused observations of bright sources with Athena/X-IFU. Astron. & Astrophys. **664**, A29 (2022). https://doi.org/10.1051/0004-6361/202243606. arXiv:2205.01126

36. C.A. Kilbourne, W.B. Doriese, S.R. Bandler, R.P. Brekosky, A.D. Brown, J.A. Chervenak, M.E. Eckart, F.M. Finkbeiner, G.C. Hilton, K.D. Irwin, N. Iyomoto, R.L. Kelley, F.S. Porter, C.D. Reintsema, S.J. Smith, J.N. Ullom, Multiplexed readout of uniform arrays of TES x-ray microcalorimeters suitable for Constellation-X, in *Space Telescopes and Instrumentation 2008: Ultraviolet to Gamma Ray, Society of Photo-Optical Instrumentation Engineers (SPIE) Conference Series*, ed. by M.J.L. Turner, K.A. Flanagan, vol. 7011 (2008), p. 701104. https://doi.org/10.1117/12.790027
37. M. Kiviranta, et al., SQUID-based readout schemes for microcalorimeter arrays, in *Low Temperature Detectors, American Institute of Physics Conference Series*, ed. by F.S. Porter, D. McCammon, M. Galeazzi, C.K. Stahle, vol, 605 (2002), pp. 295–300. https://doi.org/10.1063/1.1457649
38. A. Kozorezov, A.A. Golubov, D.D.E. Martin, P.A.J. de Korte, M.A. Lindeman, R.A. Hijmering, J. van der Kuur, H.F.C. Hoevers, L. Gottardi, M.Y. Kupriyanov, J.K. Wigmore, Modelling the resistive state in a transition edge sensor. Appl. Phys. Lett. **99**(6), 063503 (2011). https://doi.org/10.1063/1.3621829
39. R. Kraft, M. Markevitch, C. Kilbourne, J.S. Adams, H. Akamatsu, M. Ayromlou, S.R. Bandler, D.A. Bennett, A. Bhardwaj, V. Biffi, D. Bodewits, A. Bogdan, M. Bonamente, S. Borgani, G. Branduardi-Raymont, J.N. Bregman, J.N. Burchett, J. Cann, J. Carter, P. Chakraborty, E. Churazov, R.A. Crain, R. Cumbee, R. Dave, M. DiPirro, K. Dolag, W. Bertrand Doriese, J. Drake, W. Dunn, M. Eckart, D. Eckert, S. Ettori, W. Forman, M. Galeazzi, A. Gall, E. Gatuzz, N. Hell, E. Hodges-Kluck, C. Jackman, A. Jahromi, F. Jennings, C. Jones, P. Kaaret, P.J. Kavanagh, R.L. Kelley, I. Khabibullin, C.G. Kim, D. Koutroumpa, O. Kovacs, K.D. Kuntz, S.C. Lin, E. Lau, S.H. Lee, M. Leutenegger, C. Lisse, L. Lovisari, D. McCammon, S. McEntee, F. Mernier, E.D. Miller, D. Nagai, M. Negro, D. Nelson, J.U. Ness, P. Nulsen, A. Ogorzalek, B.D. Oppenheimer, L. Oskinova, D. Patnaude, R.W. Pfeifle, A. Pillepich, P. Plucinsky, D. Pooley, F.S. Porter, S. Randall, E. Rasia, J. Raymond, M. Ruszkowski, K. Sakai, A. Sarkar, M. Sasaki, K. Sato, G. Schellenberger, J. Schaye, A. Simionescu, S.J. Smith, J.F. Steiner, J. Stern, Y. Su, M. Sun, G. Tremblay, N. Truong, J. Tutt, S. Veilleux, A. Vikhlinin, S. Vladutescu-Zopp, M. Vogelsberger, S.A. Walker, K. Weaver, D.M. Weigt, J. Werk, N. Werner, S.J. Wolk, C. Zhang, W.W. Zhang, I. Zhuravleva, J. ZuHone, Line Emission Mapper (LEM): probing the physics of cosmic ecosystems (2022). https://doi.org/10.48550/arXiv.2211.09827. arXiv:2211.09827
40. C.J. Lambert, R. Raimondi, REVIEW ARTICLE: Phase-coherent transport in hybrid superconducting nanostructures. J. Phys. Conden. Matter **10**(5), 901–941 (1998). https://doi.org/10.1088/0953-8984/10/5/003. arXiv:cond-mat/9708056
41. LiteBIRD Collaboration, E. Allys, K. Arnold, J. Aumont, R. Aurlien, S. Azzoni, C. Baccigalupi, A.J. Banday, R. Banerji, R.B. Barreiro, N. Bartolo, L. Bautista, D. Beck, S. Beckman, M. Bersanelli, F. Boulanger, M. Brilenkov, M. Bucher, E. Calabrese, P. Campeti, A. Carones, F.J. Casas, A. Catalano, V. Chan, K. Cheung, Y. Chinone, S.E. Clark, F. Columbro, G. D'Alessandro, P. de Bernardis, T. de Haan, E. de la Hoz, M. De Petris, S. Della Torre, P. Diego-Palazuelos, M. Dobbs, T. Dotani, J.M. Duval, T. Elleflot, H.K. Eriksen, J. Errard, T. Essinger-Hileman, F. Finelli, R. Flauger, C. Franceschet, U. Fuskeland, M. Galloway, K. Ganga, M. Gerbino, M. Gervasi, R.T. Génova-Santos, T. Ghigna, S. Giardiello, E. Gjerløw, J. Grain, F. Grupp, A. Gruppuso, J.E. Gudmundsson, N.W. Halverson, P. Hargrave, T. Hasebe, M. Hasegawa, M. Hazumi, S. Henrot-Versillé, B. Hensley, L.T. Hergt, D. Herman, E. Hivon, R.A. Hlozek, A.L. Hornsby, Y. Hoshino, J. Hubmayr, K. Ichiki, T. Iida, H. Imada, H. Ishino, G. Jaehnig, N. Katayama, A. Kato, R. Keskitalo, T. Kisner, Y. Kobayashi, A. Kogut, K. Kohri, E. Komatsu, K. Komatsu, K. Konishi, N. Krachmalnicoff, C.L. Kuo, L. Lamagna, M. Lattanzi, A.T. Lee, C. Leloup, F. Levrier, E. Linder, G. Luzzi, J. Macias-Perez, T. Maciaszek, B. Maffei, D. Maino, S. Mandelli, E. Martínez-González, S. Masi, M. Massa, S. Matarrese, F.T. Matsuda, T. Matsumura, L. Mele, M. Migliaccio, Y. Minami, A. Moggi, J. Montgomery, L. Montier, G. Morgante, B. Mot, Y. Nagano, T. Nagasaki, R. Nagata, R. Nakano, T. Namikawa, F. Nati, P. Natoli, S. Nerval, F. Noviello, K. Odagiri, S. Oguri, H. Ohsaki, L. Pagano, A. Paiella, D. Paoletti, A. Passerini, G. Patanchon, F. Piacentini, M. Piat, G. Pisano, G. Polenta, D. Poletti, T. Prouvé, G. Puglisi, D. Rambaud, C. Raum, S. Realini, M. Reinecke, M. Remazeilles, A. Ritacco, G. Roudil,

J.A. Rubino-Martin, M. Russell, H. Sakurai, Y. Sakurai, M. Sasaki, D. Scott, Y. Sekimoto, K. Shinozaki, M. Shiraishi, P. Shirron, G. Signorelli, F. Spinella, S. Stever, R. Stompor, S. Sugiyama, R.M. Sullivan, A. Suzuki, T.L. Svalheim, E. Switzer, R. Takaku, H. Takakura, Y. Takase, A. Tartari, Y. Terao, J. Thermeau, H. Thommesen, K.L. Thompson, M. Tomasi, M. Tominaga, M. Tristram, M. Tsuji, M. Tsujimoto, L. Vacher, P. Vielva, N. Vittorio, W. Wang, K. Watanuki, I.K. Wehus, J. Weller, B. Westbrook, J. Wilms, B. Winter, E.J. Wollack, J. Yumoto, M. Zannoni, Probing Cosmic Inflation with the LiteBIRD Cosmic Microwave Background Polarization Survey (2022). https://doi.org/10.48550/arXiv.2202.02773. arXiv:2202.02773
42. D. McCammon, *Semiconductor Thermistors* (Springer, 2005), pp. 35–62
43. D. McCammon, *Thermal Equilibrium Calorimeters - An Introduction*, vol. 99 (Springer, 2005), p. 1
44. A.R. Miniussi, J.S. Adams, S.R. Bandler, J.A. Chervenak, A.M. Datesman, M.E. Eckart, A.J. Ewin, F.M. Finkbeiner, R.L. Kelley, C.A. Kilbourne, F.S. Porter, J.E. Sadleir, K. Sakai, S.J. Smith, N.A. Wakeham, E.J. Wassell, W. Yoon, Performance of an X-ray microcalorimeter with a 240 μm absorber and a 50 μm TES bilayer. J. Low Temper. Phys. **193**(3–4), 337–343 (2018). https://doi.org/10.1007/s10909-018-1974-4
45. S.H. Moseley, J.C. Mather, D. McCammon, Thermal detectors as x-ray spectrometers. J. Appl. Phys. **56**(5), 1257–1262 (1984). https://doi.org/10.1063/1.334129
46. K. Nagayoshi, et al., Development of a Ti/Au TES microcalorimeter array as a backup sensor for the athena/X-IFU instrument. J. Low Temper. Phys. **199**(3–4), 943–948 (2020). https://doi.org/10.1007/s10909-019-02282-8
47. LiteBIRDCollaborationand others, Probing Cosmic Inflation with the LiteBIRD Cosmic Microwave Background Polarization Survey (2022). arXiv:2202.02773
48. P. Peille, T. Dauser, C. Kirsch, R. den Hartog, E. Cucchetti, J. Wilms, D. Barret, J.W. den Herder, L. Piro, The performance of the athena X-ray integral field unit at very high count rates. J. Low Temper. Phys. **193**(5–6), 940–948 (2018). https://doi.org/10.1007/s10909-018-1964-6
49. F.S. Porter, et al., The astro-H soft X-ray spectrometer (SXS), in *The Thirteenth International Workshop on Low Temperature Detectors - LTD13*, ed. by B. Young, B. Cabrera, A. Miller. American Institute of Physics Conference Series, vol. 1185 (2009), pp. 91–94. https://doi.org/10.1063/1.3292564
50. F.S. Porter, et al., The detector subsystem for the SXS instrument on the ASTRO-H observatory, in *Space Telescopes and Instrumentation 2010: Ultraviolet to Gamma Ray, Society of Photo-Optical Instrumentation Engineers (SPIE) Conference Series*, ed. by M. Arnaud, S.S. Murray, T. Takahashi, vol. 7732 (2010), p. 77323J. https://doi.org/10.1117/12.857888
51. K. Sakai, J.S. Adams, S.R. Bandler, D.A. Bennett, K.D. Irwin, J.A.B. Mates, Development of space-flight compatible room-temperature electronics for the Lynx x-ray microcalorimeter. J. Astron. Telesc. Instrum. Syst. **5**(2), 021013 (2019). https://doi.org/10.1117/1.JATIS.5.2.021013
52. K. Sato, et al., Super DIOS mission for exploring "dark baryon", in *Society of Photo-Optical Instrumentation Engineers (SPIE) Conference Series*, vol. 11444 (2020), p. 114445O. https://doi.org/10.1117/12.2561681
53. N.O. Sehgal, CMB-HD: an ultra-deep, high-resolution millimeter-wave survey over half the sky. Bull. Amer. Astron. Soc. **51**, 6 (2019). arXiv:1906.10134
54. S.J. Smith, et al., Transition-edge sensor pixel parameter design of the microcalorimeter array for the x-ray integral field unit on Athena, in *Space Telescopes and Instrumentation 2016: Ultraviolet to Gamma Ray, Society of Photo-Optical Instrumentation Engineers (SPIE) Conference Series*, ed. by J.W.A. den Herder, T. Takahashi, M. Bautz, vol. 9905 (2016), p. 99052H. https://doi.org/10.1117/12.2231749
55. A.E. Szymkowiak, R.L. Kelley, S.H. Moseley, C.K. Stahle, Signal processing for microcalorimeters. J. Low Temp. Phys. **93**(3–4), 281–285 (1993). https://doi.org/10.1007/BF00693433
56. J.N. Ullom, D.A. Bennett, Review of superconducting transition-edge sensors for x-ray and gamma-ray spectroscopy. Supercond. Sci. Technol. **28**(8), 084003 (2015). https://doi.org/10.1088/0953-2048/28/8/084003

57. J. van der Kuur, et al., Optimising the multiplex factor of the frequency domain multiplexed readout of the TES-based microcalorimeter imaging array for the X-IFU instrument on the Athena x-ray observatory, in *Space Telescopes and Instrumentation 2016: Ultraviolet to Gamma Ray, Society of Photo-Optical Instrumentation Engineers (SPIE) Conference Series*, ed. by J.W.A. den Herder, T. Takahashi, M. Bautz, vol. 9905 (2016), p. 99055R. https://doi.org/10.1117/12.2232830
58. J. Vink, Supernova remnants: the X-ray perspective. Astron Astrophy Rev **20**, 49 (2012). https://doi.org/10.1007/s00159-011-0049-1. arXiv:1112.0576
59. T. Yoshino, K. Mukai, Y. Ezoe, N.Y. Yamasaki, K. Mitsuda, H. Kurabayashi, Y. Ishisaki, T. Takano, R. Maeda, The Noise and Energy Resolution of the Ti/Au Bilayer X-ray TES Calorimeter with an Au Absorber. Journal of Low Temperature Physics **151**(1–2), 185–189 (2008). https://doi.org/10.1007/s10909-007-9617-1

Chapter 5
Hitomi/XRISM Micro-Calorimeter

Kosuke Sato, Yuusuke Uchida, and Kumi Ishikawa

Scope

This chapter describes the micro-calorimeter systems, SXS and *Resolve* onboard ASTRO-H (Hitomi) and X-Ray Imaging and Spectroscopy Mission (XRISM), respectively. The design of these instruments is basically the same with a few changes. This chapter is also intended to help *Resolve* users understand the basic principles and configurations of this instrument for data analysis. This also serves as a gateway to a collection of references for more technical details on specific topics. The structure of this chapter is as follows. An introduction to and motivation for the micro-calorimeter are described in Sect. 5.1. We explain the configuration of the instruments such as the detector, onboard event processing, and cooling systems in Sect. 5.2. The data processing, screening, and calibrations on the ground tests are described in Sect. 5.3. In Sect. 5.4, we report the Hitomi/SXS performance in orbit and the observations for the Perseus cluster. Finally, we describe the performance of XRISM/*Resolve* achieved in the ground test in Sect. 5.5.

Kumi Ishikawa—on behalf of the Resolve team.

K. Sato (✉)
Saitama University, Saitama, Japan
e-mail: ksksato@phy.saitama-u.ac.jp

Y. Uchida
Tokyo University of Science, Tokyo, Japan
e-mail: yuuchida@rs.tus.ac.jp

K. Ishikawa
Tokyo Metropolitan University, Hachioji, Japan
e-mail: kumi@tmu.ac.jp

C. Bambi and J. Jiang (eds.), *High-Resolution X-ray Spectroscopy*, Springer Series in Astrophysics and Cosmology, https://doi.org/10.1007/978-981-99-4409-5_5

5.1 Introduction

X-ray micro-calorimeters detect the energy of incident X-rays by converting the energy of each photon into heat in an absorber with small heat capacity attached to a highly sensitive thermometer [27, 28, 31, 40]. In principle, a high resolution of $E/\Delta E$ >1000 can be obtained, and the high resolution can be achieved in soft X-ray to hard X-ray ranges. X-ray micro-calorimeters are non-dispersive detectors. Compared to dispersive detectors such as grating spectrometers, which provide high energy resolution below about 1 keV, their energy resolution does not deteriorate over a whole energy range and their performance does not degrade for diffuse objects such as clusters of galaxies. Thus, they have the potential to investigate many phenomena in high-energy astrophysics.

X-ray micro-calorimeters are thermal sensors that detect the energy of incident X-ray photons as a small temperature rise in the sensor. An X-ray micro-calorimeter consists of an absorber, thermometer, thermal link, and heat sink as shown in Fig. 5.1. An absorbed X-ray photon transfers energy to a photo-electron that then imparts energy to electrons and phonons in a process called thermalization. Ideally, a quasi-equilibrium state is achieved, and the resulting distribution of energy can be characterized by a temperature. This small temperature change ΔT can be expressed approximately by $\Delta T = E/C$, where the incident energy is E and the heat capacity of the sensor is C. To avoid position-dependent effects, equilibration in the absorber must be faster than the equilibration time of the thermometer with the absorber. The heat generated in the absorber escapes through the thermal link to the heat sink. The temperature returns to the steady state by the time constant of $\tau \sim C/G$, where G is thermal conductance. The energy resolution of the micro-calorimeter is expressed [21] as

$$\Delta E \propto \sqrt{k_\mathrm{B} T^2 C/\alpha},$$

where k_B is the Boltzmann constant, T is a temperature of absorber, and $\alpha = \mathrm{d}\ln R/\mathrm{d}\ln T$ is the logarithmic sensitivity of the thermometer [29]. The equation indicates that ΔE is constant over the energy, in contrast to the grating spectrometer with constant $\Delta\lambda$ over wavelengths.

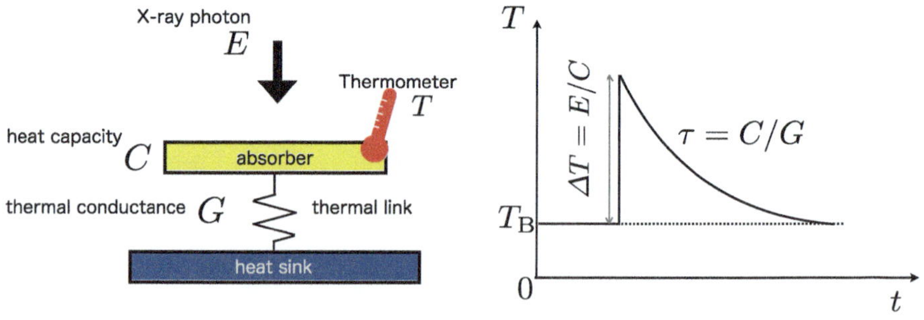

Fig. 5.1 Schematics of micro-calorimeter

Historically, the idea of the X-ray micro-calorimeter originated as a search for a material with semiconductor band-gap smaller than Si to enable higher spectral resolution in semiconductor devices. This led to the identification of candidate materials that are commonly used in infrared bolometers, and this eventually lead to the suggestion [40] of using a bolometer-type detector to thermally detect the energies of individual X-ray photons. This concept was initially proposed for the AXAF mission (now named Chandra) by NASA (National Aeronautics and Space Administration), but restructuring of that program led to the adoption of a micro-calorimeter spectrometer, called the X-ray Spectrometer (XRS), on the ASTRO-E mission [27], which was the launch of ASTRO-E in 2000 by ISAS (Institute of Space and Astronautical Science) and NASA, but it was not put into orbit due to a rocket malfunction. On the other hand, this XRS-type sensor has been used in many ground-based (e.g., [10]) and rocket experiments with success such as the X-Ray Quantum Calorimeter (XQC) [36]. The recovery mission of ASTRO-E, Suzaku [38], was successfully launched, and the in-orbit performance (e.g., the energy resolution of the calibration pixel) basically achieved in the early phase after launch [28]. However, the He cryogen was lost and the operations of XRS were terminated before observing celestial objects.

The ASTRO-H (Hitomi) was launched on February 17, 2016, as the 6th Japanese X-ray astronomy satellite and developed under the international collaboration of JAXA (Japan Aerospace Exploration Agency), NASA, ESA (European Space Agency), and CSA (Canadian Space Agency) [51]. The Hitomi SXS is a system that combines an X-ray micro-calorimeter spectrometer with a Soft X-ray Telescope (SXT) to cover a $3' \times 3'$ field of view (FOV) with an angular resolution of $1.7'$ (half power diameter) [26, 41] as shown in Table 5.1. The micro-calorimeter detector is an array of 6×6 pixels, arranged at an 832 μm pitch [31]. It was operated at 50 mK under a stable environment inside a dewar with a multi-stage cooling system as described in Sect. 5.2.4 in detail. The SXS was designed to achieve an energy resolu-

Table 5.1 Requirements for Hitomi/SXS and XRISM/*Resolve* as shown in [23]. [Reproduced with permission from Ishisaki, Y., et al., Proc. SPIE Int. Soc. Opt. Eng., 12181, 121811S (2022)]

Item	Requirement
Energy resolution (FWHM)	$\leq$7 eV
Energy range	0.3–12 keV
Non-X-ray background per array	$\leq 2 \times 10^{-3} \mathrm{cs}^{-1} \mathrm{keV}^{-1}$
Absolute energy scale	$\leq$2 eV
Field of view	$\geq 2.9' \times 2.9'$
Angular resolution (Half power diameter)	$\leq 1.7'$
Effective area	$\geq 160\,\mathrm{cm}^2$ at 1 keV, $\geq 210\,\mathrm{cm}^2$ at 6 keV
Count rate per array	$\geq 150\,\mathrm{cs}^{-1}$
Absolute time tagging accuracy	$\leq$1 ms
Lifetime	$\geq$ 3 years (nominal), $\geq$9 months (contingency)
Instrument duty cycle	$\geq$90% (helium mode)

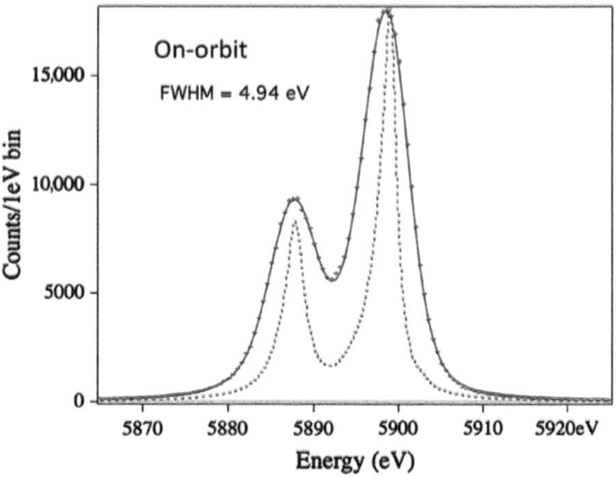

Fig. 5.2 The SXS performance in orbit [44]. The spectra of Mn Kα at 5.9 keV for all the pixels were added together. [Reproduced with permission from Porter, F. S., Boyce, K. R., Chiao, M. P., et al., Journal of Astronomical Telescopes, Instruments, and Systems, 4, 011218 (2018). Copyright 2018 Author(s), licensed under a Creative Commons Attribution 4.0 License]

tion of better than 7 eV (FWHM) at 6 keV and has actually achieved 5 eV (FWHM) at 6 keV in orbit as shown in Figure 5.2. However, the Hitomi had an accident with the attitude control system in March 2016, about a month after the launch. Thus, only the X-ray spectra from a few celestial objects in relatively higher energy bands could be observed because the gate valve on the SXS dewar, which consists of a Be window and its support structure had not been opened.

The design of the *Resolve* onboard XRISM is basically the same as the Hitomi SXS[1] with a few changes. See Sect. 3 in [23] for the details. Figure 5.3 shows a block diagram of the *Resolve*. The dewar for the *Resolve* is located on the base panel, while the X-ray Mirror Assembly (XMA) is mounted on top of the optical bench, as shown in Fig. 5.4 [23].

5.2 Instruments of Hitomi/SXS and XRISM/*Resolve*

This section describes each component in the *Resolve* instrument. The detector system, micro-calorimeter, and anti-coincidence detector are described in Sect. 5.2.1. We explain the signal processing chain to read X-ray events with the analog and digital electronics box in Sect. 5.2.2, and the characteristics, such as the read-out system and energy spectrum, of the micro-calorimeter in Sect. 5.2.3. The cooling chain to cool the detector to 50 mK with He cryogen, mechanical coolers, and adiabatic demagnetization refrigerators (ADR) in the He dewar is shown in Sect. 5.2.4. We describe the optical chain through the X-ray path in the He dewar including filter wheel (FW), aperture assembly (ApA), blocking filters, and gate valve (GV) in Sect. 5.2.5. Section 5.2.6 describes the onboard calibration sources, to correct temporal energy gain variation later on the ground.

[1] https://xrism.isas.jaxa.jp/research/analysis/manuals/xrqr_v1.pdf.

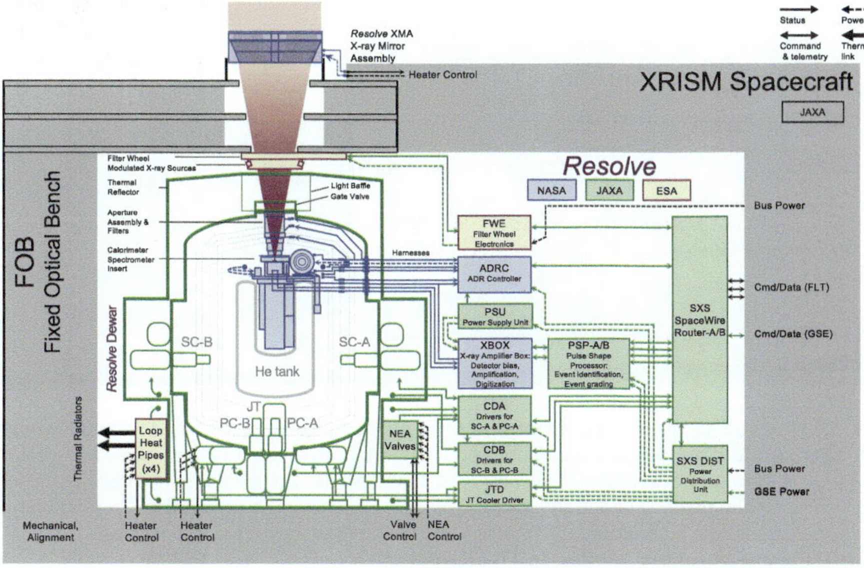

Fig. 5.3 Block diagram of the *Resolve* instrument [23]. Details are found in the text. [Reproduced with permission from Ishisaki, Y., et al., Proc. SPIE Int. Soc. Opt. Eng., 12181, 121811S (2022)]

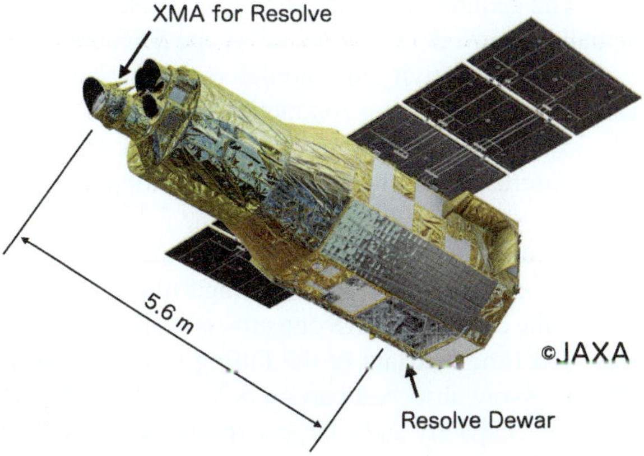

Fig. 5.4 Schematic view of the XRISM spacecraft

5.2.1 Detector System

The SXS and *Resolve* detector assembly is basically the same: a 36-pixel X-ray micro-calorimeter array whose pixels consist of ion-implanted Si thermistors with HgTe absorbers at an 832 μm pitch as shown in the left panel in Fig. 5.5 [11, 31]. One upper left corner of the array, outside the FOV of the X-ray mirror in Fig. 5.5, is a calibration pixel. This pixel is illuminated by a collimated ^{55}Fe source which continuously provides monitoring of the gain scale and the line spread function

Fig. 5.5 (Left): SXS spare array. An array of 6 × 6 pixels comprises a field of view, and the upper left pixel outside of the FOV is a calibration pixel. (Right): Portion of a flight-candidate array before completing the attachment of the X-ray absorbers to the pixels. The thermally isolated thermistors are visible on the pixels without absorbers [31]. [Reproduced with permission from Kilbourne et al., J. Astron. Telesc., Instrum., Syst. 4(1), 011214 (2018). Copyright 2018 Author(s), licensed under a Creative Commons Attribution 4.0 License]

(LSF). The array covers a field of view of $3' \times 3'$ and an energy range of 0.3 – 12 keV. The requirements are shown in Table 5.1, and the energy resolution in orbit actually achieved 5 eV at 6 keV by the SXS onboard Hitomi, as mentioned above.

As high-sensitivity thermometers, semiconductor thermistors that operate in the variable-range-hopping conduction regime have been often used in astrophysics [28, 31, 35]. Particularly, the resistance, R, of an ion-implanted Si thermistor at low temperatures, T, in the variable-range-hopping regime with the Coulomb gap can be expressed as $R(T) = R_0 \exp[(T_0/T)^{0.5}]$, where R_0 and T_0 are constants. Ion-implanted Si often deviates from this equation, but performance can be predicted in the small-signal regime from the logarithmic sensitivity α at the operating temperature. In the case of the SXS detector, α is about -6.3 at the detector temperature under bias. The time constant of the falling time (i.e., the time to return to a steady state) after an X-ray absorbed photon is 3.5 ms [31]. The absorber material should have a small heat capacity and a large stopping power of X-rays to increase the temperature change during energy injection. A semimetal, HgTe, has been used as an absorber as a material that satisfies those conditions since the XRS. The HgTe absorbers are attached with epoxy on each pixel as shown in the right panel in Fig. 5.5. Both the designed quantum efficiency at 6 keV and a filling factor of the SXS were greater than 95%. This type of micro-calorimeter typically has high electrical impedance. Due to the coupling capacitance of the long harnesses, the high impedance makes the system extremely sensitive to microphonics. Thus, a junction field-effect transistor (JFET) source follower circuit has been applied to convert the high impedance of the detector to the low output impedance of the JFET [28].

The SXS and the *Resolve* also have the same anti-coincidence (anti-co) detector as the XRS to exclude cosmic-ray events and to monitor the particle environment

[28, 31]. The cosmic rays hit both the detector pixel and anti-co and hence the background events due to cosmic rays can be removed by anti-coincidence. The silicon ionization detector is used for the anti-co detector, which is located behind the micro-calorimeter detector array. The anti-co detector consists of a 1 cm^2 × 0.5 mm high purity silicon configured as a p-i-n diode, and the anti-co detector covers a larger area than the micro-calorimeter detector. In the case of the SXS, the pulse falling time and the dead time were 0.15 ms and $<$ 1 ms, respectively, and the signal was much faster than the micro-calorimeter signal (3.5 ms).

5.2.2 Event Processing System

Event processing, using a filter to maximize the signal-to-noise ratio, is essential to achieve the high spectral performance of the micro-calorimeters. The optimal filtering [9] is thus adopted. In the optimal filtering, a template is created for each pixel from the responsivity calculated by an average pulse and the noise spectrum. The pulse height is calculated by cross-correlation between the optimal filter template and the waveform of each X-ray event. If another X-ray event (pulse) contaminates the waveform, the pulse height would not be calculated correctly. Therefore, the contamination of another event should be detected accurately. The derivatives of time series data are examined to detect the contamination of another pulse because they are more sensitive than the time series data. It is necessary to discriminate grades according to the time interval to the preceding and following pulses and to change the pulse processing for each grade. The pulse shape processor (PSP) of the SXS and the *Resolve* instruments is designed to implement these functions.

The signals of the micro-calorimeter pixels and the anti-co detector are amplified and digitized by the analog signal processing unit called the X-ray amplifier BOX (XBOX). The digitized data sampled with a sampling rate of 12.5 kHz by the analog-to-digital converter (ADC) in the XBOX, hereafter called `adc_sample`, is transferred to the PSP for event detection and pulse height calculation. The PSP applies a boxcar filter to the `adc_sample` and calculates the `derivative`. Figure 5.6a, b show the `adc_sample` of the pulse shape and the `derivative` [42]. Both the `adc_sample` and the `derivative` are stored in a buffer memory named waveform ring buffer (WFRB) in the PSP [24].

The PSP triggers X-ray events, assigns an event grade, and performs optimal filtering to calculate pulse heights. An X-ray event is triggered when the `derivative` exceeds the event threshold determined for each pixel. Once triggered, the PSP searches for the following pulse by subtracting average `derivative` scaled to the primary pulse, and by checking if the residual exceeds the event threshold again. This process is needed because another pulse, if exists, contaminates the preceding pulse signal to introduce an error in the pulse height calculation. If there is no other pulse within ±884 samples (70.72 ms) with respect to the triggered time (see the top panel of Fig. 5.7), the triggered event is classified as high-resolution primary (Hp) grade. Regarding Hp grade events, the PSP adopts the record of 1024 (81.92

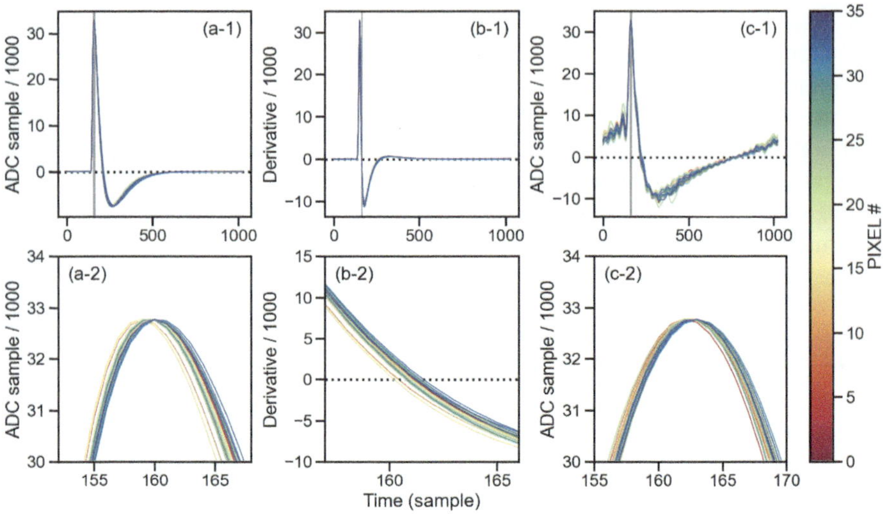

Fig. 5.6 Pulse profiles (**a-1**), profiles of its time derivatives (**b-1**), and the templates of high-resolution grade (Hp) (**c-1**) for each pixel in [42]. Figures **a-2**, **b-2**, and **c-2** show figures zoomed in around the peaking time (sample) of the pulse profiles for **a-1**, **b-1**, and **c-1**, respectively. There are time offsets among pixels, which are compensated by the template for each pixel. The pulse overshoots the baseline as shown in the upper panels due to the AC coupling in the XBOX. [Reproduced with permission from Omama, T., Tsujimoto, M., Sawada, M., et al., Proc. SPIE Int. Soc. Opt. Eng., 12181, 121861 (2022)]

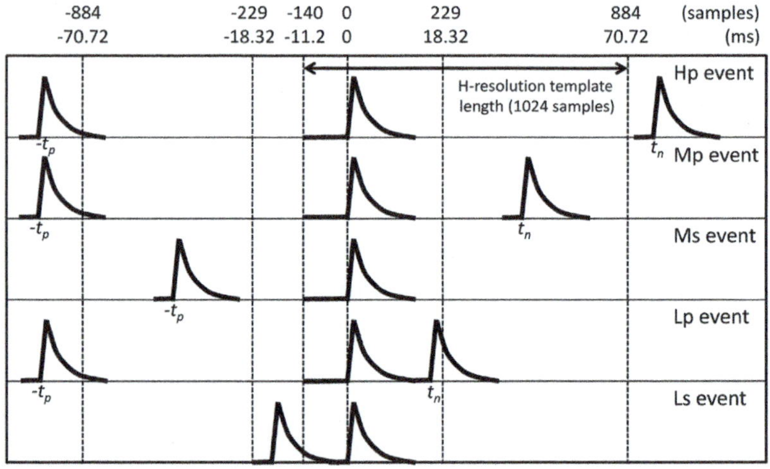

Fig. 5.7 Schematic view of grades for pulse arriving at $t = 0$ [24]. [Reproduced with permission from Ishisaki, Y., et al., Journal of Astronomical Telescopes, Instruments, and Systems, 4(1), 011217 (2018). Copyright 2018 Author(s), licensed under a Creative Commons Attribution 4.0 License]

ms) including 140 preceding samples (see Fig. 5.6a-1), and calculates the cross-correlation between this pulse record and the optimal filter template of the same length (Fig. 5.6c-1). The maximum of the cross-correlation gives an estimate of the pulse height. The PSP searches for the maximum cross-correlation by shifting the pulse record. The final pulse height amplitude (PHA) and the photon arrival time are obtained at a sub-integer sample (1/16, which corresponds to 5 μs resolution) of the sampling frequency, by using the cross-correlation values at −1, 0, and +1 shifts with respect to the shift that gives the maximum, and assuming the cross-correlation is locally represented by the second order polynomial of time near the maximum. See [9, 24] for further details on optimal filter processing.

If there is no preceding pulse from the −884 sample, and if there is a second pulse within the following 884 samples, but not within the following 229 samples (18.32 ms, see the second panel of Fig. 5.7), the triggered event is classified as medium-resolution primary (Mp) grade. As shown in Fig. 5.7, the time intervals to the preceding pulse and to the following pulse determine the grade. Regarding the event of Mp or medium-resolution secondary (Ms) grade, the PSP adopts the record of 256 samples (20.48 ms) and determines its pulse height using the short template.

In a similar way, using the interval of ±229 samples (18.32 ms), the triggered event is classified as low-resolution primary (Lp) grade or low-resolution secondary (Ls) grade. The PSP does not apply the optimal filter to the events of Lp or Ls grades but calculates the pulse height by subtracting the baseline value from the maximum value of the `adc_sample`.

Table 5.2 summarizes the relationship between the event grade and the time interval of the two pulses. See [24] for further details on the PSP implementation. The energy resolution of the Mp events is close to that of the Hp events, and hence, Hp and Mp events will be used in the standard data analysis.

The PSP has the capability to generate the templates onboard, as well as to upload pre-calculated templates on the ground. Templates that are used with the *Resolve* PSP in the spacecraft-level test, as well as the launch time, are those generated on the ground using the instrument-level test data (version of 2021 Dec 13), where the flight dewar and the flight electronics were used. High-frequency weight was removed for the flight templates with a cut-off frequency 366 Hz because otherwise, the instrument could be vulnerable to increases in high-frequency noise that might occur after the optimal filter was calculated (for details, see Sect. 5.5 in [23]).

Table 5.2 The event grade definition [24]. t_n is the interval to the next pulse, while t_p is the interval to the previous pulse. [Reproduced with permission from Ishisaki, Y., et al., Journal of Astronomical Telescopes, Instruments, and Systems, 4(1), 011217 (2018). Copyright 2018 Author(s), licensed under a Creative Commons Attribution 4.0 License]

	$t_p \leq 18.32$	$18.32 < t_p \leq 70.72$	$70.72 < t_p$
$t_n \leq 18.32$	Ls	Ls	Lp
$18.32 < t_n \leq 70.72$	Ls	Ms	Mp
$70.72 < t_n$	Ls	Ms	Hp

5.2.3 Characteristics of the Micro-Calorimeters

In this section, we describe the evaluation of an absolute energy scale and energy resolution of incident X-ray photons to characterize the micro-calorimeter array. We also explain the effects of noise in the readout system and high-count observations, which affect the evaluation of these energy scales and energy resolutions.

The energy scale, i.e., the relation between the measured pulse height and the incident photon energy, is a key parameter to calculate accurate absolute measurements of photon energy. The SXS and the *Resolve* employ very sensitive thermometers and the energy scale is affected by these instruments' thermal environments. Different detector temperatures and radiative loads lead to changes in the energy scale. Therefore, it is necessary to remove temporal variations in the gain. The SXS and the *Resolve* detector energy scale consists of two components: 1. all the pixels have a common gain that varies together with the detector's effective temperature which depends on the calorimeter thermal sink (CTS) temperature, 2. a differential gain among the pixels due to large temperature excursions or changes in radiative loading [13, 34, 44]. Details of the gain correction are described in Sect. 5.3.5.

The high energy resolution is a fundamental science requirement of the SXS and the *Resolve* and is reflected in the LSF. The LSF for each pixel across the science bandpass was characterized by nearly monochromatic X-rays generated by several monochromators. An example of the LSF is shown in Fig. 5.8. The LSF is described with two components; a Gaussian core and an extended LSF which includes an exponential tail with e-folding of ∼12 eV, an electron loss continuum,

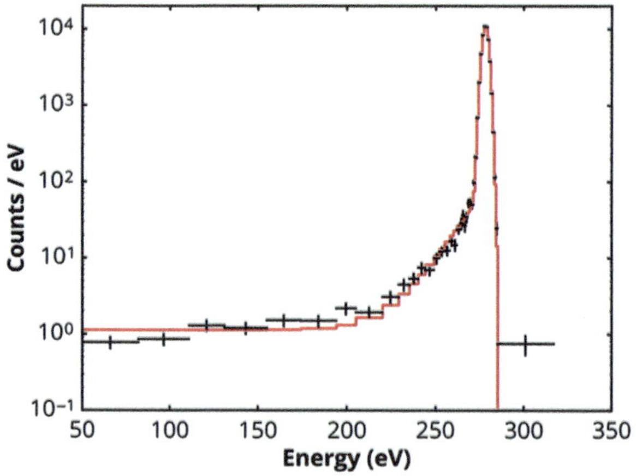

Fig. 5.8 The line-spread function (LSF) for a single SXS pixel in response to monochromatic X-rays at 282 eV [13]. The black points show data and the red curve shows the LSF model, illustrating three components: the Gaussian core which dominates the line shape, an exponential tail to low energies, and the electron loss continuum. [Reproduced with permission from Eckart, M. E., Adams, J. S., Boyce, K. R., et al., Journal of Astronomical Telescopes, Instruments, and Systems, 4, 021406 (2018). Copyright 2018 Author(s), licensed under a Creative Commons Attribution 4.0 License]

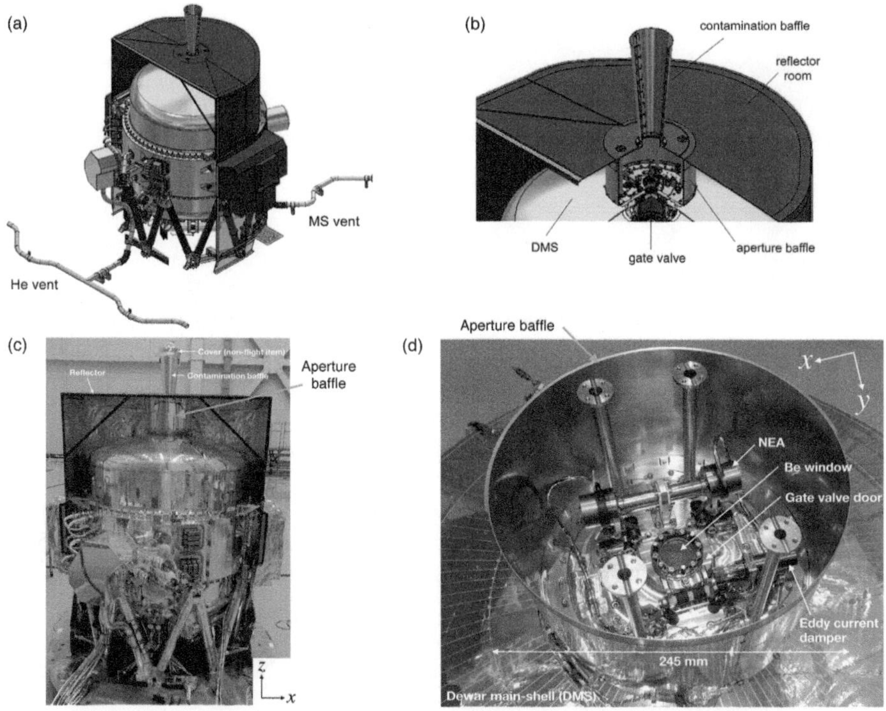

Fig. 5.12 Drawings of **a** the *Resolve* dewar and **b** the dewar topside [15]. Pictures of **c** the *Resolve* flight dewar and **d** inside the aperture baffle [23]. [Reproduced with permissions from Ezoe, Y., Ishisaki, Y., Fujimoto, R., et al., Cryogenics, 108, 103016 (2020), and Ishisaki, Y., et al., Proc. SPIE Int. Soc. Opt. Eng., 12181, 121811S (2022)]

5.2.5 Optical Chain

X-rays collected by the mirror assembly go through some components. First, X-rays pass the FW to control the incident X-ray, then they reach the detector by passing through a GV, an ApA, and several filters.

As shown in Fig. 5.13, the FW has six positions. It rotates to set a filter position that a user selects. The positions are assigned to "Open", "Be" filter, "ND" filter, "Polyimide (OBF)" filter, and "^{55}Fe". The two positions of "Open", which mean blank, are our default position. "Be" filter consists of beryllium of a thickness of 27 μm and blocks the lower energy X-rays. The neutral density (ND) filter is also blocking X-rays for bright objects. "ND" filter is made of a molybdenum sheet whose thickness is 0.25 mm with a lot of 1.1 mm diameter holes. The opening ratio is computed as about 24.5%. The filter reduced the photons for the whole energy band. "Polyimide" filter is utilized to block contamination of the optical light. For calibrations of the energy gain scale of the detector array, ^{55}Fe sources are installed in the position "Fe-55".

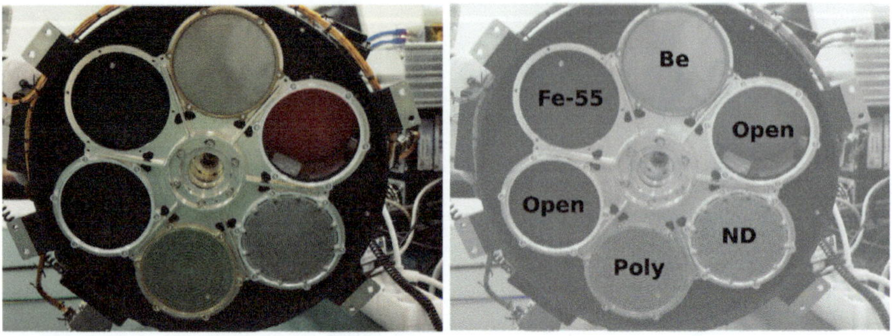

Fig. 5.13 Pictures of the SXS FW [12]. [Reproduced with permission from de Vries, C. P., Haas, D., Yamasaki, N. Y., et al., Journal of Astronomical Telescopes, Instruments, and Systems, 4, 011204 (2018). Copyright 2018 Author(s), licensed under a Creative Commons Attribution 4.0 License]

The primary purpose of these filters is to reduce the photon count rate in an appropriate dynamic range of the intensity of bright sources. As described in Sect. 5.2.3, there is a limitation for processing the signals so-called PSP limit. In addition, the incoming photon rate changes the event grade which is related to energy resolution. Therefore, it is needed to suppress the number of incoming photons to the detector array. Guest observers of XRISM can choose two filters, "ND" and "Be" for blocking X-rays, in place of the "Open" position.

The GV is attached to the DMS at the X-ray aperture to seal the dewar guard vacuum on the ground. The GV is planned to be kept closed for about one month after launch to avoid contamination on the DMS filter which is one of five thin filters installed in the X-ray incident path to prevent optical loading by photons larger than X-rays. For the ground test and observations in orbit prior to opening the GV, a beryllium window [37] with $\sim 270\ \mu$m thickness is installed in the GV door.

The SXS and the *Resolve* need to operate under the stable environment of 50 mK in orbit in order to keep the high energy resolution [30]. Thus, the ApA protects the coldest stage from the thermal radiation in the higher-temperature stages in the dewar and also prevents radiation from the outside. The ApA provides five thin-film radiation-blocking filters for the thermal radiation from the equipment, and the optical/UV photons from the sky. The ApA also includes the baffle and their support structures as shown in Fig. 5.14.

Since the area around the GV was open to space in the SXS design, there was a risk of damaging the DMS filter, the outermost filter, due to micro-meteoroid and orbital debris (MMOD) after opening the GV. If MMOD hits the DMS filter and makes a hole, radiation would penetrate inside and degrade the detector's performance. Therefore, a cylindrical aperture baffle made of aluminum is newly introduced around the GV area, which can reduce the probability of MMOD hit to the DMS filter to <0.03 per year at an altitude of 570 km [23]. The aperture baffle can also protect the DMS filter from the Earth albedo (optical light leak) and atomic oxygen other than MMOD.

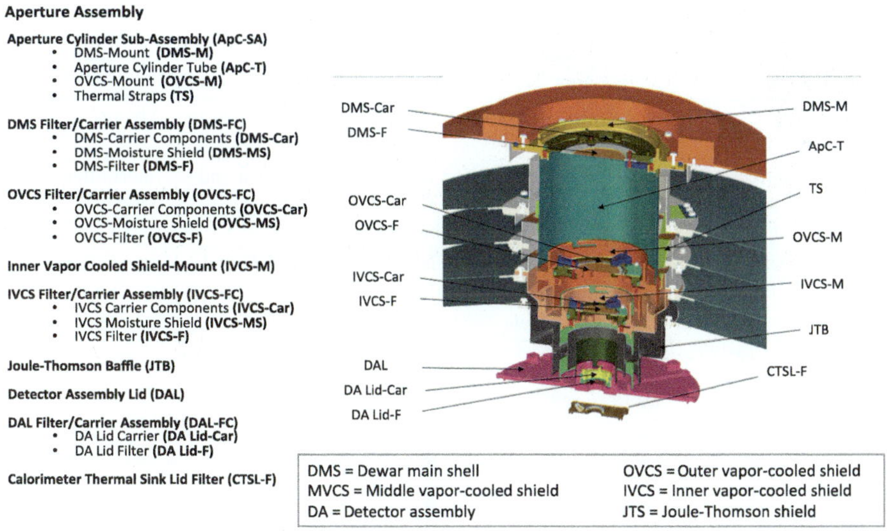

Fig. 5.14 Schematic view of the SXS aperture assembly and blocking filters [30]. [Reproduced with permission from Kilbourne, C. A., Adams, J. S., Arsenovic, P., et al., Journal of Astronomical Telescopes, Instruments, and Systems, 4, 011215 (2018). Copyright 2018 Author(s), licensed under a Creative Commons Attribution 4.0 License]

Figure 5.12b–d show drawings of the *Resolve* dewar top and the aperture baffle, and pictures of the *Resolve* flight dewar and the inside of the aperture baffle, respectively [15, 23].

5.2.6 Calibration Sources

In micro-calorimeter spectrometers, the calibration of the detector system is important because it is the characteristics of the detector response that provide the spectral information [13, 34, 44]. Particularly, the energy gain scale and LSF are essential calibration items to represent the spectral feature of the incident X-ray photons for astrophysical investigations. Therefore, each pixel required its own gain scale calibration and the gain also needs to be characterized independently for each event grade.

The SXS and the *Resolve* detector system has three types of calibration sources as shown in Table 5.3. First, a calibration pixel is mounted as described in Sect. 5.2.1, which is continually illuminated by a collimated ^{55}Fe radioactive source. The calibration pixel allows for continuous monitoring of short-term gain drift and for correcting common drift among the pixels. Second, ^{55}Fe source installed in the FW can illuminate the whole calorimeter array. Third, the SXS and the *Resolve* adopt a modulated X-ray source (MXS) [12, 13]. MXS can generate characteristic X-rays using pho-

Table 5.3 Calibration sources

Source Type	Irradiated area	Line	Operation
^{55}Fe source	Pixel 12	Decay by electron capture to Mn	Always
"Direct" sources of MXS	All pixels	Fluorescent lines of Cu and Cr, and bremsstrahlung radiation	Intermittent operation
"Indirect" sources of MXS	All pixels	Fluorescent lines of Al and Mg	Intermittent operation
Fe-55 filter	All pixels	Fluorescent lines of Mn	Supplemental use

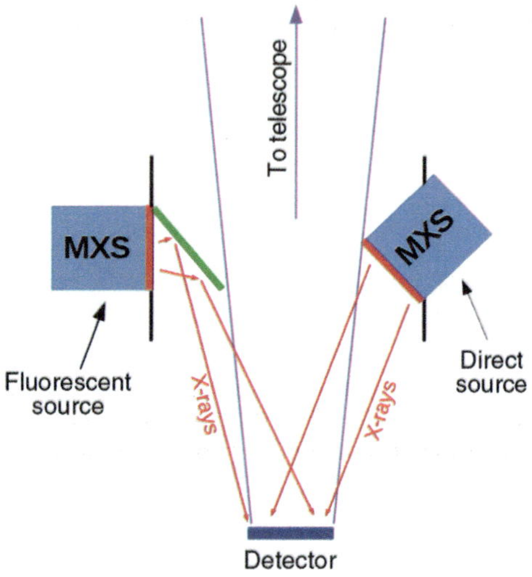

Fig. 5.15 Schematic view of MXS [12]. Characteristic X-ray lines of Cu and Cr and a continuum of the bremsstrahlung radiation are emitted from the direct source. On the other hand, X-rays from the indirect source include Al and Mg fluorescent lines. [Reproduced with permission from de Vries, C. P., Haas, D., Yamasaki, N. Y., et al., Journal of Astronomical Telescopes, Instruments, and Systems, 4, 011204 (2018). Copyright 2018 Author(s), licensed under a Creative Commons Attribution 4.0 License]

toelectrons, which are created by UV lights from an LED and accelerated to targets with a high voltage, with an adjustable duty cycle [47].

The MXS consists of two sources, called direct and indirect sources (see Fig. 5.15). Each source has a redundant system: nominal (NOM) and redundant (RED) sides. The X-rays from direct and indirect MXS exhibit different line spectra. The direct source generates Cu Kα and Cr Kα as well as the bremsstrahlung continuum, which provides the calibration references in the 5–8 keV band. The indirect source illuminates a secondary target to generate Mg Kα and Al Kα fluorescence line providing the references to trace occasionally low-energy gain scale in the 1–2 keV band [47].

The MXS irradiation can cover all pixels and pulsed X-ray emission from the MXS can be performed by switching the LED periodically. During observations of celestial targets, correction of the gain scale can be carried out using the MXS lines. X-ray events caused by the MXS can be separated from the X-rays of the celestial target using the time information of the MXS pulses.

5.3 Data Processing and Calibrations

5.3.1 Offline Processing of XRISM/Resolve

XRISM data analysis starts by converting raw spacecraft packet telemetry received on the ground into the standard FITS format. This conversion of space packets to FITS format, along with the assignment of time, is performed in the pre-pipeline processing stage. This is followed by the pipeline processing stage where the data are calibrated and screened.

The pre-pipeline (PPL) represents the first step of the processing [54]. Raw telemetry data are divided into First Fits Files (FFFs) that organize the data for distribution into the unfiltered event (uf). For the *Resolve*, the PPL (1) creates housekeeping (HK), calorimeter pixel event and anti-co event FFFs, (2) assigns time (with no barycentric correction), (3) divides files by filter wheel setting, and (4) constructs good time interval (GTI) files for excluding the lost event duration. The *Resolve* HK data such as temperatures, currents, and voltages are stored in the HK1 FITS file. The micro-calorimeter pixel event (but not the anti-co) data are divided into separate satellite slew and pointing files. Additional HK data including the detector templates, average pulses, and noise spectra at 1 k and 8 k resolutions are stored in the HK2 FITS file.

In the PPL, there are also common processes for all the instruments. One of these is the creation of the observation mode GTI. This divides the data into the time intervals before and after the attitude stabilization for each maneuver to an observation target, which are called slew and pointing durations, respectively. For the *Resolve*, the micro-calorimeter pixel event data are divided into slew and pointing in the PPL, while the anti-co event is not divided.

The pipeline (PL) processing applies event calibration and screening to FFFs, merging a number of GTI files for the latter purpose. Some of these, such as the GTI used to exclude times of ADR recycling are created in the PL. Users may fully reprocess an entire sequence of XRISM data using the HEASoft ftools, `xapipeline` or `rslpipeline`. The `rslpipeline` ftools applies all individual *Resolve* calibration and screening steps in the proper order and creates preview products. Reprocessing is recommended whenever a calibration or software update occurs, or when any non-standard processing is called for. The calibration steps include the assignment of coordinates (using the ftools `coordevt`) and energy (`rslpha2pi`), as well as the calculation of the gain history (see Sect. 5.3.5 below).

5.3.2 Calibration Files

In order to derive accurate information from observation data of celestial objects of interest, the response of the instrument at the time of observation must be accounted for. The XRISM team (and the Hitomi team[2]) provides this in the form of calibration files that are used directly by the data processing tools.

5.3.3 Pixel Event Data

After PL processing, the *Resolve* event FITS file columns for time, gain calibrated invariant pulse height, event grade, event status, lost event indicators, and so on are filled. Energy gain for each pixel is corrected based on gain histories created by the `rslgain` tool. To avoid contamination of photons from celestial objects, it is important that flags recorded in an event file indicate coincidence with the anti-co detector, the MXS pulse timing, and cross-talk events. The column bits of the event `STATUS` column, populated by the `rslflagpix` tool, record these flags as indicated in Table 5.4. `STATUS` bits are set by `rslflagpix` based on temporal coincidence within timescales defined in a calibration file (by default), or set by the user. Two types of electrical cross-talk are flagged based on the relative timing of events in other pixels with wiring proximity by applying relatively long and short intervals.

5.3.4 Time Assignment

The time of the micro-calorimeter events and anti-co events is first assigned onboard by the PSP based on the event trigger times. For X-ray grades of the micro-calorimeter pixel events (H, Mp, Ms, Lp, and Ls), the trigger times are further converted into photon-arrival times (calibrated event times) in the offline PPL process using the timing coefficients stored in a designated calibration file with the `rslsamcnt` and `xatime` tasks. There is no such offline event time conversion for the anti-co events, but the relative offset from the micro-calorimeter pixel events is calibrated and stored in the same calibration file. It is taken into account in the event flagging processes based on time-coincidence (see Sect. 5.3.3). The purpose of the timing calibration can be divided into two: one is for the relative timing accuracy to perform the time-coincidence-based event flagging and screening and the other is for the absolute timing accuracy (requirement of 1 ms; Table 5.1) desired from the scientific needs for time-domain astrophysical data analysis.

---

[2] Calibration files for Hitomi are available on the web page (https://heasarc.gsfc.nasa.gov/docs/hitomi/calib/). There are detailed description documents for each calibration file.

Table 5.4 List of `STATUS` bit[a]

`STATUS` bit	Description
0	Within general GTI
1	Within individual GTI for each PIXEL
2	Coincidence with anti-coincidence detector signal
3	Temporal proximity with another pixel
4	Coincidence with calibration pixel (12) within temporal proximity
5	"cross-talk"; coincidence with calibration pixel (12), and the energy is recoil energy
6	"cross-talk"; coincidence with adjacent wire pixel within short time scale
7	"cross-talk"; the pixel has maximum PHA in the events group within a short time scale
8	"MXS"; within MXS GTI (direct mode, including afterglow)
9	"MXS"; within MXS GTI (direct mode, considering only afterglow)
10	"MXS"; within MXS GTI (indirect mode, including afterglow)
11	"MXS"; within MXS GTI (indirect mode, considering only afterglow)
12	"cross-talk"; coincidence with adjacent wire pixel within long time scale
13	"cross-talk"; the pixel has maximum PHA in the events group within a long time scale
14	Reserved
15	Reserved

[a]This table is written in fhelp of `rslflagpix`

In the onboard time assignment, the trigger time for the H and M grades is set at which the time derivative of a pulse record exceeds a threshold, while for the L grade, it is set at which the pulse derivative reaches its maximum. This causes two major offsets in the trigger times: (1) a systematic delay for the L with respect to the H and M due to the different definitions of the trigger point and (2) an earlier trigger for higher-energy events for the H and M due to applying a uniform threshold for any X-ray energies. The latter is particularly large for low energy events below $\sim$1 keV because the trigger point is closer to the maximum of the pulse derivative rather than its onset. This large energy dependence is mostly corrected in the onboard optimal filter processing. However, there is still a residual, smaller energy dependence caused by the non-linearity of the detector pulse. An actual detector pulse rises slightly faster for higher energy X-ray events, but the optimal filter uses a template evaluated at single energy (5.9 keV for the SXS and the *Resolve*). Thus, (2'-a) the onboard optimal filter overcorrects (undercorrects) the energy dependence for higher (lower) energy events, predicting slightly earlier (later) trigger times. For the L, although the time at the derivative maximum does not depend on much, (2'-b) there is also a small energy dependence on this trigger time. The relative offsets (1), (2'-a), and (2'-b) require us to perform the relative timing calibration. In addition, (3) there is also a common delay to all the grades and energies originating from various factors such as a group delay in the analog detector pulse processing at the XBOX, which needs to be corrected as well to achieve the required absolute timing accuracy.

In the offline process, the relative and absolute timing offsets described above are corrected using the following equation:

$$\begin{aligned}&[\text{Calibrated Arrival Time}]\\&= [\text{Trigger Time}] - (a \times (0.25 \times \texttt{RISE\_TIME}) + b \times \texttt{DERIV\_MAX} + c)\,,\end{aligned}$$

where RISE_TIME is the time separation from the (original) trigger time to the time at which the pulse derivative reaches the maximum and DERIV_MAX is the derivative value at the maximum. The coefficients a, b, and c are defined separately for the different event grades and the micro-calorimeter pixels and stored in a designated calibration file. Both RISE_TIME and DERIV_MAX can be used as a measurement of the X-ray energy. Thus, the coefficients a and b can be used to correct the residual energy dependence (2'-a) and (2'-b). Because of the overall pulse linearity (except a small non-linearity mentioned above), RISE_TIME is expected to change nearly proportional to DERIV_MAX, and using either one of the coefficients a and b is sufficient to correct the residual energy dependence. For the SXS and the Resolve, only the coefficient b is used and the other one is kept at 0. The other relative offset, (1), is corrected by having different values of the c coefficient for the different grades. The absolute offset, (3), is corrected by having a common offset in c for all the grades and pixels. The above formalization of the timing conversion and coefficients also allows us to calibrate possible pixel-by-pixel offsets, although this type of offset should be negligibly small if changes in the trigger threshold between the template generation and actual observations are common to all the pixels.

5.3.5 Gain Correction and Line Spread Function

The energy gain scale and LSF are calibrated on the ground and in orbit [13, 34]. The calibration file for energy gain scale correction is prepared for three operation temperatures, 49 mK, 50 mK and 51 mK, for each channel based on the calibration on the ground. Thus this indicates that the gain is different by the effective temperature of CSIs [44]. In order to compensate for the energy drift during the observation due to the fluctuation of the temperature, it is necessary to correct the gain. The *Resolve* has three types of radiation sources for gain correction shown in Table 5.3.

The tool to estimate the gain drift is available as `rslgain` for the *Resolve* and `sxsgain` for the SXS. The output file is called the gain history file. By reading it as an option to the tool `rslpha2pi` for the *Resolve* (`sxspha2pi` for the SXS) to convert pulse height to energy, users can obtain event files with the energy corrected for temperature fluctuation.

As shown in Fig. 5.16, there is the possibility to underestimate the pulse height for the Ms grade pulse with an optimum filter. The magnitude of the underestimation is

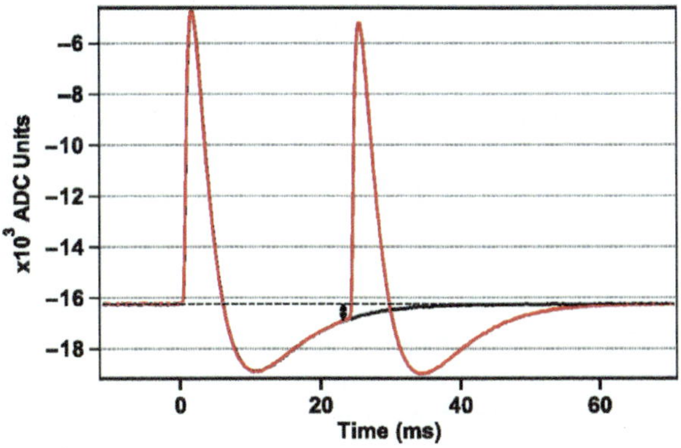

Fig. 5.16 Pulse profile of a Medium resolution event [13]. PSP gives the grades, Mp and Ms, to the first pulse and the second pulse, respectively. If the Ms grade pulse occurs before the Mp grade pulse could not be recovered completely to baseline, the evaluated pulse height of the Ms will underestimate. [Reproduced with permission from Eckart, M. E., Adams, J. S., Boyce, K. R., et al., Journal of Astronomical Telescopes, Instruments, and Systems, 4, 021406 (2018). Copyright 2018 Author(s), licensed under a Creative Commons Attribution 4.0 License]

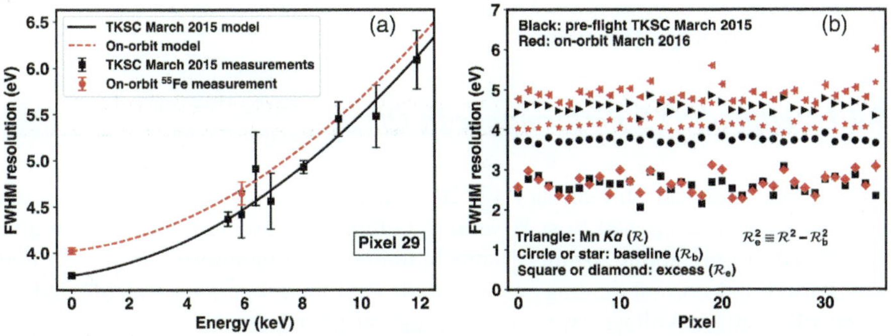

Fig. 5.17 The energy resolution of the Gaussian core for SXS [34]. Black points and lines show the data and model on the ground tests at Tsukuba Space Center (TKSC), respectively. The red ones show the data and model in orbit. These data were obtained by Hitomi/SXS. [Reproduced with permission from Leutenegger, M. A., Audard, M., Boyce, K. R., et al., Journal of Astronomical Telescopes, Instruments, and Systems, 4, 021407 (2018). Copyright 2018 Author(s), licensed under a Creative Commons Attribution 4.0 License]

different by the arrival time of the second pulse. The effect is that the energy resolution is getting worse for the Ms grade event. The pulse height can be recalculated by using the tool `rslseccor` (`sxsseccor`) with calibration files after reassigning the grade for the event with `rslsecid` (`sxssecid`).

The LSF is described with two components; Gaussian "core" and exponential "tail" as described in Sect. 5.2.3 [13, 34]. The Gaussian core is related to the detector noise, which depends on the operating environment whether on the ground or in orbit. For SXS, Fig. 5.17 shows the differences in the energy resolution of the core

Gaussian between data taken on the ground and in orbit. The Gaussian core should be calibrated in orbit by using the calibration sources, the MXS and/or ^{55}Fe. The energy resolution of the Gaussian core is modeled as

$$\mathscr{R}(E) = \sqrt{\mathscr{R}_{\mathrm{b}} + \mathscr{R}_{\mathrm{e}}(E_{\mathrm{ref}}) \left(\frac{E}{E_{\mathrm{ref}}} \right)},$$

where $\mathscr{R}_{\mathrm{b}}$ is "baseline" resolution which is obtained by using the optimal filter for noise event and $\mathscr{R}_{\mathrm{e}}(E_{\mathrm{ref}})$ is the excess energy resolution at reference energy E_{ref}. The Gaussian LSF is evaluated for each pixel and contained in a calibration file, `rmfparam`. It would be changed over time. On the other hand, the extended LSF is dependent on two different mechanisms and described by common parameters for all pixels. The mechanisms are escapes and electron loss continuum in the absorber (see Sect. 4.2.2 in [13]). The features are empirically modeled, which is contained in the calibration file. The tool `rslrmf` (`sxsrmf`) calculates the response matrix with considering LSF.

5.4 Hitomi/SXS Results

5.4.1 Hitomi/SXS Performance in Orbit

Hitomi was launched on February 17, 2016. During the rocket's acceleration, the He vent valve was opened to start venting helium gas through the PP 5 min after launch. After only about 20 min of launch, temperature monitoring of the He tank was started and normal venting was verified. And then, operations of SC and PC cryocoolers at low voltages were started, and an hour later the SC power was set at the nominal power (50 W×2). On the next day (February 18), the PC was set at the nominal power, and the operation of the JT cryocooler started. The JT power was increased gradually, and the JTS temperature reached 4.5 K on February 21. The ADR first recycle was carried out on February 21, and then the ADR recycles were performed periodically and the detector temperature was kept at 50 mK. Figure 5.18 shows temperature profiles of the dewar until the last telemetry on March 26 [17]. The He tank temperature was slowly decreasing and reached ~1.12 K at the end. A mass flow rate of the vented helium gas was estimated based on the ground test results to be about 34 μg s^{-1}, which corresponded to 0.70 mW heat load on the He tank [14]. Temperature increases during ADR recycling can also be used to estimate the heat load on the He tank, and they were consistent [49]. Based on the estimated heat load, the lifetime requirement of 3 years would have been satisfied, even if degradation of the cryocoolers is considered.

The SXS was still in the commissioning phase when Hitomi lost attitude control. The in-orbit operation of the SXS, including the operations of the power system, cooling system, and signal acquisition system, was a complete success [19]. For the

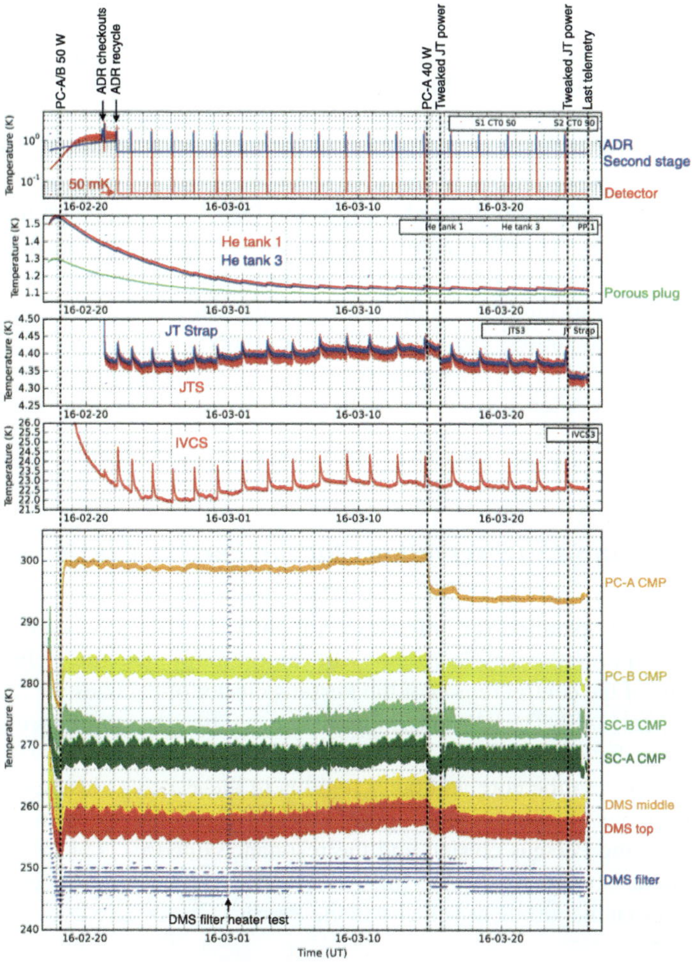

Fig. 5.18 Temperature profiles of the SXS dewar in orbit [17]. [Reproduced with permission from Fujimoto, R., Takei, Y., Mitsuda, K., et al., Journal of Astronomical Telescopes, Instruments, and Systems, 4, 011208 (2018). Copyright 2018 Author(s), licensed under a Creative Commons Attribution 4.0 License]

cooling system to run detectors at 50 mK, the cooling chain worked successfully as mentioned above, and cooling is performed by the ADR and the temperature is kept within a few μ Kelvin by Proportional-Integral-Differential (PID) control in orbit [17]. The vibration isolation system (VIS) for micro-vibration due to the mechanical cryocooler operations also worked as expected and the in-orbit detector spectral performance and cryocooler cooling performance were consistent with that on the ground [52]. As a result, the SXS was confirmed to have comparable performance in the ground tests and in orbit, including the performance for data readout, and achieved a high energy resolution of 5 eV at 6 keV in orbit as shown in Fig. 5.2 [44]. On the other hand, because the SXS operations ended early in the commissioning phase, the on-orbit detector could not be fully calibrated. The GV was not opened and the performance below 2 keV could not be verified, nor could the MXS be operated. The measurement of the gain scale, which determines the absolute measurement of photon energy, and the measurement of the LSF were also insufficient because the entire SXS instrument had not in thermal equilibrium [34]. Thus, although the SXS was not fully verified for on-orbit performance, it achieved high energy resolution even in orbit and showed the potential for future missions using X-ray micro-calorimeters. The initial results of the SXS reaffirmed the importance of high energy resolution spectroscopic observations and emphasized the importance of the SXS recovery as described in Sect. 5.4.2. Therefore, the recovery mission, XRISM has been proposed to achieve the key science goal of Hitomi as a joint project among JAXA, NASA, and ESA [53].

5.4.2 Hitomi/SXS Observations for the Perseus Cluster

Despite being in operation for only about a month, the SXS produced a very large number of important scientific results with high energy resolution. In particular, observations of the central region of the Perseus cluster of galaxies are reported in a series of papers [1–8]. The SXS observation revealed fine structures such as Fe He-α emission lines for the first time. The observed velocity dispersion of 187 ± 13 and 164 ± 10 km s^{-1} in the core and outer regions in the Perseus cluster, respectively, was much quieter than expected. Figure 5.19 left shows the spectrum of Fe He-α in the core of the Perseus cluster, and the resonance line (w) is clearly wider than the instrumental broadening (blue line) and thermal broadening (black line). This result shows that the turbulent velocity in the Perseus cluster core does not affect the

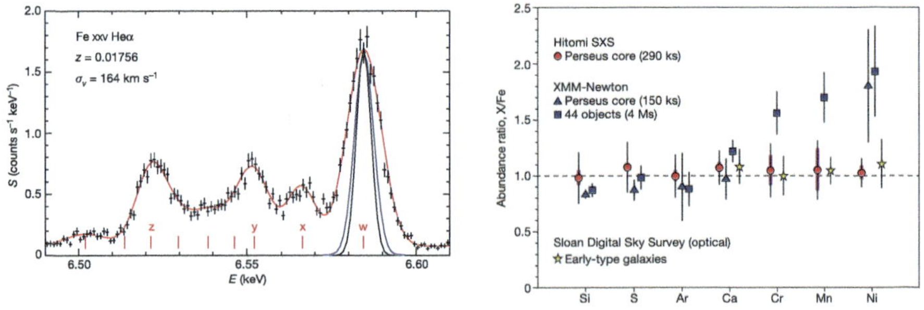

Fig. 5.19 (Left): SXS spectra of Fe$_{\mathrm{XXV}}$ He-α lines in the Perseus cluster core region [2]. (Right): Metal abundance rations to Fe of the Perseus cluster observed by SXS [3]. [Reproduced with permissions from Aharonian, F., et al., [Hitomi], Nature 535, 117–121 (2016), and Aharonian, F., et al., [Hitomi], Nature 551, 478–480 (2017)]

estimation of non-thermal effects on the mass estimates of galaxy clusters, which also affects the determination of cosmological parameters. The fine structure of the emission lines can now be clearly distinguished, and the ratio of resonant to forbidden lines in the iron emission lines due to resonant scattering has also been clarified, demonstrating for the first time that resonant scattering occurs at the center of the Perseus cluster. The SXS observation also gave the detection of not only the main strong lines from Si, S, Ar, Cr, and Fe but the weak lines from Cr, Mn, and Ni with high significance levels. These metal abundance ratios to Fe are similar to the Solar ratios as shown in Fig. 5.19, and it constrained the metal enrichment mechanism of the Perseus cluster core. This result indicates that the elemental enrichment processes in our Galaxy and the Perseus Cluster are similar to each other, suggesting that the elemental cycles in the Universe would be universal. It is particularly important to note that observations of the Perseus cluster with the micro-calorimeter have achieved higher abundance determination accuracy than longer-duration observations with CCDs, despite the short observation time. The SXS demonstrated the power of high-energy spectroscopic observations and the importance of micro-calorimeter in high-energy astrophysics.

5.5 XRISM/*Resolve* Performance on Ground Test

Table 5.1 shows the requirements of the *Resolve* as same as the SXS. As for the *Resolve* performance, in the instrument level test in March 2022, the energy resolution for all the pixels met the requirement with a margin for high and medium resolution grade as shown in Fig. 5.20. Figure 5.21 shows a number of characteristic X-ray lines for the calibrations of the energy scale in which the pulse height is converted to the energy scale with a fit using a sixth-order polynomial function for each pixel. The absolute energy scale over 0.3–9 keV energy range also met the requirement with margin. The calibration of the gain scale and line spread function in the 4–25 keV energy range were also done both in cryogen mode and in cryogen-free mode, and those in the 0.3–10 keV range in cryogen mode with the gate valve open.

For the purpose of simulations of in-orbit conditions, the shield coolers were operated with a higher power to cool the IVCS, and the heat load of 0.55 mW on the He tank was obtained. Considering 10% uncertainty and a conservative average heat load from the ADR, the heat load on the He tank at the beginning of life is expected to be 0.75 mW. Even if degradation of the cryocoolers (30%) is included, the heat load is expected to satisfy the requirement of < 1 mW at the end of life. Based on these results, the lifetime is roughly estimated to be 4.0 years for 35 L liquid helium. After launch, the heat load on the He tank can be estimated in the same way as the SXS, and the estimation of the life would be verified.

The ADR plays a key role in cooling the detector to 50 mK, and its performance directly affects the detector's performance. One of the critical functions of the ADR is to achieve high observing efficiency, which means that keeping 50 mK for an extended period of time in temperature-stable conditions is required. The require-

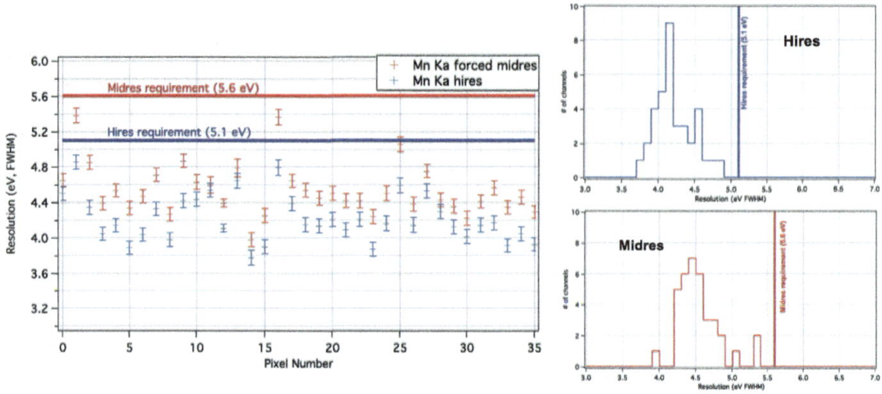

Fig. 5.20 (Left): Energy resolution (FWHM) of each pixel at 5.9 keV. (Right) Histograms of high and medium-resolution events for 36 pixels. [23]. [Reproduced with permission from Ishisaki, Y., et al., Proc. SPIE Int. Soc. Opt. Eng., 12181, 121811S (2022)]

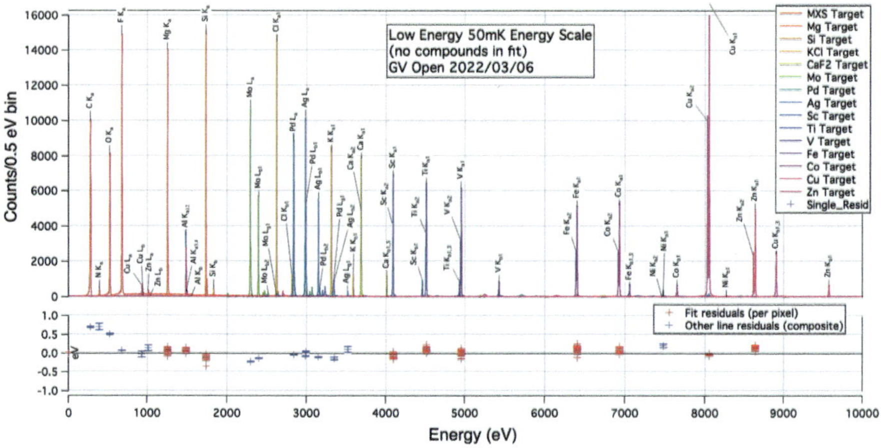

Fig. 5.21 Calibration of absolute energy scale accuracy for a number of characteristic X-rays with the gate valve open [23]. [Reproduced with permission from Ishisaki, Y., et al., Proc. SPIE Int. Soc. Opt. Eng., 12181, 121811S (2022)]

ment for temperature stability is 2.5 μK RMS (root mean square) or better during a 10 min period, in cryogen and cryogen-free modes. During the ground test, the operation algorithm and parameters were optimized. As a result, the performance of the temperature stability was 0.6 μK RMS on average in cryogen mode and 0.7 μK RMS in cryogen-free mode [23].

In cryogen mode, the hold time which is the time to keep at 50 mK by one duty ADR cycle, was about 37.8 h with the LHe temperature of typically about 1.24 K, then an operational duty cycle was >97%. In cryogen-free mode, the hold time was about 16.7 h. During the cryogen-free mode, the He tank was cooled to a stable 1.4 K using the stage-3 ADR. The operational duty cycle was obtained to be >93% [23]. In the ground tests, the DMS temperature was higher than that expected in orbit as mentioned above. Therefore, the He tank temperature is expected to become lower to 1.12 K in orbit, and the smaller heat loads on the ADR will increase the hold time in cryogen mode. On the other hand, since in cryogen-free mode, the He tank temperature is controlled to keep the same at 1.4 K, the hold time is not affected.

In order to keep tracking the gain drift in orbit, the MXS operation mounted on the FW is needed to illuminate the detector array in a pulsed mode at a duty cycle to be ~1%. With the MXS operation, the MXS pulse-on time intervals are excluded in a GTI file, and it causes a small amount of decrease in observation efficiency. The MXS pulses also cause an increase in the instrumental (non-X-ray) background due to their exponential tails by each pulse and the loss of the throughput due to changes in the branching ratio of the event grade. Therefore, the calibrations for the pulse parameters on the ground has carried out to optimize the MXS operation in orbit, and the optimal solution for the MXS operation has been established [47].

Also, the high count rate data sets are taken to evaluate the *Resolve* performance change [39]. There are three possible causes of spectral performance degradation: 1. CPU limit, 2. pile-up, and 3. electrical cross-talk. As for the CPU limit, the requirement of pulse processing is >200 s^{-1} $array^{-1}$, including background and spurious events without event losses. The high count rate data set in the instrument level test on the ground showed that this requirement should be met. The pile-up distorts the spectrum and count rate due to the arrival times of the primary and secondary pulse events being too close (see also Sect. 5.2.2). Electrical cross-talk is another possible cause of energy resolution degradation. When one pulse arrives, a part of its energy is deposited into adjacent pixels, which in turn affects another pulse that would normally be unrelated. The behavior for the high count rate event is well studied and modeled with the data set in the instrument level tests on the ground [39].

After the instrumental level test, the *Resolve* was integrated into the XRISM satellite in April 2022. The XRISM has conducted initial electrical and thermal vacuum tests and has confirmed that there is no change in the *Resolve* performance [19, 20, 33]. The data sets for calibrations are also taken to develop a plan for MXS operation in orbit and to estimate the influence of high count rate observations. The XRISM will be transported to the Tanegashima Space Center for launch after mechanical environmental tests and final electrical tests.

Acknowledgements This work was made possible by the collaborative efforts of all members of the Resolve team, SHI, and NEC engineers, which we greatly appreciate. Particularly, the authors are grateful to M. Eckart, C. Kilbourne, R. Kelley, M. Lowenstein, R. Shipman, P. Shirron, Y. Ishisaki, R. Fujimoto, M. Tsujimoto, M. Sawada, and Y. Takei for valuable comments and suggestions to our manuscript.

References

1. F. Aharonian et al., [Hitomi], Astrophys. J. Lett. **837**(1), L15 (2017). https://doi.org/10.3847/2041-8213/aa61fa, arXiv:1607.07420 [astro-ph.HE]
2. F. Aharonian et al., [Hitomi], Nature **535**, 117–121 (2016). https://doi.org/10.1038/nature18627, arXiv:1607.04487 [astro-ph.GA]
3. F. Aharonian et al., [Hitomi], Nature **551**, 478–480 (2017). https://doi.org/10.1038/nature24301, arXiv:1711.10035 [astro-ph.HE]
4. F. Aharonian et al., [Hitomi], Publ. Astron. Soc. Jpn. **70**(2), 10 (2018). https://doi.org/10.1093/pasj/psx127, arXiv:1710.04648 [astro-ph.HE]
5. F. Aharonian et al., [Hitomi], Publ. Astron. Soc. Jpn. **70**(2), 12 (2018). https://doi.org/10.1093/pasj/psx156, arXiv:1712.05407 [astro-ph.HE]
6. F. Aharonian et al., [Hitomi], Publ. Astron. Soc. Jpn. **70**(2), 13 (2018). https://doi.org/10.1093/pasj/psx147, arXiv:1711.06289 [astro-ph.HE]
7. F. Aharonian et al., [Hitomi], Publ. Astron. Soc. Jpn. **70**(2), 9 (2018). https://doi.org/10.1093/pasj/psx138, arXiv:1711.00240 [astro-ph.HE]
8. F. Aharonian et al., [Hitomi], Publ. Astron. Soc. Jpn. **70**(2), 11 (2018). https://doi.org/10.1093/pasj/psy004, arXiv:1712.06612 [astro-ph.HE]
9. K.R. Boyce, M.D. Audley, R.G. Baker et al., Proc. SPIE Int. Soc. Opt. Eng. **3765**, 741 (1999). https://doi.org/10.1117/12.366557
10. G.V. Brown, P. Beiersdorfer, H. Chen et al., Phys. Rev. Lett. **96**, 253201 (2006). https://doi.org/10.1103/PhysRevLett.96.253201
11. M.P. Chiao, J. Adams, P. Goodwin et al., J. Astron. Telesc. Instrum. Syst. **4**, 021404 (2018). https://doi.org/10.1117/1.JATIS.4.2.021404
12. C.P. de Vries, D. Haas, N.Y. Yamasaki et al., J. Astron. Telesc. Instrum. Syst. **4**, 011204 (2018). https://doi.org/10.1117/1.JATIS.4.1.011204
13. M.E. Eckart, J.S. Adams, K.R. Boyce et al., J. Astron. Telesc. Instrum. Syst. **4**, 021406 (2018). https://doi.org/10.1117/1.JATIS.4.2.021406
14. Y. Ezoe, M. DiPirro, R. Fujimoto et al., J. Astron. Telesc. Instrum. Syst. **1**, 011203 (2017). https://doi.org/10.1117/1.JATIS.4.1.011203
15. Y. Ezoe, Y. Ishisaki, R. Fujimoto et al., Cryogenics **108**, 103016 (2020). https://doi.org/10.1016/j.cryogenics.2019.103016
16. R. Fujimoto, K. Mitsuda, N. Yamasaki et al., Cryogenics **50**, 488 (2010). https://doi.org/10.1016/j.cryogenics.2010.02.004
17. R. Fujimoto, Y. Takei, K. Mitsuda et al., J. Astron. Telesc. Instrum. Syst. **4**, 011208 (2018). https://doi.org/10.1117/1.JATIS.4.1.011208
18. T. Hasebe, R. Imamura, M. Tsujimoto et al., J Astron. Telescopes. Instrum. Syst. **9**, 014003 (2023). https://doi.org/10.1117/1.JATIS.9.1.014003
19. T. Hasebe, R. Imamura, M. Tsujimoto et al., J Astron. Telescopes. Instrum. Syst. **9**, 014003 (2023). https://doi.org/10.1117/1.JATIS.9.1.014003
20. R. Imamura, Awaki, H., M. Tsujimoto et al., J. Low. Temp. Phy. **211**, 426 (2023). https://doi.org/10.1007/s10909-022-02935-1
21. K.D. Irwin, Appl. Phys. Lett. **66**, 1995 (1998). https://doi.org/10.1063/1.113674
22. K. Ishikawa, Y. Ezoe, H. Yamaguchi et al., Cryogenics **50**, 507–511 (2010). https://doi.org/10.1016/j.cryogenics.2010.02.008
23. Y. Ishisaki et al., Proc. SPIE Int. Soc. Opt. Eng. **12181**, 121811S (2022)
24. Y. Ishisaki et al., J. Astron. Telesc. Instrum. Syst. **4**(1), 011217 (2018). https://doi.org/10.1117/1.JATIS.4.1.011217
25. K. Kanao, S. Yoshida, M. Miyaoka et al., Cryogenics **88**, 143 (2017). https://doi.org/10.1016/j.cryogenics.2017.10.018
26. R.L. Kelley, H. Akamatsu, P. Azzarello et al., Proc. SPIE Int. Soc. Opt. Eng. **9905**, 99050V (2016). https://doi.org/10.1117/12.2232509
27. R.L. Kelley, M.D. Audley, K.R. Boyce et al., Proc. SPIE Int. Soc. Opt. Eng. **3765**, 114 (1999). https://doi.org/10.1117/12.366493

28. R.L. Kelley, K. Mitsuda, C.A. Allen et al., Publ. Astron. Soc. Jpn. **59**, 77 (2007). https://doi.org/10.1093/pasj/59.sp1.S77
29. M. Kilbourne, Irwin, Phys. Today **52**(8), 32 (1999). https://doi.org/10.1063/1.882776
30. C.A. Kilbourne, J.S. Adams, P. Arsenovic et al., J. Astron. Telesc. Instrum. Syst. **4**, 011215 (2018). https://doi.org/10.1117/1.JATIS.4.1.011215
31. C.A. Kilbourne, J.S. Adams, R.P. Brekosky et al., J. Astron. Telesc. Instrum. Syst. **4**, 011214 (2018). https://doi.org/10.1117/1.JATIS.4.1.011214
32. C.A. Kilbourne, M. Sawada, M. Tsujimoto et al., Publ. Astron. Soc. Jpn. **70**(18) (2018). https://doi.org/10.1093/pasj/psx139
33. M. Kurihara, M. Tsujimoto, M. E. Eckart et al., J. Astron. Telescopes. Instrum. Syst. **9**, 018004 (2023). https://doi.org/10.1117/1.JATIS.9.1.018004
34. M.A. Leutenegger, M. Audard, K.R. Boyce et al., J. Astron. Telesc. Instrum. Syst. **4**, 021407 (2018). https://doi.org/10.1117/1.JATIS.4.2.021407
35. D. McCammon, Cryogenic Particle Detect. **35** (2005). https://doi.org/10.1007/10933596_2
36. D. McCammon, R. Almy, E. Apodaca et al., Astrophys. J. **576**, 188 (2002). https://doi.org/10.1086/341727
37. T. Midooka, M. Tsujimoto, S. Kitamoto et al., J. Astron. Telesc. Instrum. Syst. **7**, 028005 (2021). https://doi.org/10.1117/1.JATIS.7.2.028005
38. K. Mitsuda, M. Bautz, H. Inoue et al., Publ. Astron. Soc. Jpn. **59**, S1 (2007). https://doi.org/10.1093/pasj/59.sp1.S1
39. M. Mizumoto, M. Tsujimoto, R.S. Cumbee et al., Proc. SPIE Int. Soc. Opt. Eng. **12181**, 121815Z (2022). https://doi.org/10.1117/12.2628784
40. S.H. Moseley, J.C. Mather, D. McCammon, J. Appl. Phys. **56**, 1257 (1984). https://doi.org/10.1063/1.334129
41. T. Okajima, Y. Soong, P. Serlemitsos et al., Proc. SPIE Int. Soc. Opt. Eng. **9905**, 99050Z (2016). https://doi.org/10.1117/12.2231705
42. T. Omama, M. Tsujimoto, M. Sawada et al., Proc. SPIE Int. Soc. Opt. Eng. **12181**, 121861 (2022). https://doi.org/10.1117/12.2629753
43. F.S. Porter, J.S. Adams, G.V. Brown et al., Proc. SPIE Int. Soc. Opt. Eng. **7732**, 77323J (2010). https://doi.org/10.1117/12.857888
44. F.S. Porter, K.R. Boyce, M.P. Chiao et al., J. Astron. Telesc. Instrum. Syst. **4**, 011218 (2018). https://doi.org/10.1117/1.JATIS.4.1.011218
45. Y. Sato, K. Sawada, K. Shinozaki et al., Cryogenics **64**, 182–188 (2014). https://doi.org/10.1016/j.cryogenics.2014.04.022
46. Y. Sato, K. Tanaka, H. Sugita et al., Cryogenics **116**, 103306 (2021). https://doi.org/10.1016/j.cryogenics.2021.103306
47. M. Sawada et al., Proc. SPIE Int. Soc. Opt. Eng. **12181**, 1218164 (2022)
48. P.J. Shirron, M.J. DiPirro, *Advances in Cryogenic Engineering* (Springer, Boston, MA, 1998), pp. 949–956. https://doi.org/10.1007/978-1-4757-9047-4_119
49. P.J. Shirron, M.O. Kimball, B.L. James et al., J. Astron. Telesc. Instrum. Syst. **4**, 021403 (2018). https://doi.org/10.1117/1.JATIS.4.2.021403
50. G.A. Sneiderman, P.J. Shirron, R. Fujimoto et al., J. Astron. Telesc. Instrum. Syst. **4**, 021408 (2018). https://doi.org/10.1117/1.JATIS.4.2.021408
51. T. Takahashi, M. Kokubun, K. Mitsuda et al., J. Astron. Telesc. Instrum. Syst. **4**, 021402 (2018). https://doi.org/10.1117/1.JATIS.4.2.021402
52. Y. Takei, S. Yasuda, K. Ishimura et al., J. Astron. Telesc. Instrum. Syst. **4**, 011216 (2018). https://doi.org/10.1117/1.JATIS.4.1.011216
53. M. Tashiro, H. Maejima, K. Toda et al., Proc. SPIE Int. Soc. Opt. Eng. **10699**, 1069922 (2018). https://doi.org/10.1117/12.2309455
54. Y. Terada, M. Holland, M. Loewenstein et al., J. Astron. Telesc. Instrum. Syst. **7**, 037001 (2021). https://doi.org/10.1117/1.JATIS.7.3.037001

Chapter 6
High-Resolution Spectral Analysis

Yijun Wang and Jelle de Plaa

Note: The goal of this chapter is to setup the `pion` model (a photoionisation equilibrium model) for the soft X-ray absorption features in a nearby (z = 0.00386) Seyfert 1 galaxy observed with XMM-Newton observations (EPIC-pn and RGS). A simulated spectrum was used because this thread merely intends to show the setup of the `pion` model.

6.1 Introduction

Active galactic nuclei (AGN) accrete matter onto a central supermassive black hole (SMBH) to produce intense broadband radiation, which can ionise and drive away the surrounding matter in form of outflows, such as warm ionised and cold molecular outflows [6]. Many observational proofs have implied that outflows might play an important role in affecting the star formation, evolution, and even environment of their host galaxies, also known as AGN feedback [9]. Therefore, investigating properties of outflows might significantly help us to understand the formation of AGN outflows and their feedback efficiency to the host galaxy.

Warm absorbers (WAs) [1, 5] are one type of warm ionised outflows [13], which can be detected via absorption features along the line of sight in the ultraviolet (UV) and soft X-rays [14]. WAs have the outflow velocities from about one hundred to several thousand km s^{-1} [3, 7] and they might originate in the accretion disk [4, 11],

Y. Wang (✉)
Nanjing University, Nanjing, China
e-mail: wangyijun@nju.edu.cn

J. de Plaa
SRON, Leiden, The Netherlands
e-mail: j.de.plaa@sron.nl

C. Bambi and J. Jiang (eds.), *High-Resolution X-ray Spectroscopy*, Springer Series in Astrophysics and Cosmology, https://doi.org/10.1007/978-981-99-4409-5_6

broad line region (BLR) [16, 20], or dusty torus [2, 10]. Until now, WAs have been found in about 50% of nearby AGN [12, 17, 19].

SPEX[1] is an X-ray spectral fitting package that is optimised for analysing high-resolution astrophysical X-ray spectra observed by the current and future X-ray observatories. Currently, SPEX is the only code that allows for the spectral-energy-distribution (SED) and the ionisation balance to be fitted simultaneously, while all other codes have to pre-calculate the ionization balance on a given SED. The pion model of SPEX is a robust and self-consistent photoionisation code that can simultaneously calculate the thermal/ionisation balance and the plasma spectrum in the photoionisation equilibrium. In addition, SPEX has a large atomic database that has been continuously developed at SRON[2] Netherlands Institute for Space Research since 1970s. These characteristics make SPEX a powerful tool to study the physical properties of plasma in different astrophysical conditions. In this chapter, for the high-resolution X-ray spectra of a nearby Seyfert 1 galaxy, we aim to use SPEX to simultaneously fit the continuum spectra and the absorption features caused by WAs (described by the pion model) to obtain relevant physical parameters.

6.2 Preparation

The user can use the following two methods to convert OGIP spectra and response files into SPEX format:

- The trafo[3] program in SPEX software, which asks questions interactively;
- The ogip2spex[4] script of the SPEX Python tools (pyspextools[5]), which works with command-line arguments to gather all the input and is easy for scripting.

To follow the thread in this chapter, the user can download the example files (simulated_PN.spo, simulated_PN.res, simulated_RGS.spo, simulated_RGS.res) here.[6] The user can also make all the following analysis through Pyspex[7] module which offers a Python interface to the SPEX program.

[1] The user can download different versions of SPEX software from SPEX Zenodo page (https://doi.org/10.5281/zenodo.1924563) and install the entire program following the instructions in SPEX GitHub page (https://spex-xray.github.io/spex-help/getstarted/install.html).

[2] https://www.sron.nl.

[3] https://spex-xray.github.io/spex-help/getstarted/runtrafo.html?highlight=trafo.

[4] https://spex-xray.github.io/pyspextools/tutorials/ogip2spex.html.

[5] https://spex-xray.github.io/pyspextools/.

[6] https://doi.org/10.5281/zenodo.7241267.

[7] https://spex-xray.github.io/spex-help/pyspex.html.

6.3 Start SPEX

Start SPEX in a linux terminal window:

```
user@linux: > spex
Welcome user to SPEX version 3.07.01

NEW in this version of SPEX:
02-08-2022 Fixed issue with optimal binning

Currently using SPEXACT version 2.07.00. Type 'help var
calc' for details.
SPEX>
```

6.4 Load Data

Use data[8] command to load the spectra and response files of EPIC-pn and RGS data for a nearby (z = 0.00386) Seyfert 1 galaxy:

```
SPEX> data simulated_PN simulated_PN
SPEX> data simulated_RGS simulated_RGS
```

6.5 Plot Data and Save the Plotting

Use plot[9] command to open the graphic device xs (xserver) and plot the data that are loaded in Sect. 6.4 on this device:

```
SPEX> plot dev xs
SPEX> plot type data
```

Set the x-axis plot in the unit (ux) of keV, on a log scale (log), and ranging (rx) from 0.3 to 13 keV:

```
SPEX> plot ux kev
SPEX> plot x log
SPEX> plot rx 0.3:13
```

[8] https://spex-xray.github.io/spex-help/reference/commands/data.html.

[9] https://spex-xray.github.io/spex-help/reference/commands/plot.html.

Set the y-axis plot in the unit (`uy`) of Counts s^{-1} $Å^{-1}$, on a linear scale (`lin`), and ranging (`ry`) from 0 to 25 Counts s^{-1} $Å^{-1}$:

```
SPEX> plot uy fa
SPEX> plot y lin
SPEX> plot ry 0:25
```

Use `ignore`[10] command to ignore the data set below 2 keV and above 12 keV for instrument 1 (EPIC-pn here), and ignore the data set above 2 keV for instrument 2 (RGS here):

```
SPEX> ignore instrument 1 0:2 unit kev
SPEX> ignore instrument 1 12:100 unit kev
SPEX> ignore instrument 2 2:100 unit kev
```

Optically bins (`obin`[11] command) the data channels 1:10000 for instrument 1 (EPIC-pn here), and bins (`bin`[12] command) the data channels 1:10000 by a factor of 4 for instrument 2 (RGS here):

```
SPEX> obin instrument 1 1:10000
SPEX> bin instrument 2 1:10000 4
```

Set the plot with data line weight (`lw`) of 3 and without displaying subtracted background (`back`), the plot identification (`cap id`), the upper title text (`cap ut`), and the lower title text (`cap lt`):

```
SPEX> plot set all
SPEX> plot data lw 3
SPEX> plot back disp f
SPEX> plot cap id disp f
SPEX> plot cap ut disp f
SPEX> plot cap lt disp f
```

Set the data colour (`data col`) of instrument 1 (EPIC-pn here) to blue:

```
SPEX> plot set 1
SPEX> plot data col 4
```

Finally, refresh the plot (see Fig. 6.1):

```
SPEX> plot
```

[10] https://spex-xray.github.io/spex-help/reference/commands/ignore.html.
[11] https://spex-xray.github.io/spex-help/reference/commands/obin.html.
[12] https://spex-xray.github.io/spex-help/reference/commands/bin.html.

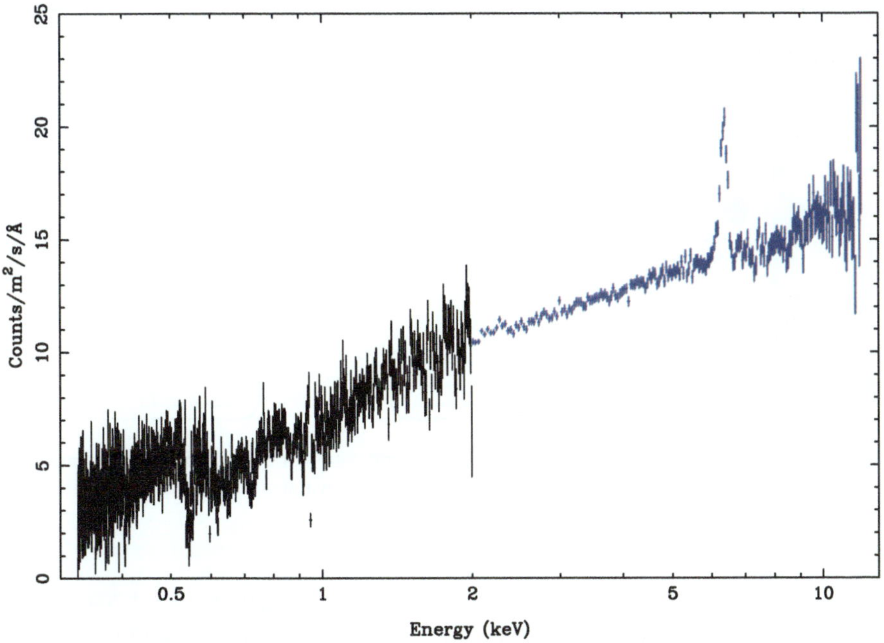

Fig. 6.1 The data spectra. *Black points:* XMM-Newton/RGS. *Blue points:* XMM-Newton/EPIC-pn

The user can open a colour postscript graphics device and save the output file to data.ps:

```
SPEX> plot dev cps data.ps
SPEX> plot
SPEX> plot close 2
```

6.6 Define Model Components and Component Relations (Step-by-Step)

Here we are looking at the components in this Seyfert 1 galaxy. For a typical Seyfert 1 galaxy, the SED of the intrinsic continuum above the Lyman limit along our line-of-sight has three components [15, 20] (see Fig. 6.2):

- A Comptonised disk component (`comt`: comptonisation model) for optical to soft X-ray data, which is produced by the accretion disk radiation,
- A power-law component (`pow`: power law model) for X-ray data, which is produced by Comptonisation of optical/UV disk photons by a corona of hot electrons,

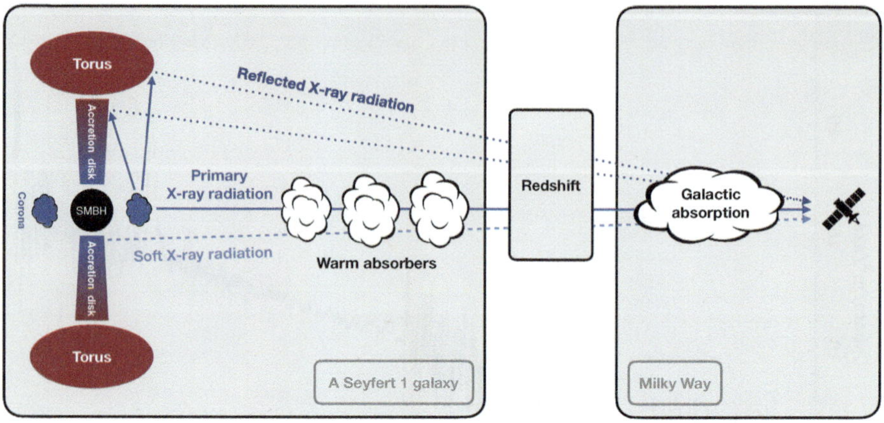

Fig. 6.2 Cartoon of the X-ray radiation from the central region of a typical Seyfert 1 galaxy to the observer (not to scale). Here three WA components are shown and numbers of them will vary in different cases

- A neutral reflection component (`refl`: reflection model) for hard X-rays data, which is produced by the reprocessing of the X-ray photons from the corona by the molecular torus, the BLR, and the disk.

This intrinsic continuum represents the central radiation of this galaxy. However, we cannot directly observe this intrinsic continuum. Before arriving at the observer, the intrinsic continuum will pass through various types of gases in itself (such as WAs), be redshifted, and then pass through the interstellar medium in our Milky Way (see Fig. 6.2; in some cases, it will experience more complex physical processes). This long journey will be represented as the multiple absorption features on the continuum. Next we will define the model components to describe the spectra in order to trace these physical processes.

6.6.1 Set the Distance of the Source

The distance of the source must be set in order to calculate the flux received at Earth. The user can specify the distance in a number of different units (please see details here[13]). Here we use redshift (parameter `z` in the following):

[13] https://spex-xray.github.io/spex-help/reference/commands/distance.html.

```
SPEX> distance 0.00386 z

Distances assuming H0 = 70.0 km/s/Mpc, Omega_m = 0.300 Omega_Lambda = 0.700 Omega_r = 0.000
Sector         m          A.U.          ly          pc          kpc         Mpc  redshift       cz      age(yr)
----------------------------------------------------
1      5.116E+23  3.420E+12  5.408E+07  1.658E+07  1.658E+04  16.5808    0.0039  1157.2  5.377E+07
----------------------------------------------------
```

6.6.2 Set the Redshift Component

The user can use `com`[14] command to add `reds`[15] model in order to apply a redshift effect. Next we use `par`[16] command to set its redshift (`z`) value (`val`) to be 0.00386.

```
SPEX> com reds
You have defined    1 component.
SPEX> par 1 1 z val 0.00386
```

6.6.3 Set the Galactic Absorption

The radiation from the source will pass through the neutral gas in our Milky Way, which will produce absorption features in the X-ray spectra. In SPEX, these absorption features can be described by the `hot`[17] model. This model calculates the transmission of a plasma in the collisional ionisation equilibrium with cosmic abundances.

We set the hydrogen column density (`nh`) value of the plasma to be 2.07×10^{-4} (in the unit of 10^{28} m^{-2}) and set its electron temperature (`t`) value to be 5×10^{-4} (in the unit of keV). We make the state (`stat`) of `nh` and `t` being frozen (`f`) in the fit, which means that they will not be fitted in the fit:

```
SPEX> com hot
You have defined 2 components.
SPEX> par 1 2 nh val 2.07e-4
SPEX> par 1 2 t val 5e-4
SPEX> par 1 2 t stat f
SPEX> par 1 2 nh stat f
```

[14] https://spex-xray.github.io/spex-help/reference/commands/component.html.
[15] https://spex-xray.github.io/spex-help/models/reds.html.
[16] https://spex-xray.github.io/spex-help/reference/commands/par.html.
[17] https://spex-xray.github.io/spex-help/models/hot.html.

6.6.4 Set the SED

As we mentioned in Sect. 6.6, the SED of this source has three components: a Comptonised disk component (`comt`[18] model), a power-law component (`pow`[19] model), and a neutral reflection component (`refl`[20] model)

We set the normalisation (`norm`) value of the power-law component to be 4.5×10^6 (in the unit of 10^{44} photon s^{-1} keV^{-1}) and set the photon index (`gamm`) value to be 1.88. We make the state of `norm` and `gamm` being free (`t`) in the fit:

```
SPEX> com pow
You have defined 3 components.
SPEX> par 1 3 norm val 4.5e6
SPEX> par 1 3 gamm val 1.88
SPEX> par 1 3 norm stat t
SPEX> par 1 3 gamm stat t
```

We set the cut-off energy (`ecut`) value of the ionising spectrum to be 300 (in the unit of keV), and set the scale for reflection (`scal`) value to be 0.68. We make the `ecut` state being frozen and make the `scal` state being free in the fit. We couple (`couple`) the `norm` and `gamm` of the `refl` component to those of the `pow` component, which means that these two parameters of these two components will be simultaneously fitted. Here we set `pow` to be 0 in order to ensure only the reflected spectrum being given, set `disk` to be 0 in order to make the spectrum not being convolved with an accretion disk profile, and set `fgr` to be 0 in order to not use general relativity.

```
SPEX> com refl
You have defined 4 components.
SPEX> par 1 4 ecut val 300
SPEX> par 1 4 scal val 0.68
SPEX> par 1 4 ecut stat f
SPEX> par 1 4 scal stat t
SPEX> par 1 4 norm couple 1 3 norm
SPEX> par 1 4 gamm couple 1 3 gamm
SPEX> par 1 4 pow val 0
SPEX> par 1 4 disk val 0
SPEX> par 1 4 fgr val 0
```

[18] https://spex-xray.github.io/spex-help/models/comt.html.

[19] https://spex-xray.github.io/spex-help/models/pow.html.

[20] https://spex-xray.github.io/spex-help/models/refl.html.

We set the normalisation (`norm`) value of the `comt` component to be 10^{10} (in the unit of 10^{44} photon s^{-1} keV^{-1}), set the seed photons temperature (`t0`) value to be 0.01 (in the unit of keV), set the plasma temperature (`t1`) value to be 0.06 (in the unit of keV), and set the optical depth (`tau`) value to be 30. We make the state of these parameters being free in the fit:

```
SPEX> com comt
You have defined 5 components.
SPEX> par 1 5 norm val 1.0e10
SPEX> par 1 5 t0 val 0.01
SPEX> par 1 5 t1 val 0.06
SPEX> par 1 5 tau val 30
SPEX> par 1 5 norm stat t
SPEX> par 1 5 t0 stat t
SPEX> par 1 5 t1 stat t
SPEX> par 1 5 tau stat t
```

6.6.5 Apply an Exponential Cut-Off to the Power-Law

Use `etau`[21] model to apply exponential cut-offs to the power-law component of the SED both below the low-energy cut-off and above the high-energy cut-off. The low-energy cut-off is related to the Lyman limit, while the high-energy cut-off is related to the temperature and optical depth of the plasma of hot electrons in corona. For `etau` model, the spectrum has a high-energy cut-off with $a > 0$, while it has a low-energy cut-off with $a < 0$, and for $a = 0$ the transmission is flat. The larger the value of a, the sharper the cut-off is.

We set the index (`a`) value of the first `etau` component to be 1 in order to make this component becoming a high-energy cut-off. We set the optical depth (`tau0`) value to be 3.2361×10^{-3} in order to make the high-energy cut-off energy being around 300 keV. We make the state of `a` and `tau0` being frozen:

```
SPEX> com etau
You have defined 6 components.
SPEX> par 1 6 a val 1
SPEX> par 1 6 tau0 val 3.2362e-3
SPEX> par 1 6 a stat f
SPEX> par 1 6 tau0 stat f
```

[21] https://spex-xray.github.io/spex-help/models/etau.html.

We set the index (a) value of the second etau component to be -1 in order to make this component becoming a low-energy cut-off. We set the tau0 value to be 1.3605×10^{-2} in order to make the low-energy cut-off energy being 0.0136 keV (Lyman limit). We make the state of a and tau0 being frozen:

```
SPEX> com etau
You have defined 7 components.
SPEX> par 1 7 a val -1
SPEX> par 1 7 tau0 val 1.3605e-2
SPEX> par 1 7 a stat f
SPEX> par 1 7 tau0 stat f
```

6.6.6 Set the pion *(Absorption) Components*

We think that there are three WA components in this system, therefore we introduce three pion[22] components (pion: SPEX photoionised plasma model) to model the warm absorber outflows. The pion model calculates the transmission and emission of a slab of photo-ionised plasma, where all ionic column densities are linked through a photo-ionisation model. The pion model can self-consistently calculate the photo-ionisation equilibrium using the available plasma routines of SPEX. The relevant parameter is the ionisation parameter $\xi = L/nr^2$ [18], where L is the source luminosity, n is the hydrogen density of the plasma and r is the distance of the plasma from the ionising source.

For the first WA component, we set the hydrogen column density (nh) value to be 2×10^{-2} (in the unit of 10^{28} m^{-2}), set the ionisation parameter (xil) value to be 3.3 (in the log scale; in the unit of W m), set the average systematic velocity (zv) value to be -1300 (in the unit of km s^{-1}), and set the root mean square velocity (v) value to be 20 (in the unit of km s^{-1}):

```
SPEX> com pion
You have defined 8 components.
** Pion model: take care about proper COM REL use!
** Check the manual in case of errors at: https://spex-xray.github.
io/spex-help/models/pion.html
SPEX> par 1 8 nh val 2.0e-2
SPEX> par 1 8 xil val 3.3
SPEX> par 1 8 zv val -1300
SPEX> par 1 8 v val 20
```

[22] https://spex-xray.github.io/spex-help/models/pion.html.

For the second WA component, we set the nh value to be 2.5×10^{-3} (in the unit of 10^{28} m^{-2}), set the xil value to be 2.5 (in the log scale; in the unit of W m), set the zv value to be -500 (in the unit of km s^{-1}), and set the v value to be 100 (in the unit of km s^{-1}):

```
SPEX> com pion
You have defined 9 components.
** Pion model: take care about proper COM REL use!
** Check the manual in case of errors at: https://spex-xray.github.
io/spex-help/models/pion.html
SPEX> par 1 9 nh val 2.5e-3
SPEX> par 1 9 xil val 2.5
SPEX> par 1 9 zv val -500
SPEX> par 1 9 v val 100
```

For the third WA component, we set the nh value to be 2×10^{-3} (in the unit of 10^{28} m^{-2}), set the xil value to be -1.0 (in the log scale; in the unit of W m), set the zv value to be -100 (in the unit of km s^{-1}), and set the v value to be 200 (in the unit of km s^{-1}):

```
SPEX> com pion
You have defined 10 components.
** Pion model: take care about proper COM REL use!
** Check the manual in case of errors at: https://spex-xray.github.
io/spex-help/models/pion.html
SPEX> par 1 10 nh val 2.0e-3
SPEX> par 1 10 xil val -1.0
SPEX> par 1 10 zv val -100
SPEX> par 1 10 v val 200
```

We make the state of these four parameters of all the WA components (8:10) being free in the fit:

```
SPEX> par 1 8:10 nh stat t
SPEX> par 1 8:10 xil stat t
SPEX> par 1 8:10 zv stat t
SPEX> par 1 8:10 v stat t
```

6.6.7 Set the Component Relations for Radiation Along Line-of-sight

As Fig. 6.2 shows, photons from both the Comptonised disk and power-law components are screened by the warm absorber components at the redshift of the target, as well as the galactic absorption before reaching the detector. Photons from the neutral reflection component is assumed not to be screened by the warm absorber for simplicity. It is still redshifted and requires the galactic absorption.

We define the component relations (`com rel`[23]) according to the order from the source to the observer. For example, from the source to the observer, the power-law continuum will experience the following journey (see Fig. 6.2): power-law continuum (component 3 defined before) → high-energy cut-off (component 6) → low-energy cut-off (component 7) → high-ionisation WA in this source (component 8) → mid-ionisation WA in this source (component 9) → low-ionisation WA in this source (component 10) → being redshifted (component 1) → interstellar medium in our Milky Way (component 2):

```
SPEX> com rel 3 6,7,8,9,10,1,2
SPEX> com rel 4 1,2
SPEX> com rel 5 8,9,10,1,2
```

Assuming that the WA components closer to the central engine are defined first (with a smaller component index), photons transmitted from the inner `pion` components are screened by all the outer `pion` components at the redshift of the target, as well as the galactic absorption before reaching the detector. The ionisation parameter of the WA component from the inner to the outer region usually decreases, which is the basis of defining the relations between different WA components:

```
SPEX> com rel 8 9,10,1,2
SPEX> com rel 9 10,1,2
SPEX> com rel 10 1,2
```

[23] https://spex-xray.github.io/spex-help/reference/commands/component.html.

6.6.8 Check the Model Settings and Calculate

We check the setting of the component relations. The following `model show`[24] command prints the currently-used spectral models to the screen that include all additive and multiplicative components.

```
SPEX> model show
________________

Number of sectors : 1
Sector: 1 Number of model components: 10
Nr. 1: reds
Nr. 2: hot
Nr. 3: pow [6,7,8,9,10,1,2 ]
Nr. 4: refl[1,2 ]
Nr. 5: comt[8,9,10,1,2 ]
Nr. 6: etau
Nr. 7: etau
Nr. 8: pion[9,10,1,2 ]
Nr. 9: pion[10,1,2 ]
Nr. 10: pion[1,2 ]
```

We check the setting of the free parameters and calculate the 1–1000 Ryd ionising luminosity which is usually used to calculate the ionisation state of plasma. For this, we need to set the energy band using `elim`[25]:

[24] https://spex-xray.github.io/spex-help/reference/commands/model.html.

[25] https://spex-xray.github.io/spex-help/reference/commands/elim.html.

```
SPEX> elim 1:1000 Ryd
Fluxes and luminosities will be calculated between
1.360570E-02 and 13.6057 keV
SPEX> calculate
SPEX> par show free

--------------------------------------------------
sect comp mod  acro parameter with unit     value          status  minimum    maximum  lsec lcom lpar

   1    3 pow  norm Norm (1E44 ph/s/keV)    4500000.       thawn   0.0        1.00E+20
   1    3 pow  gamm Photon index            1.880000       thawn  -10.        10.

   1    4 refl scal Scale for reflection    0.6800000      thawn   0.0        1.00E+10

   1    5 comt norm Norm (1E44 ph/s/keV)    1.0000000E+10  thawn   0.0        1.00E+20
   1    5 comt t0   Wien temp (keV)         9.9999998E-03  thawn   1.00E-05   1.00E+10
   1    5 comt t1   Plasma temp (keV)       5.9999999E-02  thawn   1.00E-05   1.00E+10
   1    5 comt tau  Optical depth           30.00000       thawn   1.00E-03   1.00E+03

   1    8 pion nh   X-Column (1E28/m**2)    2.0000000E-02  thawn   1.00E-14   1.00E+20
   1    8 pion xil  Log xi (1E-9 Wm)        3.300000       thawn  -7.0        10.
   1    8 pion v    RMS Velocity (km/s)     20.00000       thawn   0.0        3.00E+05
   1    8 pion zv   Average vel. (km/s)    -1300.000       thawn  -1.00E+05   1.00E+05

   1    9 pion nh   X-Column (1E28/m**2)    2.4999999E-03  thawn   1.00E-14   1.00E+20
   1    9 pion xil  Log xi (1E-9 Wm)        2.500000       thawn  -7.0        10.
   1    9 pion v    RMS Velocity (km/s)     100.0000       thawn   0.0        3.00E+05
   1    9 pion zv   Average vel. (km/s)    -500.0000       thawn  -1.00E+05   1.00E+05

   1   10 pion nh   X-Column (1E28/m**2)    2.0000001E-03  thawn   1.00E-14   1.00E+20
   1   10 pion xil  Log xi (1E-9 Wm)       -1.000000       thawn  -7.0        10.
   1   10 pion v    RMS Velocity (km/s)     200.0000       thawn   0.0        3.00E+05
   1   10 pion zv   Average vel. (km/s)    -100.0000       thawn  -1.00E+05   1.00E+05

Instrument     1 region     1 has norm    1.00000E+00 and is frozen
Instrument     2 region     1 has norm    1.00000E+00 and is frozen
----------------------------------------
Fluxes and rest frame luminosities between  1.36057E-02 and    13.606     keV

sect comp mod     photon flux   energy flux    nr of photons luminosity
                 (phot/m**2/s)   (W/m**2)      (photons/s)         (W)
   1    3 pow     250.552       7.133011E-14  2.239087E+52  4.630977E+35
   1    4 refl    4.11401       4.715256E-15  1.381814E+49  1.542371E+34
   1    5 comt    24.2510       1.208219E-15  4.796796E+52  3.859213E+35
   1    8 pion    0.00000       0.00000       0.00000       0.00000
   1    9 pion    0.00000       0.00000       0.00000       0.00000
   1   10 pion    0.00000       0.00000       0.00000       0.00000

----------------------------------------
Fit method        : Classical Levenberg-Marquardt
Fit statistic     : C-statistic
C-statistic       :       972.73
Expected C-stat   :       996.07 +/-        44.71
Chi-squared value :      1047.12
Degrees of freedom:         0
W-statistic       :       914.11
Contributions of instruments and regions:
  Ins    Reg     Bins       C-stat  Exp C-stat  Rms C-stat       chi**2       W-stat
    1      1      205       187.43      205.05       20.25       188.68       186.71
    2      1      788       785.30      791.02       39.86       858.44       727.40
```

6.10 Check the Model

When we set the plot type to data, the observed spectrum and the model can be shown simultaneously (see Sect. 6.7 and Fig. 6.4). However, the model that is shown in this plot type is folded through the response matrix. Therefore, in order to show the model spectrum that is not convolved by any instrumental profile, the user can set the plot type to model (`plot type model` command) (Fig. 6.6).

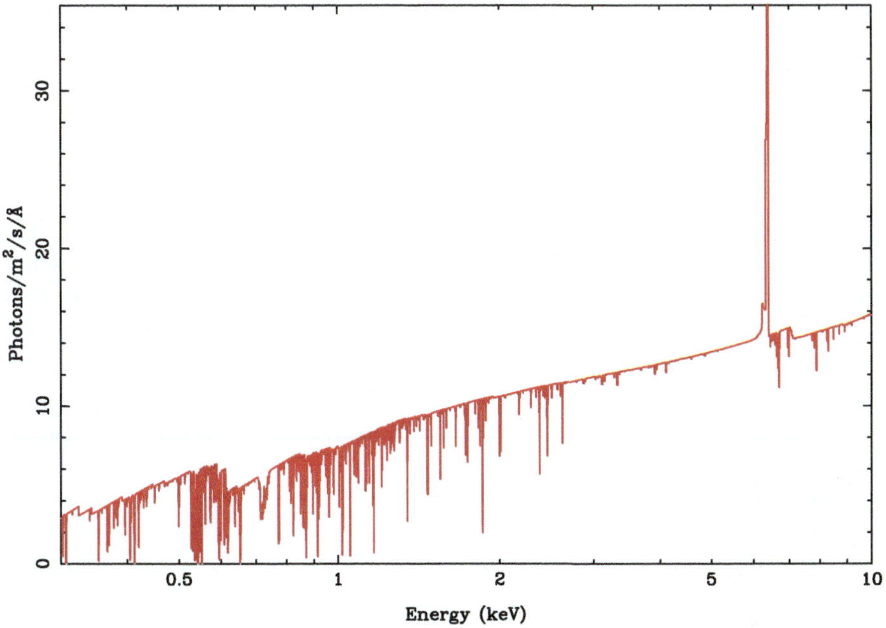

Fig. 6.6 The best-fit model in the plot type of model

6.10.1 Check the Best-Fit Model

This best-fit model can be shown using the following commands:

```
SPEX> plot type model
SPEX> plot ux kev
SPEX> plot uy a
SPEX> plot x log
SPEX> plot y lin
SPEX> plot rx 0.3:10
SPEX> plot ry 0:40
SPEX> plot fill disp f
SPEX> plot
```

6.10.2 Check the Intrinsic Continuum Model

In order to show the intrinsic continuum model, it is necessary to remove the effect from the warm absorbers and neutral gas in our Milky Way. Therefore, the user can set the "`fcov`" parameter to 0.0 (`fcov val 0.0`), which means that the plasma gas does not cover the central radiation region (`fcov=1` means a full covering). It is more convenient than changing the normalisation of spectral models (Fig. 6.7):

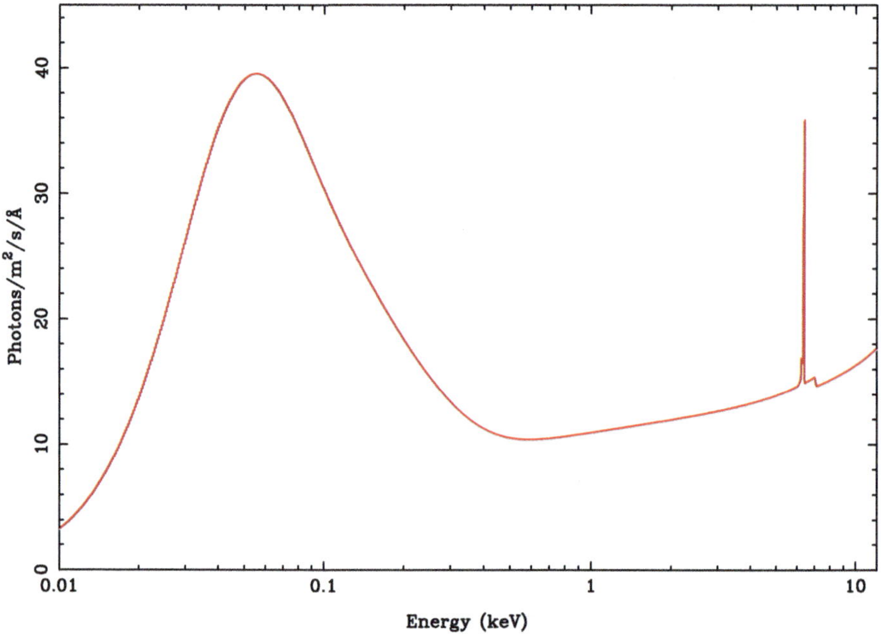

Fig. 6.7 The intrinsic continuum model

```
SPEX> par 1 2 fcov val 0.0
SPEX> par 1 8:10 fcov val 0.0
SPEX> calculate
SPEX> plot rx 0.01:12
SPEX> plot ry 0:45
SPEX> plot
```

6.11 Check the Warm Absorber Components with Different Outflowing Velocities

Sometimes the user needs to the check how much potential components with different outflowing velocities at a reference energy. It can be achieved in SPEX through plotting the x-axis in velocity units (km s^{-1}) with a reference energy provided (here is 20.062 Å that corresponds to the O IV absorption edge) (Fig. 6.8):

```
SPEX> plot type data
SPEX> plot ux vel 20.062 ang
SPEX> plot uy fa
SPEX> plot x lin
SPEX> plot y lin
SPEX> plot rx -3000:3000
SPEX> plot ry -1:9
```

6.12 Final Remarks

This is the end of this analysis thread. If you want, you can quit SPEX now:

```
SPEX> quit
Thank you for using SPEX!
```

Below, we provide a useful command file.

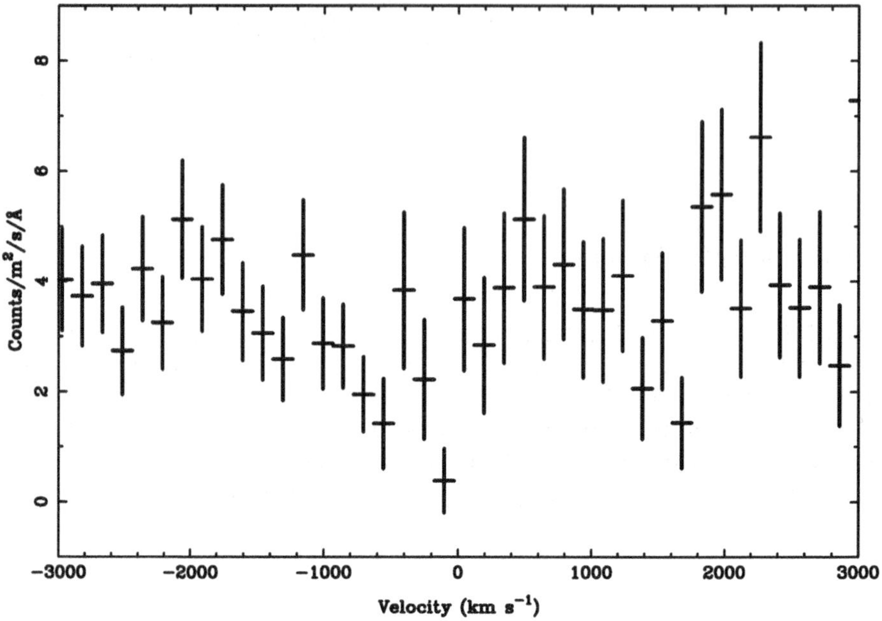

Fig. 6.8 The velocity spectrum at 20.062 Å

6.12.1 Running Scripts

The user can save all the above commands of this thread to a self-defined file named `myrunning.com`. Then use `log`[29] to load the above command file into SPEX:

```
SPEX> log exe myrunning
```

6.13 Summary

In this chapter, we provide some commands and threads to analyse the high-resolution X-ray spectra of a typical Seyfert 1 galaxy with SPEX. Using SPEX, we could simultaneously fit the continuum radiation and the absorption features caused by warm absorber outflows to obtain relevant physical parameters. SPEX also provides some methods to check the physical properties of absorption lines and edges, and obtain the physical model, and to present the different outflowing components, etc.

[29] https://spex-xray.github.io/spex-help/reference/commands/log.html.

We cannot introduce all the commands and threads of SPEX in this chapter, so the reader could refer to SPEX GitHub page (https://spex-xray.github.io/spex-help/getstarted/install.html) to know more details about SPEX.

References

1. D.M. Crenshaw, S.B. Kraemer, I.M. George, ARA&A **41**, 117 (2003). https://doi.org/10.1146/annurev.astro.41.082801.100328
2. A.J. Blustin, M.J. Page, S.V. Fuerst, G. Branduardi-Raymont, C.E. Ashton, A&A **431**, 111 (2005). https://doi.org/10.1051/0004-6361:20041775
3. J. Ebrero, J.S. Kaastra, G.A. Kriss, C.P. de Vries, E. Costantini, MNRAS **435**, 3028 (2013). https://doi.org/10.1093/mnras/stt1497
4. M. Elvis, ApJ **545**, 63 (2000). https://doi.org/10.1086/317778
5. J.P. Halpern, ApJ **281**, 90 (1984). https://doi.org/10.1086/162077
6. C.M. Harrison, T. Costa, C.N. Tadhunter et al., Nat. Astron. **2**, 198 (2018). https://doi.org/10.1038/s41550-018-0403-6
7. J.S. Kaastra, R. Mewe, D.A. Liedahl, S. Komossa, A.C. Brinkman, A&A **354**, L83 (2000). www.astro-ph/0002345
8. J.S. Kaastra, A.J.J. Raassen, J. de Plaa, Liyi Gu (2022) SPEX X-ray spectral fitting package (3.07.01). Zenodo.https://doi.org/10.5281/zenodo.7037609
9. A.R. King, K.A. Pounds, MNRAS **345**, 657 (2003). https://doi.org/10.1046/j.1365-8711.2003.06980.x
10. J.H. Krolik, G.A. Kriss, APJ **561**, 684 (2001). https://doi.org/10.1086/323442
11. Y. Krongold, F. Nicastro, M. Elvis et al., ApJ **659**, 1022 (2007). https://doi.org/10.1086/512476
12. S. Laha, M. Guainazzi, G.C. Dewangan, S. Chakravorty, A.K. Kembhavi, MNRAS **441**, 2613 (2014). https://doi.org/10.1093/mnras/stu669
13. S. Laha, C.S. Reynolds, J. Reeves et al., Nat. Astron. **5**, 13 (2021). https://doi.org/10.1038/s41550-020-01255-2
14. J. Mao, M. Mehdipour, J.S. Kaastra et al., A&A **621**, A99 (2019). https://doi.org/10.1051/0004-6361/201833191
15. M. Mehdipour, J.S. Kaastra, G.A. Kriss et al., A&A **575**, A22 (2016). https://doi.org/10.1051/0004-6361/201425373
16. C.S. Reynolds, A.C. Fabian, MNRAS **273**, 1167 (1995). https://doi.org/10.1093/mnras/273.4.1167
17. C.S. Reynolds, MNRAS **286**, 513 (1997). https://doi.org/10.1093/mnras/286.3.513
18. C.B. Tarter, W.H. Tucker, E.E. Salpeter, ApJ **156**, 943 (1969). https://doi.org/10.1086/150026
19. F. Tombesi, M. Cappi, J.N. Reeves et al., MNRAS **430**, 1102 (2013). https://doi.org/10.1093/mnras/sts692
20. Y. Wang, J. Kaastra, M. Mehdipour et al., A&A **657**, A77 (2022). https://doi.org/10.1051/0004-6361/202141599

Part II
Science

make the coronal approximation, i.e. that the gas is optically thin and that all excited ions return to their ground states in times short compared with the time between collisions. As shown in the previous section, such plasmas are expected to be far from LTE for typical coronal densities which are $\leq 10^{10}$ cm^{-3}.

Also, the characteristic times and distances for significant changes in the density and temperature are assumed to be long compared with those required to establish and maintain a steady state. This means that transient effects are unimportant and that the state of ionization is determined by the instantaneous temperature of the gas. In computing the ionization equilibrium, it is assumed that ionization by electron collisions is balanced by radiative and dielectronic recombination. It is also assumed that the field of ionizing radiation is weak, so photoionization can be neglected. The physical processes and rates for the processes which dominate in coronal plasmas have been summarized in the extensive review by [8]. The main processes which determine the charge state distribution are electron impact collisional ionization and radiative and dielectronic recombination. Many of the ingredients of such models, along with pioneering results, were presented by [43, 56].

The important free parameters describing a CIE plasma are the electron kinetic temperature, elemental abundances, and the emission measure. This last quantity is defined as $\int n^2 dV$ where n is the gas number density and the integral is over volume. The procedure for CIE spectrum calculation consists of first calculating the abundances of the ions in the gas (the charge state distribution), employing the equilibrium assumption. Then these are multiplied by the emission measure and the elemental abundances, and finally by the decay rate for the upper level for each line or free-bound transition.

One quantity which is of particular interest for CIE plasmas is the net cooling rate. This is important in modeling the properties of intergalactic and circumgalactic gas, and the early universe [59]. An example of this quantity versus temperature is shown in Fig. 7.1.

Fig. 7.1 Net cooling rate for coronal gas versus temperature. Curves of various metallicity are labeled. From [27]

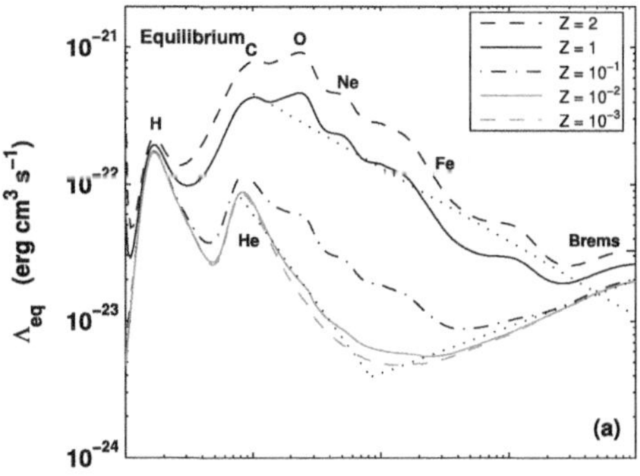

7.3.1 Stellar Coronae

The quintessential application of CIE models is to stellar coronae. However, this is complicated by the fact that stellar coronae span a range of temperatures, and also may have other physical characteristics which are not consistent with CIE models. The highest signal-to-noise non-solar coronal X-ray spectrum is that of Capella, as observed by the *Chandra* High energy transmission grating (HETG). Examples of the distribution of emission measure (differential emission measure or DEM) with temperature for fits to these spectra, using various atomic code packages, are shown in Fig. 7.2. This shows that there is a broad range of temperatures represented, from ~0.2–2 keV, and that the various models used to fit the spectra very nearly agree, when the assumptions made in the 'advanced' model are adopted [30].

7.3.2 Cooling Flows

An important subset of CIE models is the cooling flows. According to this model the temperature of the gas spans a range, and the DEM of each temperature is inversely proportional to the bolometric luminosity at that temperature. If the blobs of plasma cool in thermal isolation at constant pressure, and the dominant energy loss mechanism is via X-radiation, then the differential luminosity distribution, i.e., the luminosity radiated per unit temperature interval, must be proportional to the mass deposition rate, $\dot{M}$, i.e. $\frac{dL}{dT} = \frac{5\dot{M}k}{3\mu m_p}$. In the case of clusters of galaxies, there is a clear deficit of gas at low temperatures compared with this prediction [54], leading to the prediction

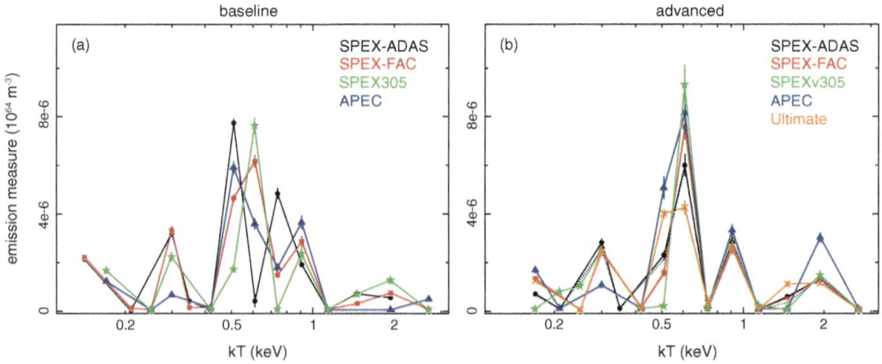

Fig. 7.2 Differential emission measure for fits to the Spectrum of the active star Capella using data from the *Chandra* High Energy Transmission Grating (HETG). Various curves correspond to different databases and codes. Panel **b** differs from panel **a** in that it allows many of the element abundances to be free parameters, and allows the temperature used in calculating the emitted spectrum to differ from the temperature used in the ionization balance. It also allows the density to be a free parameter. From [30]

that the gas is heated via feedback from a central AGN. Cooling flow models are implemented in XSPEC in the mkcflow, vmkcflow and cflow models. These have been shown to fit many of the properties of accreting white dwarf systems, eg. [35].

7.3.3 Non-Maxwellian Electron Speeds

In situations where fast suprathermal particles are introduced on timescales short compared with the electron-electron equilibration time the canonical assumption of Maxwellian velocity distribution may be wrong. This is plausible in regions where particle acceleration in shocks or by plasma waves are important, or possibly when transient photoionization is rapid.

Reference [18] have detected evidence for non-Maxwellian electron distributions in the Fe XII EUV lines emitted by solar active regions. Reference [38] have simulated the effects of non-Maxwellian electron velocity distributions in a shock by using a superposition of Maxwellians. This approach has been tested quantitatively by [31]. This showed that the satellite lines of He-like ions are affected and can be used to detect non-Maxwllian distributions. The major CIE code packages have all recently been updated to include κ electron distributions [21] and searches for these effects will be part of future high-resolution X-ray spectroscopy science.

7.3.4 Charge Transfer

When the projectile is an ion rather than an electron the dominant emission process is charge transfer, whereby the projectile transfers an electron to the target ion. This reaction proceeds rapidly when it is energetically allowed, and the electron is often captured into an excited level of the target ion, followed subsequently by line emission. Rates for this process can exceed those for electron impact excitation by large factors, corresponding to the greater geometrical cross-section of the projectile. Accurate rates are challenging to calculate, owing to the many-body effects in the quasi-molecule formed during the interaction. Collections of rates include those from [13]. A comprehensive review of this topic is [55].

Spectra from charge transfer are likely to have a characteristic signature; the capture into the target ion tends to favor higher angular momentum states compared with processes in CIE plasmas. This leads to a 'flatter' decrement in line strength versus principle quantum number. Charge transfer can only dominate over CIE if there is a supply of near-neutral projectile ions and highly ionized targets. This can occur in planetary or solar system comet atmospheres, which provide neutrals, and where the projectiles are the ions in the solar wind (eg. [45]). It can also occur in objects where there the existence of the neutral projectiles was previously unknown [14]. Computational packages for calculating the X-ray spectra from charge exchange include those from [29, 68].

7.3.5 Non-equilibrium Plasmas

The CIE assumptions are not valid in situations where the plasma conditions are changing on a timescale which is short compared with the time needed for the charge state distribution to reach equilibrium. An example is behind a young supernova remnant shock [36]. The equilibration timescale depends on the rates for recombination and for ionization (eg. [69]). If so, the distribution of gas emission measure with temperature can be derived using a similarity solution [34].

7.4 Atomic Data and Comprehensiveness

It has long been realized that comprehensive atomic data is crucial to accurately calculate the thermal properties of gases hotter than $\sim 10^4$ K [49], and that this need outweighs the need for consideration of such processes as the accurate treatment of radiation transfer for traditional nebular problems. This implies consistent treatment of various ionization/excitation processes, both radiative and collisional, and their inverses, including inner-shell processes, for all of the ions of the $\geq$10 most abundant elements. Similar arguments apply to the calculation of the X-ray opacity of partially ionized gases. Furthermore, the need to calculate synthetic spectra introduces a further requirement of accuracy on atomic data: observed X-ray spectra can have a spectral resolution of $\varepsilon/\Delta\varepsilon \simeq 1000$, and in order to fit to such data, synthetic spectra must employ wavelengths and ionization potentials which are accurate at this level. Such precision cannot be achieved by current atomic physics calculations and requires dedicated experiments.

A great deal of work has been done to calculate and measure atomic energy levels, cross sections and transition probabilities for the purposes of astrophysical X-ray spectral modeling. A detailed summary would require a dedicated paper; many areas of progress have been described by [7, 25, 41]. Notable are the measurements carried out by the electron beam ion trap (EBIT) and storage rings, calculations using the FAC, HULLAC, R-Matrix and Autostructure codes. Some of this work will be highlighted in the following sections.

There are various atomic databases: for collisional plasmas these include CHIANTI [17], ADAS [72], SPEX [16] and AtomDB [24], while for photo-ionized plasmas there are XSTAR [6, 50], CLOUDY [22] and others.

7.4.1 Continuum Emission

Processes which transfer energy from electrons to radiation include bremsstrahlung, Compton scattering, and magnetic processes such as synchrotron emission. Rates for these processes are well known, although accurate relativistic rates can necessitate the use of tabulated gaunt factors or moments of the cross section. Electron-electron bremsstrahlung is generally much less important than electron-ion bremsstrahlung, except for very high temperatures. Reference [52] have tabulated relativistic rates for this process, as well as for electron-ion bremsstrahlung.

7.4.2 *Recombination*

Radiative recombination can be calculated from photoionization cross sections along with the detailed balancing requirement. Dielectronic recombination depends on the energy level structure of the recombining ion, particularly the multiply excited autoionizing states. These rates have been calculated by [4] and succeeding papers, and these have been extensively validated using experiments (eg. [63]).

It was pointed out by [46, 51] that the ionization balance needed to fit the K lines of Si in the spectrum of NGC3783 was discrepant from that needed to fit the iron UTA lines. It was suggested that this was due to the use of inaccurate dielectronic recombination rates for iron, and it was postulated that the rates where were larger by $\sim$10. This suggestion was confirmed, qualitatively, by [3], who performed distorted wave calculations of the rates and obtained rate coefficients which were even greater than those suggested by [51]. Experimental confirmation was demonstrated for ions down to Fe^{7+} [64].

7.4.3 *Collisional Ionization*

Electron impact ionization rate coefficients are challenging to calculate accurately, owing to the large dynamic range in atomic length scales and partial waves which they can access, and also to the importance of collisional excitation to autoionizing levels. As a practical matter, most current CIE simulations make use of rates which are derived from experimental cross sections, such as those of [9, 19, 74]. These are implemented by fitting to analytic formulas derived from computations, such as those of [75]. This procedure must correct for the possibility of ionization from metastable levels in the experiment.

7.5 Photoionized Plasmas

The general problem of calculating the reprocessing of ionizing continuum radiation from a star or compact object into longer wavelength lines and diffuse continuum has broad importance in astrophysics. Photoionization modeling is generally thought to include any situation in which the dominant ionization and excitation mechanism is photons from an external source. In addition, key ingredients of such models are applicable more broadly, to calculating the photoelectric opacity and the flux transmitted through any astrophysical gas. In the context of X-ray astronomy, photoionization is important in sources which contain a compact object, notably active galactic nuclei (AGN), X-ray binaries, and cataclysmic variables.

Many of the same principles apply to modeling the properties of diffuse nebulae illuminated by UV radiation from stars, i.e. H II regions and planetary nebulae. Discussions of model line emission from nebulae date from [76], but Seaton and collaborators laid the groundwork for modern numerical modeling by calculating many atomic quantities of importance to nebular modeling [11, 65, 66]. The subject has been reviewed extensively by [53], and others (see references in [50])

For the purposes of subsequent discussion, typical photoionization models calculate the ionization, excitation, and heating of cosmic gas by an external source of photons. The gas is generally assumed to be in a time-steady balance between ionization and recombination, and between heating and cooling. If the gas is optically thin, then a useful scaling parameter is the ionization parameter, defined in terms of the ratio of the incident ionizing flux to the gas density or pressure. We adopt the definition $\xi = 4\pi F/n$ where F is the incident energy flux integrated between 1 and 1000 Ry, and n is the gas number density [73]; various other definitions are also in use. When the gas is isobaric, the appropriate ionization parameter is proportional to the ratio ionizing flux/gas pressure. We adopt the combination ξ/Tor $\Xi = \xi/(kTc)$ [48] for this purpose. This much of photoionization modeling is common to today's models for X-ray sources and to more traditional models used for nebulae.

Prior to the launch of $Chandra$ and $XMM - Newton$, it was recognized that objects such as AGN and X-ray binaries would have spectral features with the diagnostic application when observed in the X-ray band. These included $K\alpha$ fluorescence lines from iron seen from many objects [2, 28], and the warm absorbers in the soft X-ray band from AGN [58]. The quantum leap in sensitivity and spectral resolution is represented by $Chandra$ and $XMM - Newton$ revealed that other new ingredients are needed in order to make the models useful for the quantitative study of X-ray sources. This motivated a great deal of work in the treatment of physical processes previously neglected, and in the accuracy and comprehensiveness of atomic data. A summary of some of these areas will occupy the remainder of this paper.

7.5.1 Radiation Transport

Although numerical methods for radiation transfer are well established [33], their implementation has the potential to be very computationally expensive. Also, radiation transfer depends sensitively on the geometrical arrangement of the gas and the sources of illuminating radiation, so it is difficult to make a calculation which has wide applicability. A calculation must treat the transfer of the ionizing continuum into the photoionized gas and transfer of cooling or reprocessed radiation out of the gas. For the purposes of X-ray astronomy, it is also useful to calculate the entire synthetic spectrum produced by the photoionized plasma. This can then be used for direct fitting to data using tools such as XSPEC,[1] ISIS[2] or SHERPA.[3]

Traditional treatments of radiation transfer make several simplifying assumptions which allow for efficient calculation and wider applicability of model results. These include: (i) Simplified geometry, such as a plane parallel slab or a spherical shell; (ii) Use of single-stream transport of the illuminating radiation; and (iii) Use of escape probability transport of resonance lines. We note that X-ray resonance line optical depths are typically much smaller than optical or UV line depths, so the effects of line transfer approximations are reduced compared with traditional nebular photoionization calculations.

Recent progress in radiation transfer for X-ray photoionization has centered around the application of sophisticated accurate transfer treatments for certain limited specialized problems. These include the adoption of the accelerated lambda iteration method in the Titan code [61]. The Monte Carlo method has been used by [67] to treat the transfer in AGN broad absorption line (BAL) and warm absorber flows. This latter work combines radiation transfer with the detailed geometry derived from a numerical hydrodynamic model. It has led to important insights into the origin of spectral features such as high velocity lines, and X-ray emission lines. This work illustrates the importance of consideration of the detailed geometrical distribution of the gas, and the inherent limitations of traditional simplified geometrical assumptions.

7.5.2 Thermal Instability

A surprising result derived from fits of photoionized models to observed spectra from AGNs is the apparent multi-phase behavior. That is, there are two or more regions of parameter space, as described by the ionization parameter ξ, which appear to be preferred by the absorbing gas. These are separated by regions of ξ which appear to be avoided. A possible explanation for this two-phase behavior is thermal instability. This is due to the properties of the cooling function in the gas which allows for two stable and one unstable solution to the thermal equilibrium equation (or more). This has been discussed in detail by many authors beginning with [23] and, in the context of AGN, by [48]. The physical origin is the fact that the net cooling function has strong temperature dependence in some regions of temperature, and weaker dependence elsewhere, leading to a local maximum in the net cooling function as shown in Fig. 7.1.

Thermal instability is associated with regions where the net cooling function $\Lambda(T)$ is a decreasing function of temperature. A stable temperature solution is characterized by a cooling function which is an increasing function of temperature; a positive temperature perturbation about a stable equilibrium temperature will lead to increased

[1] http://heasarc.gsfc.nasa.gov/docs/xanadu/xspec.

[2] http://space.mit.edu/CXC/isis.

[3] http://cxc.harvard.edu/sherpa/.

cooling, which will restore the gas to equilibrium, with the corresponding behavior for a negative temperature perturbation. Perturbations about an unstable temperature will tend to run away toward higher or lower temperature until a stable region of the cooling curve is reached. The net cooling function of a photoionized gas is globally increasing over the temperature range from $\sim$1000–10^8K, owing to strong hydrogen cooling at low temperatures, and strong inverse Compton cooling at high temperatures. Instability is associated with a local maximum to the function, so there must always be an odd number of thermal equilibrium solutions.

The instability is more likely to occur when the gas is isobaric, rather than isochoric. Isochoric gas heated by a radiation field such as the spectral energy distribution characteristic of AGN, is predicted to be thermally stable; a radiation field which is flat, or deficient in soft photons, is more likely to produce thermal instability. The presence of the thermal instability depends on interesting things: the shape of the ionizing spectrum (SED) from IR through the X-rays, the atomic rates, elemental abundances, and (weakly) in the gas density. This suggests possible diagnostic use.

The instability criterion of [23] is:

$$\left(\frac{\partial \Lambda}{\partial S}\right)_A > 0. \tag{7.8}$$

where the subscript refers to the intensive quantity which is being held constant. which can be rewritten for isochoric gas:

$$\left(\frac{\partial \Lambda}{\partial T}\right)_\rho < 0. \tag{7.9}$$

and for isobaric gas

$$\left(\frac{\partial \Lambda}{\partial T}\right)_P = \left(\frac{\partial \Lambda}{\partial T}\right)_\rho - \frac{\rho 0}{T_0}\left(\frac{\partial \Lambda}{\partial \rho}\right)_T < 0. \tag{7.10}$$

The criterion has been revised by [5], who makes the change:

$$\left(\frac{\partial \Lambda/T}{\partial S}\right)_A > 0. \tag{7.11}$$

and points out that the quantity in the brackets is the cooling rate. This criterion has been explored by [15], who show that the unstable region of parameter space is narrower than it would be using the [23] criterion, and that this has implications for the fragmentation of the flow into clouds. Points corresponding to the maximum value of Ξ on the cold stable branch of the S-curve (i.e., for log $\Xi_{C,max} = 1.10$) are dynamically most significant, as they mark the entry into a cloud formation zone and dramatic changes in the flow profiles can occur there. Figure 7.3 illustrates the thermally unstable region.

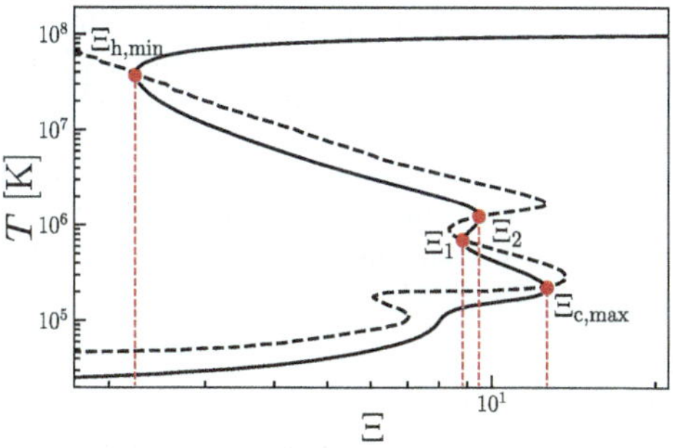

Fig. 7.3 S-curve and Balbus contour (the solid and dashed lines, respectively). Red dots mark the points $\Xi_{c,max}$, Ξ_1, Ξ_2, and $\Xi_{h,min}$. Note that $\Xi_{c,max}$ ($\Xi_{h,min}$) denotes the last (first) stable point on the "cold phase branch" ("Compton branch") of the S-curve. From [15]

The existence of thermal instability depends on the validity of the assumption of thermal (and ionization) equilibrium, and it is important to consider this when using it for quantitative work. Thermal equilibrium requires that the timescale for heating and cooling, the thermal timescale, be less than other relevant timescales. In the case of warm absorbers, which are flowing out from the AGN center, this includes the gas flow timescale.

7.5.3 *High Densities*

The accuracy of the results from all spectral fits are dependent on systematic model limitations, in addition to any statistical uncertainties. Recent theoretical work has explored the effect of higher density when computing the X-ray reflection spectra of accretion disks (still assuming a constant disk density) using the XILLVER code [26]. Calculations performed over a range of densities have demonstrated that at sufficiently high densities ($n_e > 10^{17}$ cm^{-3}), ionization effects result in a significant increase in the atmospheric temperature. However, limitations in the conventional atomic data prevented the calculation of these models at densities above $n_e = 10^{19}$ cm^{-3}.

The physics of photoionization including densities beyond those typically assumed for AGN broad line regions has been discussed previously by [57]. Density effects in coronal models leading to the suppression of dielectronic recombination (DR) have been addressed by [71]. Reference [20] have explored the effects of DR suppression and metastable populations on ionization balance and line diagnostics in the solar transition region.

The effects of high densities in photoionization models have been explored by [39]. The list of processes which can be affected by plasma effects at high densities includes: (i) The effect of collisional ionization and three-body recombination from atomic states with a large principal quantum number (high-n states) which are very close to the continuum, causes 'continuum lowering'. These processes, when their rates exceed downward radiative rates, cause the level populations to be in local thermodynamic equilibrium (LTE) with the continuum. The net result is that such states belong to the continuum rather than to the bound atom or ion, and so cannot be counted when summing to calculate the net rate coefficients for recombination or any other process where high-n states are involved. (ii) Dielectronic recombination: this process, whereby recombination occurs together with the excitation of a bound electron in the target ion, is important , particularly for photoionized plasmas. During the recombination, the doubly excited ion is particularly fragile and subject to collisional ionization. This effectively suppresses the recombination rate at high density. (iii) Three-body recombination: this process can be important for any bound level at high density. It can dominate the net recombination rate and produces a different distribution of excited states than radiative recombination. (iv) Stimulated radiative processes will become important when the radiation energy density is high at the energies of appropriate transitions, and will enhance the rates of radiative decay processes. (v) Although metastable levels are already included in state-of-the-art calculations at sufficiently high densities new levels can be collisionally populated from the ground, thereby affecting the emissivity and opacity. (vi) At sufficiently high density, the effect of neighboring ions and electrons can create plasma microfields that perturb the atomic level structure in ways which can change which decay or excitation channels are energetically allowed, and can change atomic wavefunctions and the resulting matrix elements and rate coefficients. (vii) Free-free heating is strongest at high density and for illuminating spectra with flux at soft energies.

7.5.4 Resonance Scattering

In traditional nebular modeling it was typically assumed that the gas was spherically symmetric around the continuum source and stationary. If so, resonance scattering of the incident continuum is expected to have a negligible effect on the spectrum seen by a distant observer. This is because there is a cancellation between continuum photons removed from and those added to the beam of continuum we observed from the central source (there is a small net loss of photons in the region shadowed by the central source). In this case, the region of the spectrum emergent from the cloud which includes a resonance line would show neither a deficit nor an excess at the line energy. Of course, photoionization followed by recombination leads to a redistribution of the energy of the photons absorbed by ionization, and so would produce an excess at the line energy.

It is now recognized that many photoionization-dominated sources observed in X-rays have reprocessing gas which is not spherically distributed around the continuum source. In the case of a non-spherical scattering region, the cancellation between photons scattered out of and into the observed beam is not exact, and thus scattering can have an effect on the observed spectrum. The effect depends, crudely, on the relation between the column density between the continuum source and infinity averaged over 4π steradians, as compared with the column density along the observer's line of sight to the continuum source. If the observer's line of sight traverses more gas than the spherical-average, then resonance scattering will result in a net removal of photons from the observed radiation field and the spectrum will have a deficit at the line energy. This could be termed the 'net absorption' case. If the observer's line of sight traverses less gas than the spherical-average, then resonance scattering will result in a net addition of photons to the observed radiation field and the spectrum will have an excess at the line energy. Such apparent 'net emission' can be confused with emission due to recombination or electron impact excitation, though these can be distinguished by differing dependence on atomic quantities such as oscillator strengths or collision cross sections. Furthermore, the resonance scattering cross section is much greater than the background cross sections for photoionization, so lines are expected to saturate at column densities which are small compared to the column densities where the continuum can penetrate. Recombination emission will dominate at higher column densities and scattered emission will dominate at lower column densities. Reference [44] have shown that resonance scattering has a distinct signature in the line ratios and line/continuum ratio in emitting plasmas. This is illustrated in Fig. 7.4, which shows the transition from a scattering-dominated to a recombination-dominated spectrum as a function of the column density of the plasma. The left column corresponds to what would be observed in the 'net absorption' case, and the right column corresponds to 'net emission'. At low column densities radiative pumping of resonance

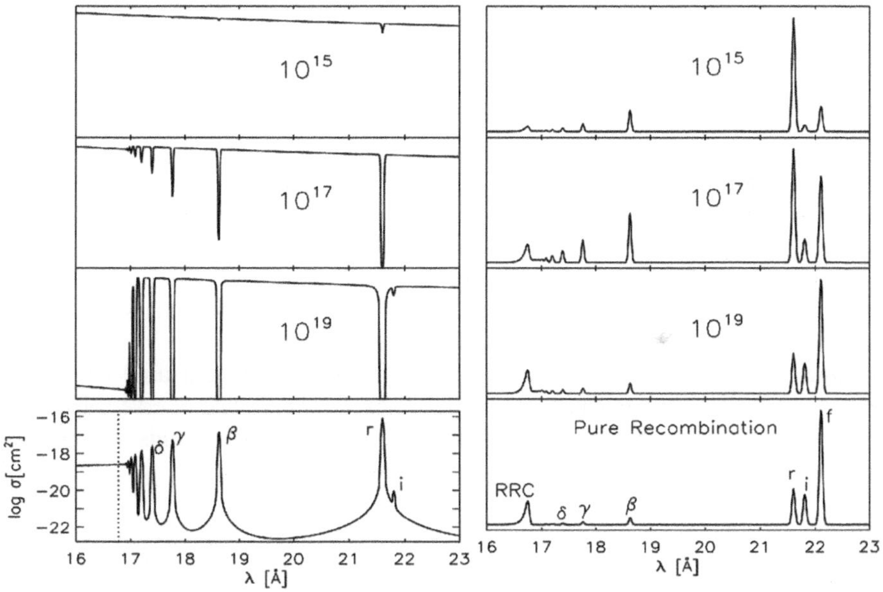

Fig. 7.4 Emission line spectra illustrating the difference between line spectra from photoionization versus resonance scattering. From [44]

lines dominates because the cross sections are larger than for photoionization, and the apparent line emission in the 'net emission' case is due to resonance scattering. At high column densities, the emission is dominated by recombination. More recent discussions of the effects of continuum photo-excitation are by [12].

7.5.5 Time Dependence

Standard photoionization models assume that the various heating, cooling, ionization, and recombination processes are in a time-steady balance. This assumption may not be valid, owing to the intrinsic time-variability of the illuminating continuum or other factors which change the cloud environment. If so, the time dependence of the changing illumination or other factors must be included. This increases the complexity of the problem significantly. It is important to include not just the time dependence of the ionization balance equations and the temperature, but also the time dependent transfer of the radiation field. This is true whenever the gas is not optically thin. Such models have only been carried out for a few situations relevant to X-ray spectroscopy so far [60, 62]. Fitting of time dependent models to variable spectra has the potential to provide significant insights, since the timescales for ionization, recombination, radiative transfer and heating or cooling depend on quantities such as the gas density which cannot be obtained in any other way.

7.5.6 Line Absorption

A key peculiarity of X-ray spectra is the importance of inner shell transitions. These are even more important often in gas dominated by photoionization, when we view the gas in absorption in front of a strong continuum source. Photoionization from inner shells can be important in this case, and emission from inner shells occurs via fluorescence following inner shell photoionization. Inner shell lines differ from valence shell lines in the importance of damping. That is, the inner shell vacancy levels can often decay via many different channels, including non-radiative Auger channels. The very short lifetime for these levels implies large natural line widths, and this affects the curve of growth for such lines, and hence the observed absorption spectrum. Examples of such strongly damped lines are shown in Figs. 7.5 and 7.6.

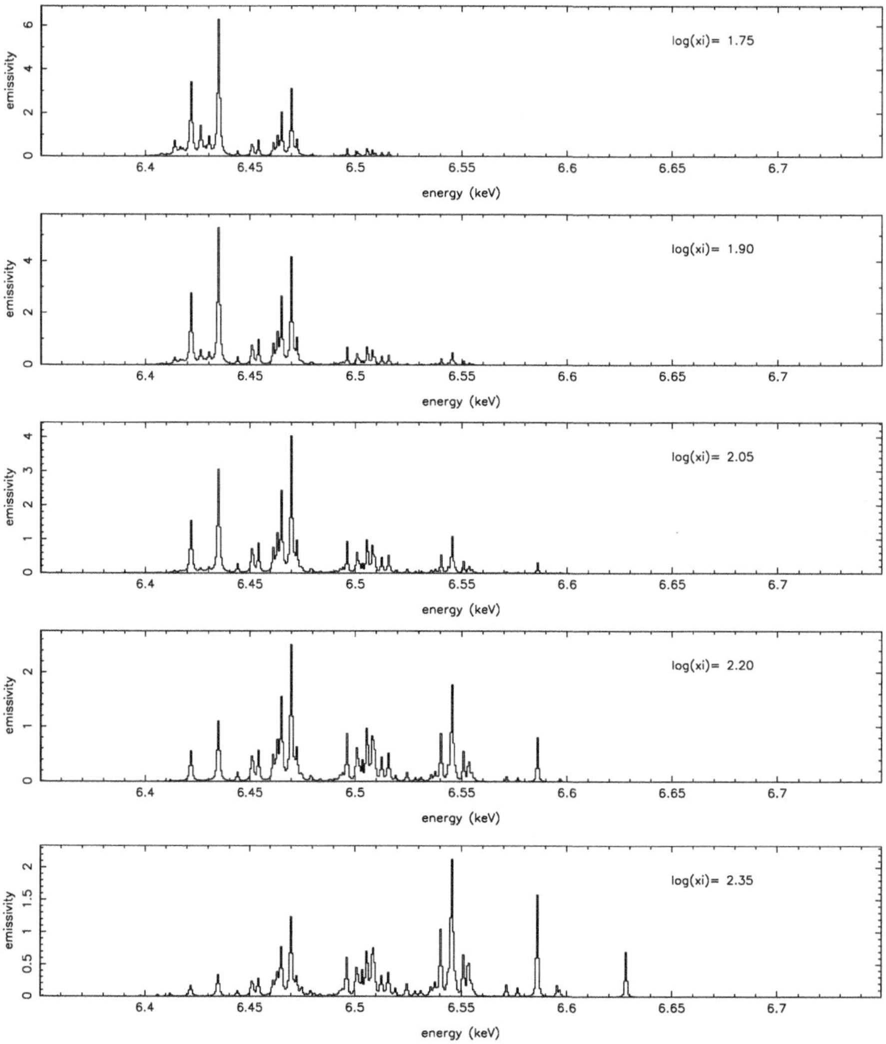

Fig. 7.5 Emission line spectra from iron in the K line region versus ionization parameter ξ. From [42]

7.6 The Future

Fitting models of plasmas to *Chandra* and *XMM – Newton* spectra provides insights into the nature of many different astrophysical sources: their degree of ionization, density, location, composition and kinematics. However, the models are likely still incomplete in important ways. For example, X-ray grating spectra with good statistics seldom give truly statistically acceptable fits to standard models. Typically,

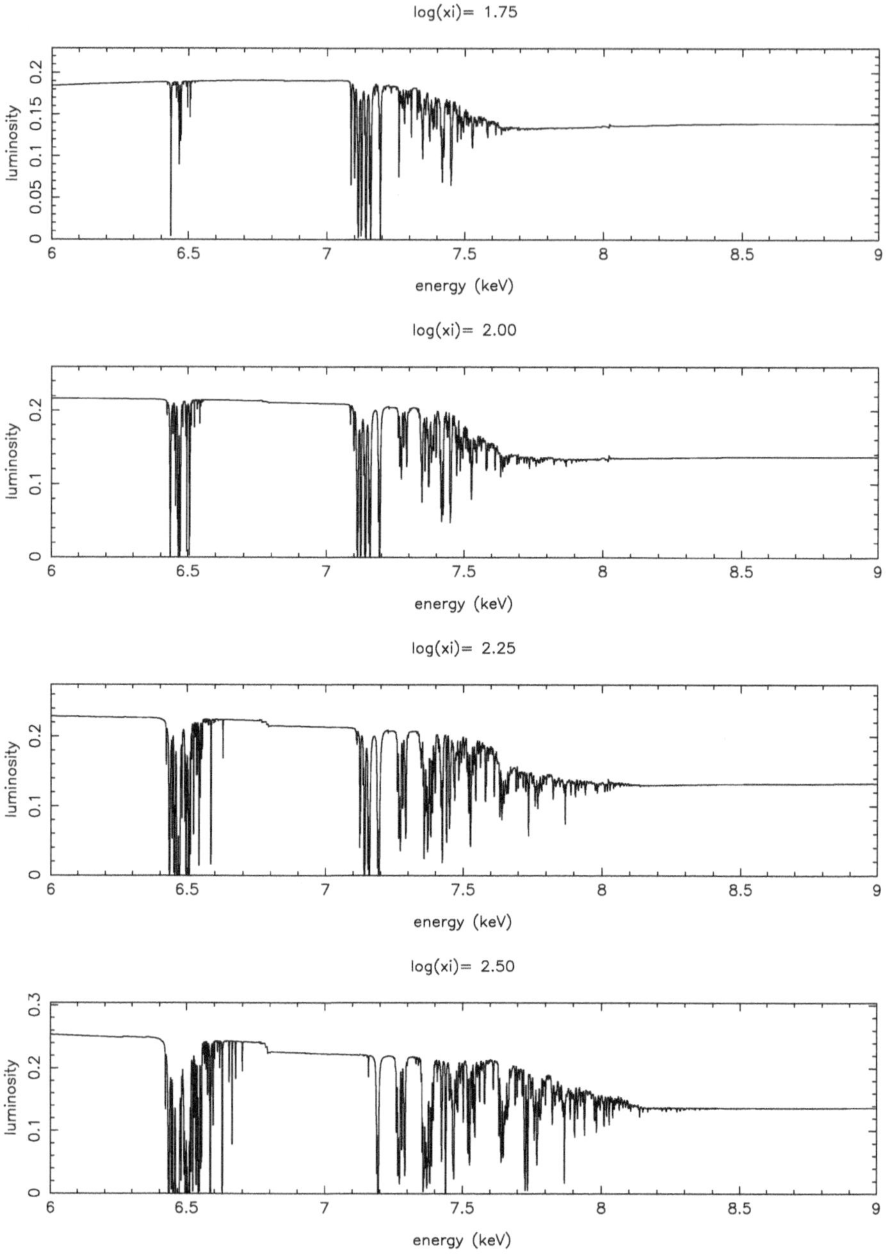

Fig. 7.6 Absorption line spectra from iron in the K line region versus ionization parameter ξ. From [42]

Chapter 8
Clusters of Galaxies

Jeremy S. Sanders

8.1 Introduction

Many galaxies in the universe are found to be gravitationally bound into objects known as groups and clusters of galaxies. The richest clusters of galaxies consist of thousands of individual galaxies, with total masses of $\sim 10^{15}\ \mathrm{M}_{\odot}$. Groups of galaxies are lower mass objects containing fewer galaxies ($\lesssim 50$), although the boundary between clusters and groups is not exact. In the hierarchical theory of the formation of structure, clusters are expected to lie at the densest regions of the cosmic web, built up by the merger of smaller structures over time (e.g. [2]).

It was noted that a number of the X-ray sources in the sky discovered by the *Uhuru* X-ray observatory were associated with groups and clusters of galaxies [15]. The discovery of an emission feature around 7 keV in an X-ray spectrum of the Perseus cluster observed using the *Ariel 5* X-ray telescope [72] provided key evidence that the X-ray emission from clusters of galaxies was thermal in nature, originating from Fe K-shell transitions within the hot plasma within the cluster. The presence of this line in Perseus and other clusters ruled out non-thermal emission mechanisms as the dominant X-ray emission processes within clusters.

This observed X-ray emission originates from the intracluster medium (ICM), the hot plasma cluster atmosphere, making up the majority of the baryonic matter in these objects. Despite this, due to their large volume, the density of ICM is relatively low. The electron density, for example, typically peaks at the centre with values between 10^{-1} and $10^{-3}\ \mathrm{cm}^{-3}$. However, the majority of mass in clusters consists of dark matter, which is only indirectly visible.

Galaxy groups range in temperature from a few 10^6 K (noting that the typical unit for cluster and group temperatures in X-ray astronomy is keV, where 1 keV is approximately 11.6 MK) to $2-3$ keV. Clusters range above these temperatures to

J. S. Sanders (✉)
Max Planck Institute for Extraterrestrial Physics, Gießenbachstraße 1, 85748 Garching, Germany
e-mail: jsanders@mpe.mpg.de

C. Bambi and J. Jiang (eds.), *High-Resolution X-ray Spectroscopy*, Springer Series in Astrophysics and Cosmology, https://doi.org/10.1007/978-981-99-4409-5_8

$\sim$10 keV. At the highest temperatures, the majority of the X-ray emission from these hydrogen-rich atmospheres is thermal bremsstrahlung (free-free) emission. At lower temperatures (below $\sim$2 keV), particularly in group scale objects, line emission becomes more important. It should be noted, however, that clusters and groups are not purely isothermal, and temperature substructure can be present. For example, as we discuss further in Sect. 8.4, these objects often contain a central cool core where the temperature drops significantly below the cluster ambient temperature. Many baryonic processes in clusters (e.g. mergers, stripping, feedback) can also affect the temperature distribution.

If the ICM entirely consisted of hydrogen and helium, then there would be little point in doing high-resolution X-ray spectroscopy in order to understand cluster physics. In fact, the metallicity of the intracluster medium is enriched to around $0.2-0.3\ Z_{\odot}$ (Sect. 8.6), meaning that the ratio of the density of metal ions compared to hydrogen, is roughly 0.2–0.3 times the value found in the sun. It is the emission lines from the metals in the ICM which enable X-ray spectroscopy and allows the detailed study of a cluster through its hot baryons.

High-resolution X-ray spectroscopy can be used in a number of different ways to study the physics of galaxy clusters, including

1. Examining the metallicity of the ICM. The majority of metals in the ICM originate from stellar processes. Studying them allows us to look at the history of star formation in cluster galaxies and constrain models of metal production, and examine the processes which transport metals through the ICM.
2. Measuring the velocity of the ICM. Given sufficient spectral resolution, the energy and width of emission lines can be used to measure the velocity distribution of the hot atmosphere. Indirect spectral probes, such as resonant scattering, can also be used to constrain the velocity in clusters. Velocities are powerful observational probes of physical processes in clusters, such as feedback or mergers.
3. Studying the temperature distribution of the ICM. In addition to the bremsstrahlung continuum emission, the relative strength of emission lines is sensitive to the temperature of the plasma. Temperature measurements can be used to study many physical processes, including whether there could be cooling of the ICM in the core of a cluster.
4. Non-thermal physical processes. Although the bulk of the cluster emission is thermal in nature, there could be other emission mechanisms taking place, such as charge exchange. In addition, more exotic particles in the cluster could produce emission lines or modulate the spectra of sources passing through the cluster.

8.2 Cluster Emission Processes and Spectral Models

The spectra from clusters of galaxies are typically assumed to be well-modelled as a collisionally-ionised optically-thin plasma. At the temperatures typically found in clusters, the continuum emission is mainly due to bremsstrahlung (free-free) emission (see Chap. 7 for a detailed discussion of astrophysical plasmas, and [53] for a review). In this case the emission is produced by scattering between electrons and ions, with ϵ_{ff}, the rate of emitted photons per unit volume and energy, scaling as

$$\epsilon_{ff} \propto n_e n_i \, T^{-1/2} E^{-1} e^{-E/kT} Z_{\mathrm{eff}}, \tag{8.1}$$

where n_e and n_i are the electron and ion number densities, respectively, T is the temperature, E is the photon energy and Z_{eff} is an effective charge of that ion species. Bremsstrahlung emission is dominated by hydrogen and helium ions. Other, smaller, continuum contributions come from free-bound and two photon emission processes. The cluster X-ray emission is heavily weighted towards the densest regions in a cluster.

For line emission, an ion is put into an excited state, usually by collisional excitation for cluster plasma, which then decays into a lower state by emitting a photon. This two-body process also scales as the particle density squared, similarly to the bremsstrahlung emission. However, as the lines are from heavier elements than helium, the number density of an ion depends on the abundance of the relevant element, and the fraction of those atoms which have the correct ionisation, which varies by temperature. Lines in the X-ray spectrum are strongest for those elements which are most abundant, but the strength also typically increases strongly according to the nuclear charge [53], which results in iron (Fe) lines being some of the strongest in the spectrum.

The assumption of collisional ionisation equilibrium (CIE) means that for a particular temperature, there is a balance between the rate at which ions and atoms are ionised to a higher ionisation state and those recombining from the higher states. In a non-equilibrium ionisation (NEI) plasma this balance in rates is not present.

The most common codes used for modelling the spectra of a collisionally-ionised plasma in equilibrium are the APEC model produced from ATOMDB [103], and the SPEXACT code in the SPEX modelling software [52] which derives from the earlier MEKAL model [56, 70, 71]. New observational data and model results are continually being incorporated into these models. For example, high resolution *Hitomi* spectra of the Perseus cluster were used to bring these models into better agreement [45]. Figure 8.1 shows some example spectra at different temperatures over the energy range typical for clusters (0.1–10 keV) and in detail the range around the Fe-L complex (0.5–1.9 keV). The results are for an XSPEC normalisation of 1 cm^{-5}, which is defined for a source with redshift z and angular diameter distance of D_A (cm) as

$$N = \frac{10^{-14}}{4\pi [D_A (1+z)]^2} \int n_{\mathrm{e}} n_{\mathrm{H}} \, \mathrm{d}V, \tag{8.2}$$

integrating the electron (n_{e} [cm^{-3}]) and Hydrogen (n_{H} [cm^{-3}]) densities over volume (dV [cm^3]).

Fe is the strongest line-producing element in the spectra of clusters of galaxies. At lower temperatures (below $\sim$ 2 keV) the Fe L-shell lines (lines from $n = 2$ in Li-like

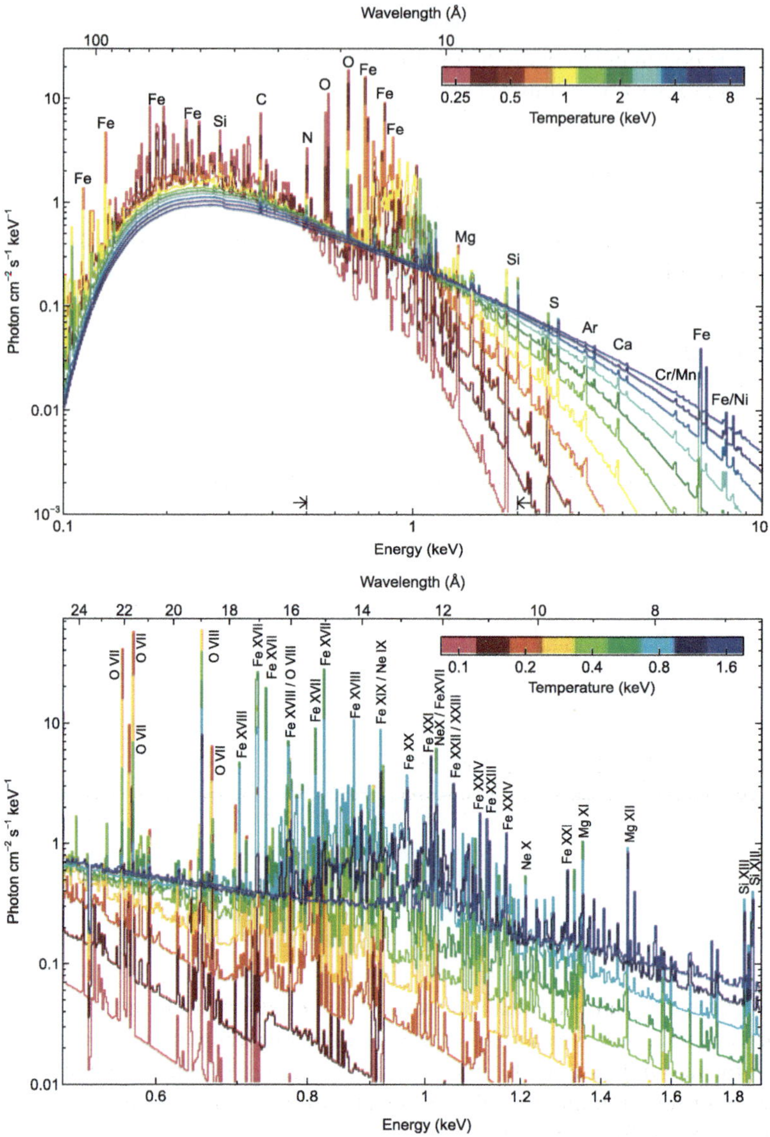

Fig. 8.1 Collisionally-ionised spectral models as a function of temperature, for a wider spectral range (top) and around Fe-L (bottom). Models were computed with APEC 3.0.9 for a 0.3 $Z_\odot$ metallicity, an XSPEC normalisation of unity, an equivalent H column absorption of 10^{20} cm^{-2} [117], in 500 spectral bins

to Ne-like iron ions) around 1 keV are dominant. At higher temperatures the He-like Fe K-shell lines at ∼6.7 keV become the main spectral feature. Other lines which are important contributors to the X-ray spectrum include those from C, N and O (in the soft sub-keV band), Ne and Mg (in the Fe-L complex range), Si and S ($\sim 2-3$ keV), Ar and Ca ($\sim 3-4$ keV) and Ni (in the Fe-L region and ∼7–8 keV).

Cluster spectra, however, are not perfectly described as perfect CIE optically thin models. Firstly, an observer has to account for absorption from material in our own Galaxy (e.g. using TBABS; [117]). The amount of absorption can be estimated from HI and dust extinction maps of the galaxy [116]. High resolution spectra can be used to model the absorption directly, although high-quality data are required (e.g. [87]). Emission from our own galaxy and the local hot bubble can also produce a significant background contribution [59, 104].

The assumption that the plasma is optically thin may also be incorrect for some resonance lines, leading to a modification of the spectral line intensity distribution over the source. We discuss this effect further, including how it can be used to infer gas motions, in Sect. 8.5.3.

There are other components which may need modelling. The X-ray background and Active Galactic Nuclei (AGN) in the cluster can introduce complexities in modelling the cluster spectrum, particularly when determining the level of the continuum. Charge exchange is another process which can give rise to emission lines and is predicted to happen at the interface between the ICM and colder gas (Sect. 8.7). Clusters are also not completely static objects, and so there are cases where there may be a lack of ionisation equilibrium (NEI). This lack of thermal equilibrium is most likely to occur in shocks in the low-density outer regions of clusters (also Sect. 8.7).

One of the largest uncertainties in spectral modelling is deciding whether more than one or how many temperature components are sufficient for describing the observed spectrum. Real clusters have temperature substructure, either due to projection effects or real structures within the studied aperture. An observer can fit either a single component, two or more components, or a continuous parametrized distribution (a differential emission measure or DEM). Given a perfect thermal model and data, it should be possible to test a number of distributions and choose the one which best fits the data, but existing data has limited statistics or spectral resolution. In addition, these components could have different metallicities; it is common for observers to assume they have the same metallicity in order to constrain the parameters. Using too few components when modelling can, in particular, lead to the determination of incorrect metallicities (the so-called Fe-bias or inverse Fe-bias, see Sect. 8.6). If a DEM is used to model the spectrum then commonly used distributions are powerlaws or a log-Gaussian (e.g. [22]).

8.3 High Spectral Resolution Data

In X-ray astronomy, one can either achieve high spectral resolution using a detector with intrinsic high spectral resolution, such as a microcalorimeter or through the use of a diffraction grating.

Extended objects, such as clusters, present some challenges when using gratings. The main issue is that gratings on X-ray telescopes are slitless, which means that the spectral resolution is degraded for an extended object when compared to a point source. The relative origin angle of an X-ray photon and its wavelength are both important for the dispersion angle. For a reflection grating instrument like the Reflection Grating Spectrometer (RGS) gratings onboard the *XMM-Newton* observatory, the wavelength (λ) is connected to the incoming angle (α), outgoing angle (β), for a spectral order (m, which is negative for this type of grating) and grating spacing (d) by

$$m\lambda = d(\cos\beta - \cos\alpha). \tag{8.3}$$

Therefore an angular form of the profile of the source along the dispersion direction is convolved with the X-ray spectrum. However, a grating does allow the spectral variation to be examined in the cross-dispersion direction, allowing a limited study of the spatial variation of the source spectrum.

For the RGS spectrometers onboard *XMM-Newton*, the spectra are broadened by approximately

$$\Delta\lambda \approx \frac{0.139}{m} \frac{\Delta\theta}{\mathrm{arcmin}} \text{ Å}, \tag{8.4}$$

where $\Delta\theta$ is the half energy width of the source and m is the (positive) spectral order (*XMM-Newton Users Handbook*).

In order to analyse the grating spectrum a measured profile of the object can be included within the spectral model (e.g. using the LPRO spectral model in the SPEX package [52, 88]), or a fitting approach can be used to fit for the line broadening caused by the spatial extent (e.g. [97]). One should note, however, that the source profile may be strongly energy dependent. For example, if a cluster has a cool core the spectrum of its centre can have strong emission lines but its emission would be continuum-dominated at a larger radius. An incorrect profile could over or underestimate the spatial broadening, particularly if measured using a low-spectral resolution instrument [96]. A measured spatial profile also includes the instrumental point spread function (PSF) which will need to be accounted for when modelling the spectral broadening.

To obtain the highest quality grating spectrum of a cluster requires an object where the line-emitting region is as compact as possible. Therefore our grating view of clusters of galaxies is biased towards those with steeply peaked surface brightness profiles. At higher redshifts, spatial line broadening becomes less of an issue, but the effective area of current grating instruments limits our ability to observe more distant clusters with reasonable exposure times.

Figure 8.2 shows the data obtained from an *XMM-Newton* RGS grating camera for an example source, the nearby Centaurus cluster of galaxies ($z = 0.010$, [97]), whose central region contains a complex structure of multiphase gas. The top panel is a high spatial resolution *Chandra* X-ray image of the source and includes the approximate RGS extraction region, showing the cross-dispersion axis direction. The centre panel plots the one-dimensional cross-dispersion image of the source as a function of the dispersion angle. The bottom panel shows the spectral orders as separated using the energy resolution of the CCD detectors. To analyse the event file from the camera, the user needs a source position and parameters for the cross-dispersion extraction (parametrized as a percentage of the PSF width) and CCD energy extraction widths. The extraction width optimises how much of the source to extract and is used to maximize the signal-to-noise ratio. Increasing the width in the cross-dispersion direction or CCD energy dimensions gives the user more photons, but includes more background. However, for extended objects, such as clusters, the cross-dispersion width also modifies the part of the source which is being examined in that spatial dimension. If a cluster has a cool core, like the example shown, a spectrum extracted from a smaller cross-dispersion extraction width will likely contain the emission lines with less background emission.

The dispersion and cross-dispersion directions depend on how the gratings are mounted in the telescope and the roll angle of the telescope on the sky. Spacecraft operational constraints also decide the rotation of the instrument at the time of the observation, meaning that the angle of the grating on the sky cannot be freely chosen.

Modelling the background components in spectra also presents a challenge for nearby clusters. For *XMM-Newton* there exists a tool which produces a background spectrum for non-X-ray-backgrounds, based on empty field observations selected using the off-axis count rate. In addition there will be cluster background components originating from radial regions beyond the radius of the cross-dispersion extraction region, in projection along the line of sight and along the dispersion axis.

A different approach to solving the source extent problem is to use a forward-modelling technique to predict the measured photons in the dispersion-cross dispersion plane and to self-consistently model the source [83, 84]. Combining observations with multiple roll angles, or also forward-modelling CCD data could help build up much more accurate 2D models of the source.

Microcalorimeters are much better suited to extended objects as they are intrinsically pixel-based instruments. If there is more than one pixel in the detector, then different spatial regions can be extracted by selecting which pixels to example. However, if the PSF of the telescope is relatively broad compared to the pixel size (as in *Hitomi* [105] or *XRISM* [120]), the pixels will not correspond to independent regions in the cluster.

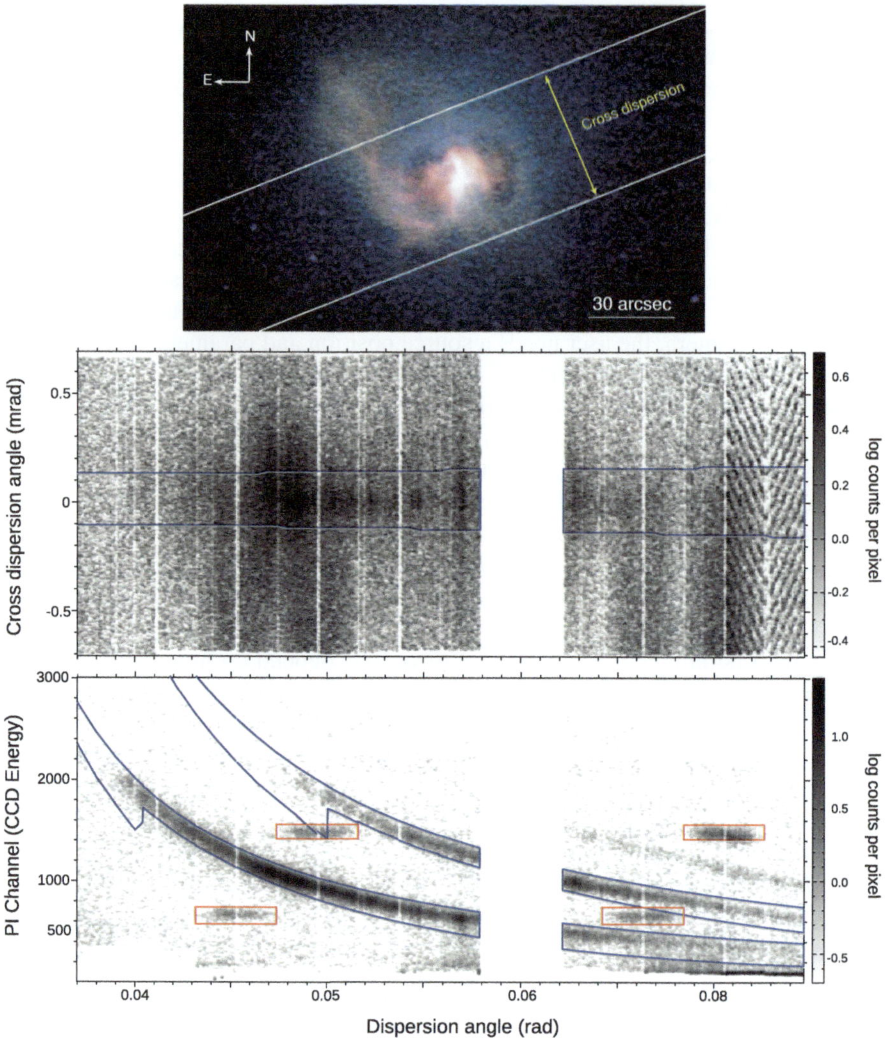

Fig. 8.2 *XMM-Newton* RGS observation of the Centaurus cluster of galaxies, taken with the RGS2 camera. (Top panel) *Chandra* RGB (0.5–1.0, 1.0–1.5 and 1.5–6.0 keV) image of Centaurus (taken from [101]) showing the RGS extraction region. (Centre panel) Distribution of the X-ray counts in cross-dispersion and dispersion-angle space, showing the 90% PSF width cross dispersion extraction region, which corresponds to around 0.8 arcmin in extraction width. (Bottom panel) Distribution of counts in the PI channel (pulse-invariant energy as measured by the CCD) as a function of dispersion angle, showing the 95% pulse-height extraction regions for the first two spectral orders. The orange boxes show the calibration sources. The *XMM* dataset is observation ID 0406200101 [97]. The empty dispersion range is due to a failed CCD in the RGS2 camera

8.4 Cool Cores

A substantial fraction of the clusters in the local universe have cores which are strong steeply-peaked emitters of X-ray radiation (e.g. [64]). This is due to the rising gas density towards the centres of these objects. As the X-ray emission scales with the square of the density, the X-ray emission is very peaked. Therefore if one calculates the mean radiative cooling timescale due to the radiation of X-rays, it can be very short, and much shorter than the age of the cluster. These clusters with dense cooler material in the centre are known as cool core clusters. Other non-cool core clusters have much flatter density profiles and isothermal cores [49], although there is a debate about whether there are two separate populations or a continuous distribution (e.g. [35]).

If there is not some form of heating, then this dense material should be rapidly cooling out of the X-ray waveband in cool core clusters. This could lead to a build-up of cold gas in the cores of clusters, and subsequently the formation of stars. The rate of material cooling out of the X-ray band is called the mass deposition rate ($\dot{M}$), within this cooling flow model [24]. $\dot{M}$ can be estimated from the excess X-ray luminosity (L) emitted in some regions where the radiative cooling timescale is shorter than the age of the object using

$$L = \frac{5}{2} \frac{\dot{M}}{\mu m} kT, \tag{8.5}$$

taking account of the gravitational work done on the gas as it flows towards the core, where T is the temperature and μm is the mean mass per particle. Typical values ranging from 10s to 100s of $\mathrm{M_\odot\,yr^{-1}}$, although there are extreme cases above $1000\,\mathrm{M_\odot\,yr^{-1}}$ [63]. Despite these expected rates of cooling, the measured star formation rates are a small fraction of the total mass deposition rate (e.g. [50, 65, 77]).

In a cooling flow, there should be a range of gas temperatures from some ambient cluster temperature down to temperature values which should not emit X-ray emission. This is typically modelled spectrally through a multi-temperature component model, where the component normalisations follow a DEM distribution. The commonly-used MKCFLOW and VMCFLOW models in the XSPEC spectral fitting package [4] follow the prescription of [73], where there is some number of temperature components between an upper and lower temperature. The differential emission measure at a temperature is then set to be inversely proportional to the bolometric luminosity at that temperature. The combined model spectrum is normalised to measure the mass deposition rate. If the lower temperature of the model is set to zero, then a full cooling flow is modelled. One limitation of this model, however, is that it does not explicitly include the gravitational work done on the gas in the cooling flow.

When high resolution *XMM*-RGS grating observations were first made of the core of nearby clusters, the spectra did not show the expected emission lines seen

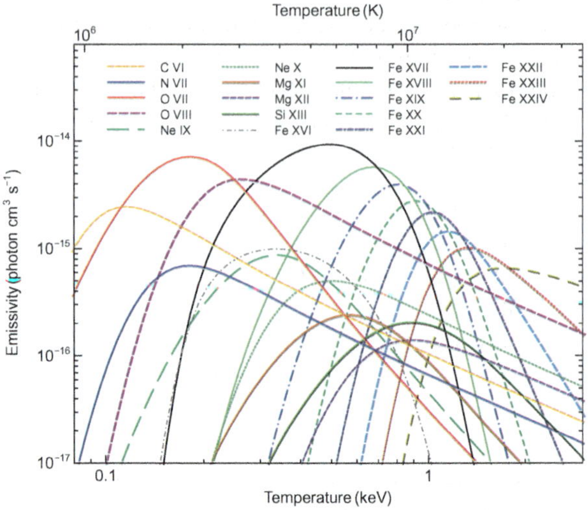

Fig. 8.3 Total line emissivities of ions in the RGS waveband (5–38Å) as a function of temperature. Emissivity here is defined to be the rate of radiative transitions divided by the product of the electron and H number densities. Therefore the relative strength of emission lines is that of a plasma with a constant density as a function of temperature. Values were calculated using APEC 3.0.9 [103] assuming solar abundances [3]

from lower temperature X-ray emitting gas, as predicted from models like the one described above. The power of these high-resolution spectra comes from a large number of temperature-sensitive emission lines present in the soft X-ray RGS band. These include Fe L-shell emission lines and others from O, Ne, Mg and Si. In Fig. 8.3 is plotted the strength of these emission lines as a function of temperature for Solar abundances. It can be seen that the presence or absence of some emission lines, or their relative strengths, are very good indicators of gas temperature. The best indicators at low X-ray emitting temperatures are the Fe XVII lines, showing gas around 0.5 keV and O VII, at around 0.2 keV.

In Abell 1835 [84] it was found that there was a lack of Fe XXIII, Fe XII and Ne X emission, implying there was little material below 2.7 keV in temperature, with less than 200 $M_\odot\ yr^{-1}$ cooling to very little temperatures, compared to the 1000 $M_\odot\ yr^{-1}$ expected from the X-ray profile. In Sérsic 159-03 [51] a two-temperature component fit was sufficient, implying less than 20% of the predicted 230 $M_\odot\ yr^{-1}$ cooling. An isothermal model was found to fit the RGS spectrum of Abell 1975 [108]. The lack of Fe XVII emission lines constrained the mass deposition rate to less than 150 $M_\odot\ yr^{-1}$, compared to the 500 $M_\odot\ yr^{-1}$ inferred from the X-ray emission. Abell 496 [106] also showed a deficit of material at lower temperatures.

Later studies examined more targets or looked at deep observations in detail. In an analysis looking at a sample of multiple clusters [82], 14 targets with high mass deposition rates were studied, finding that the emission lines indicated gas typically down to around 1/2 of the ambient temperature, but very little below 1/3 of that temperature. Figure 8.4 shows the spectra for a subset of these clusters, compared to cooling flow models with the appropriate mass deposition rates.

There is good evidence that this lack of cooling is due to the AGN feedback in these systems (for a review see [25]). In many of these clusters the central AGN are seen to be actively affecting their surroundings by the mechanical input of energy

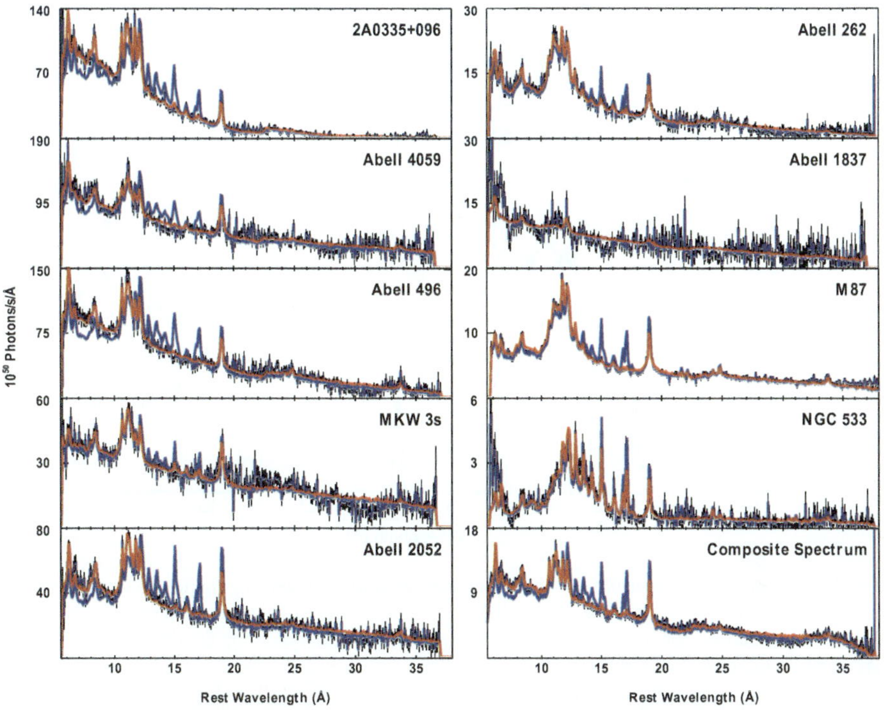

Fig. 8.4 Fluxed and de-redshifted RGS spectra of several galaxy clusters and a composite spectrum, taken from [82]. Shown on the data points is the best empirical model (red) and a cooling flow model scaled to match the soft X-ray flux (blue). It can be seen that the cooling flow model strongly overpredicts the flux of emission lines from colder gas

(as an example, see the Centaurus cluster in Fig. 8.2 top panel). The radio jets of these AGN form bubbles of radio-emitting plasma in the intracluster medium. The mechanical heating power can be estimated using both an estimate of the enthalpy of the bubbles (using the surrounding gas to measure the pressure) and a timescale for the formation of the bubbles [16]. There is a good correlation between the rate of X-ray cooling in the centre of clusters and this heating power (e.g. [48, 89]), implying that the AGN are responsible for preventing a large fraction of the cooling. The details of how the energy is dissipated from the feedback into the ICM are, however, uncertain. Studying the cool X-ray emitting gas in cluster cores and the associated multiphase material constrains models of AGN feedback. There are scenarios such as the chaotic-accretion model [32], where thermal instabilities cause cold clouds to 'rain' onto the black hole, or another where cold material can condense behind buoyantly rising AGN-inflated bubbles [66].

Deep observations of clusters allow tight constraints to be put on the level of cooling that can take place. For example the Centaurus cluster [97] shows cooling down from above 3 keV down to around 0.5 keV, as shown by the presence of

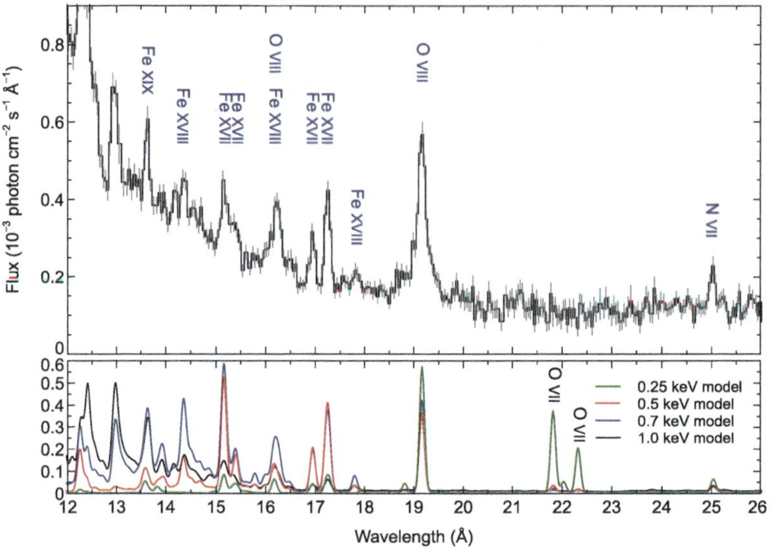

Fig. 8.5 Fluxed RGS spectrum of the core of the Centaurus cluster of galaxies, extracted from the innermost 0.8 arcmin width, taken from [97]. The bottom panel shows smoothed APEC models at various temperatures for comparison. The O VII emission lines, indicating material around 0.25 keV, are weak in the observed spectrum

the Fe XVII emission lines in the spectrum (Fig. 8.5). However, there is no strong evidence for O VII emission in the spectrum, although a reanalysis found evidence for weak emission [86], measuring the cooling levels below 0.25 keV temperature consistent with the strength of the lines. The amount of possible mass deposition as a function of temperature can be measured by fitting models consisting of multiple temperature components, approximating a DEM, or multiple cooling flow models spanning the temperature range. The results for Centaurus and three other clusters are shown in Fig. 8.6. At temperatures below 0.5 keV there are often sharp cut-offs in the temperature distribution, here constrained by the lack of O VII emission. However, in lower mass group-scale objects there is some evidence for gas at still lower temperatures, as seen by O VII emission in stacked observations [95] or in individual systems [85, 86].

The very low levels of allowed cooling found in the centres of these systems are suggestive that there is a tight connection between cooling and feedback. There cannot be strong overcooling or overheating. This conclusion can be somewhat relaxed if the coolest material can be hidden or cooled more rapidly than expected, for example by mixing. The very coolest X-ray emitting material could mix with cool gas associated with the Hα nebulae common in such systems [26], such as in the Virgo cluster [113]. In a large sample of clusters and groups, truncated cooling flow models were fit to the RGS spectra [58]. If the model was allowed to cool to zero, then the mass deposition rate was around 10–30% of the classical $\dot{M}$. However, if they terminate at

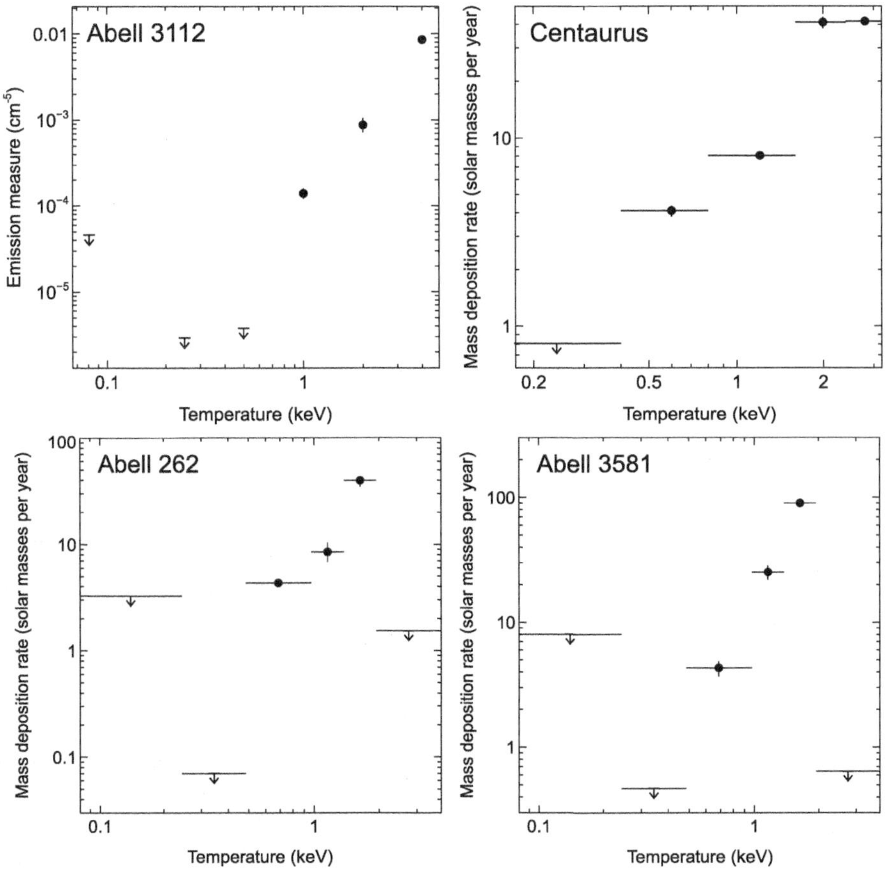

Fig. 8.6 Temperature distributions in four clusters, measured using deep *XMM*-RGS observations. For Abell 3112, the emission measure for 6 APEC components is shown (values taken from [13]). For Centaurus (values from [97]), Abell 262 and Abell 3581 (values from [98]) the data have been modelled using VMCFLOW components where the temperature range has been split into 5 or 6 sub-ranges, plotting the mass deposition rate

0.7 keV in clusters, the cooling rates are higher. Energetically the cooling gas could power the emission-line nebulae found in these systems. There also remains the possibility that absorption intrinsic within the cooling region could obscure substantial amounts of cooling material [27].

8.5 Motions in Galaxy Clusters

8.5.1 Introduction

Measuring motions in the ICM is a very important probe for many different physical phenomena in clusters. However, such measurements are very difficult to make due to the very high spectral resolution necessary to detect typical motions of $100-1000\,\mathrm{km\,s^{-1}}$, either in terms of the shifts due to bulk motions or increases in line widths due to velocity distribution. Although there are some existing results, it will not be until the launch of microcalorimeters such as those onboard *XRISM* and *Athena*, that such measurements will become more common.

One particularly important physical process which generates motions in the ICM is the AGN feedback in the cores of clusters. The feedback should induce motions of few hundred $\mathrm{km\,s^{-1}}$ in the ICM (e.g. [9, 40]). By measuring the velocity distribution around the central AGN we can measure the kinetic energy of the gas, testing feedback models. The level of energy in turbulence, shocks and sound waves can be compared to the thermal energy of the ICM and other non-thermal components. This "complete-calorimetry" of cluster cores will be very powerful for understanding AGN feedback [18]. For these purposes, ideally, is needed a telescope and detector able to resolve the spatial structures around the AGN cavities.

There are other sources for bulk motions and turbulence in a cluster. Clusters are not static objects within the hierarchical formation of structure. Merging substructures will induce shocks, flows and turbulence in the ICM. The influence of this activity is strongest in the outskirts of the cluster [55, 111]. In addition to generating turbulence, merging subclusters can cause the gas to slosh inside the potential well due to the gravitational centre moving away from the location of ICM peak during the merger [5, 124]. These sloshing motions can be several hundred $\mathrm{km\,s^{-1}}$ and last for Gyr.

Measuring the motions in clusters is not only important for our understanding of cluster evolution and AGN feedback. Non-thermal pressure support affects the assumption of hydrostatic equilibrium often made to measure cluster masses [74]. Therefore, it is useful to measure how close to equilibrium clusters are in different dynamical states. The strength of motions is also an indirect indicator of poorly measured properties of the plasma in the ICM, such as its viscosity and magnetic field substructure [122]. Flows and turbulence in clusters also affect how metals injected into the ICM are propagated through it over time [90].

8.5.2 RGS Line Widths

Even though the *XMM*-RGS have reduced spectral resolution for extended objects, it was found that they could be used to place constraints on the line of sight velocity width of the ICM. Although the spectra are broadened by the spatial extent of the

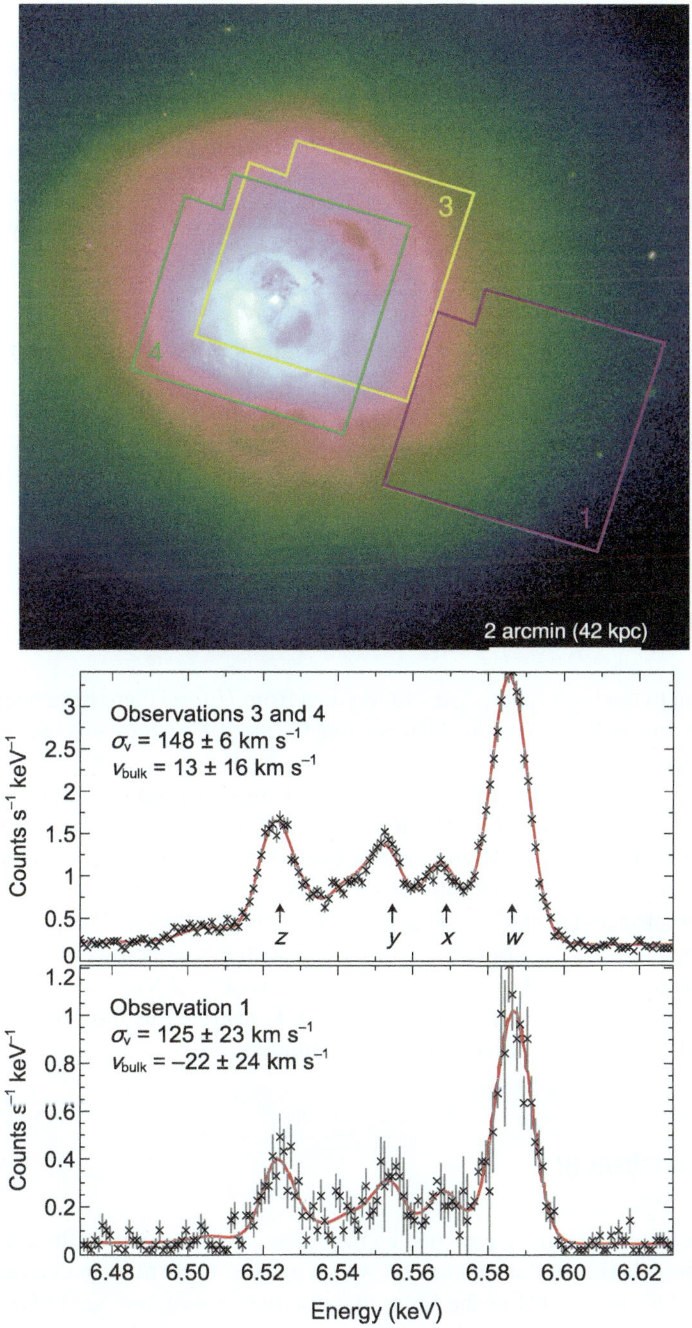
3
4
1
2 arcmin (42 kpc)
Observations 3 and 4
σv = 148 ± 6 km s−1
vbulk = 13 ± 16 km s−1
z
y
x
w
Observation 1
σv = 125 ± 23 km s−1
vbulk = −22 ± 24 km s−1
Counts s−1 keV−1
Energy (keV)

◀**Fig. 8.9** *Hitomi* measurements of the velocity line broadening (σ_v) and bulk velocity (v_{bulk}) for different regions of the Perseus cluster (data points from [44]). The bulk velocity is measured relative to the optical redshift of NGC 1275. (Top panel) 0.5–7.0 keV *Chandra* image of the cluster showing the observation footprints. (Centre panel) Spectrum around the He-like Fe XXV triplet extracted from two observations (3 and 4) pointed near the innermost region of the cluster, containing the two central cavities and the cavity to the northwest. The plot shows the data points and best-fitting model. The labelled energies are at those of the resonance (w), forbidden (z) and intercombination (x and y) lines. (Bottom panel) A spectrum taken from an offset pointing around 80 kpc to the southwest of the cluster centre, close to a cold front edge which may be generated by the sloshing of gas in the cluster potential

velocities, although there remain significant statistical and systematic uncertainties, including the uncertainties in model line emissivities.

These low values of velocity width put constraints on our models of how AGN feedback works. For example, turbulence does not effectively radially propagate if line widths are low [7, 30, 57]. Some other mechanism is required to radially transport energy in order to prevent the formation of a cooling flow.

8.5.5 CCD Measurements of Bulk Flows

In addition to the high spectral resolution data from *Hitomi* there have been previous measurements of bulk flows in clusters using CCD-type detectors. If the energy scale of the detector is known sufficiently well, the Fe-K lines can be used to measure velocities. The accurate energy calibration of *Suzaku* produced hints of motions in samples of clusters [79, 109]. Recently, *XMM-Newton* pn detector data been examined, making use of instrumental background lines to improve the understanding of the energy scale [93]. The initial results for Perseus agreed with *Hitomi* for a common region and found evidence for sloshing in the cluster. The gas in the Coma cluster appears to have a bimodal distribution with the velocities of its subclusters. The technique was similarly applied to the Centaurus and Virgo clusters to look for the effect of feedback on motions [33, 34]. CCD instruments have the advantage of large fields of view compared to microcalorimeters.

8.6 Enrichment

The intracluster medium is enriched by the stellar processes within the cluster galaxies and the galaxies in the subclusters which merged to form it. In clusters the ICM is enriched to around 1/3 of the Solar metallicity (for a review see [67]). Cool core clusters show a peak towards Solar values in their cores [68] and a flat distribution at $0.3\,Z_\odot$ in the outskirts [110, 114]. Non cool core clusters do not show this central peak [20]. The radial profiles are best measured in the Fe element, but profiles in several other elements (e.g. O, S, Si, Ar, Ca and Ni) show similar shapes [68]. In the central part of some cool core clusters there is evidence for a drop in metallicity (e.g. [68, 80, 94]).

The different stellar processes which make up the abundance pattern in a cluster produce different distributions of elements. C and N are produced in asymptotic giant branch (AGB) stars [41]. Core collapse supernovae (SNcc) produce other lighter metals ranging from O to Mg [76]. In contrast, Type Ia supernovae (SNIa), although their progenitor objects are not confidently identified, nor their explosion mechanism pinpointed exactly, are responsible for heavier elements between Cr to Ni (e.g. [54]). Intermediate mass elements between Si and Ca could be generated from either SNIa or SNcc supernovae.

The majority of baryonic mass in a cluster is in the form of the ICM and this ICM ends up of a large repository of metals that have been ejected or stripped from the member galaxies. Therefore studying the metallicity of the ICM is an important tool for understanding these stellar metal-generating processes. By obtaining the metallicity of different elements in the cluster and comparing the results to models produced by different mixtures of SNIa and SNcc models, these models can be tested. High-resolution X-ray spectroscopy is the best tool for measuring the ICM abundances.

However, most of the analyses of the radial profiles have been conducted using relatively low spectral resolution CCD instruments, with higher resolution spectra mostly concentrating on the cluster centres. The measurement of metallicity is made difficult by a number of systematic issues. One potential problem which is particularly important for lower spectral resolution data is the so-called Fe-bias (for details see [67]), which occurs in cooler group scale systems, and inverse Fe-bias, which is seen in intermediate temperature systems. In both of these cases, the problem is that the total observed spectrum originates from more than one temperature component. If only a single thermal model is fitted to the spectrum, some average temperature is obtained. In the case of Fe-bias, the average best-fitting temperature causes the best-fitting model metallicity to be reduced to fit the flatter spectral shape around the Fe-L complex. The inverse Fe-bias similarly increases the best-fitting metallicity to compensate for the inadequate modelling.

This problem is much easier to avoid when using higher spectral resolution, but it still remains to some degree in RGS-quality spectra. As seen in Fig. 8.2 RGS spectra can contain a considerable amount of multitemperature structure due to the large extraction size. The exact shape of the distribution of temperatures in the extraction region is difficult to predict. Figure 8.10 shows an example spectrum where the data are fit either by one or two thermal (CIE) components or by a log-Gaussian DEM distribution. The residual panels show no clear choice of model based on the emission lines. Simulations predict, for example, biases of 10–20% on the O/Fe ratio from the uncertainties in the temperature distribution modelling [21].

With RGS spectra the continuum can also be uncertain. This is important as metallicity measurements need both an accurate line and continuum strength. Lines from the hotter gas at larger radius in of the cluster can become very broad (due to its large spatial extent), leading to difficulty in modelling it. In addition, there can

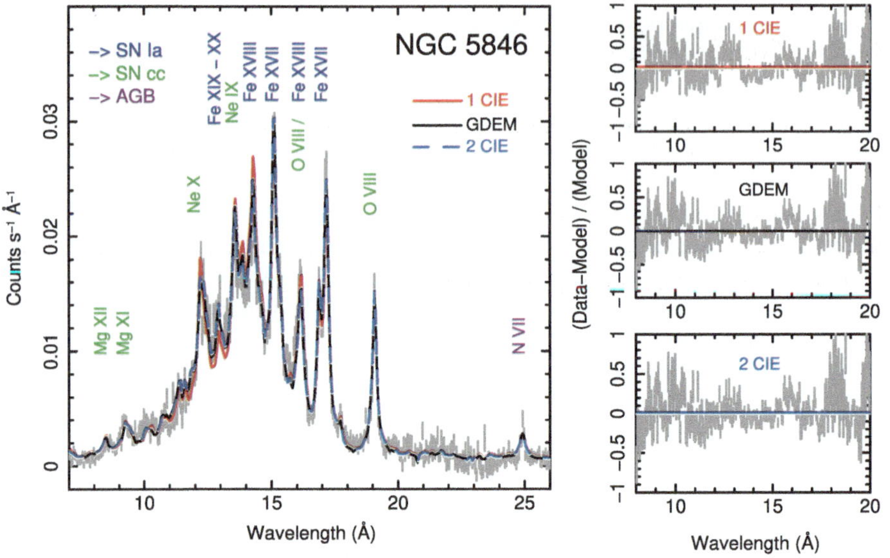

Fig. 8.10 RGS spectrum of the NGC 5846 galaxy group, taken from [21]. (Left panel) Data with models of three different temperature distributions, a single component CIE model, a Gaussian a log Gaussian distribution (GDEM) and a double CIE model. The spectral lines are labelled, and coloured according to the most likely origin of their element, i.e. SNIa, SNcc or AGB stars. (Right panels) Residuals of the three different models

be AGN contributions, both from AGN in the cluster and in the X-ray background which should be accounted for in modelling. Another effect which may bias the level of the continuum is the uncertainties of modelling the absorption towards the cluster, affecting the continuum at lower energies. A physical effect which could change the continuum is He-sedimentation, which can occur if He were to sediment to the cluster centre over its lifetime [28]. Although this build-up can be suppressed by various processes, it could affect X-ray metallicity measurements and other cluster properties (e.g. [81]).

As we have previously discussed in Sect. 8.5.3 the spectral models are not perfect, which introduces uncertainties in the line strengths and therefore the obtained metallicities. Furthermore, resonant scattering could affect abundance measurements, as shown in Perseus where its effect is 11% for Fe [46].

Microcalorimeters, such as the one that was onboard *Hitomi* will make large advances in the measurement of metallicities, due to the lack of spatial broadening, ability to study individual regions and spectral resolution. Figure 8.11 shows a combination of the spectra of Perseus using the microcalorimeter data at higher energies (soft energies are missing due to a valve in-place during these observations), RGS data in soft energies and lower resolution *Suzaku* XIS data over the whole band. The extreme improvement of the microcalorimeter over the CCD data is immediately apparent.

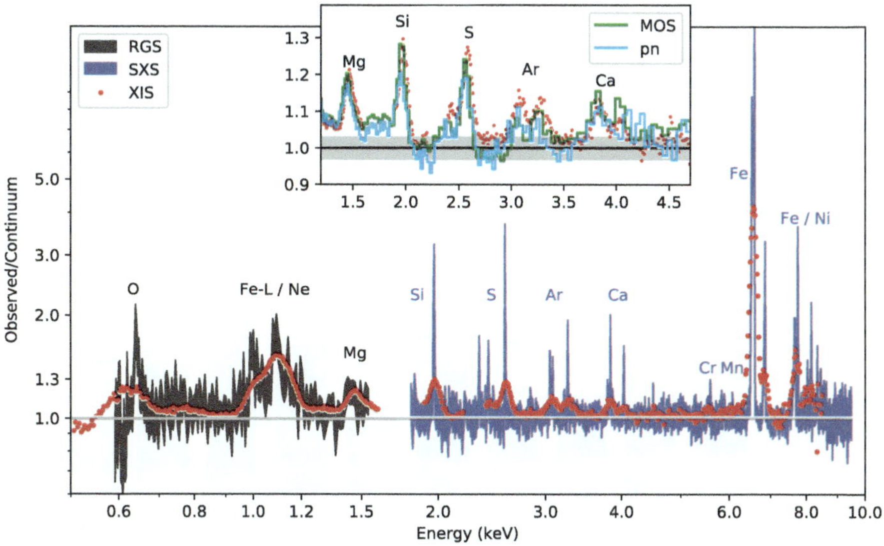

Fig. 8.11 Spectra of the core of the Perseus cluster, after having divided by continuum models, taken from [102]. Results are shown for the *XMM-Newton* RGS, *Hitomi* SXS microcalorimeter and *Suzaku* XIS CCD imaging spectrometer

Supernova models can be fitted to the ratios obtained. Different supernova models can be compared by seeing which matches the data better. For example, Fig. 8.12 shows two sets of ratios and model comparisons. The top panel is from a sample of 44 clusters, groups and giant elliptical galaxies called CHEERS [21], which were optimised for RGS measurements. The contributions of SNIa and SNcc to the best-fitting model are shown by coloured bars. It can be seen that the measured ratios are fairly similar to Solar (X/Fe=1) for many of the elements, except for Ca, Cr and Ni. Another interesting finding is that stacked RGS spectra [95] and individual analyses [60] show enhanced N ratios compared to Solar values.

In Fig. 8.12 (bottom panel) is shown the same set of ratios, obtained from high spectral resolution *Hitomi* data from the Perseus cluster, but using RGS for the light elements which could not be measured in this early spectrum [102]. In this case, all of the metals are extremely similar to Solar abundances ratios, with a simple exactly-Solar model giving a $\chi^2 = 10.7$ for 10 degrees if freedom. These ratios are remarkably similar to those found in the protosolar nebula, low-mass early-type galaxies and in near-solar metallicity Milky Way stars. However, this abundance pattern is difficult to reproduce with linear combinations of supernova models, as seen by the mismatch between the data and the models in Fig. 8.12. When comparing the data with CCD quality lower spectra, there is considerable scatter in the CCD results, depending on how the spectra were analysed.

The *Hitomi* results are unfortunately only for one cluster. Making progress in understanding nucleosynthesis and the enrichment of clusters requires observations

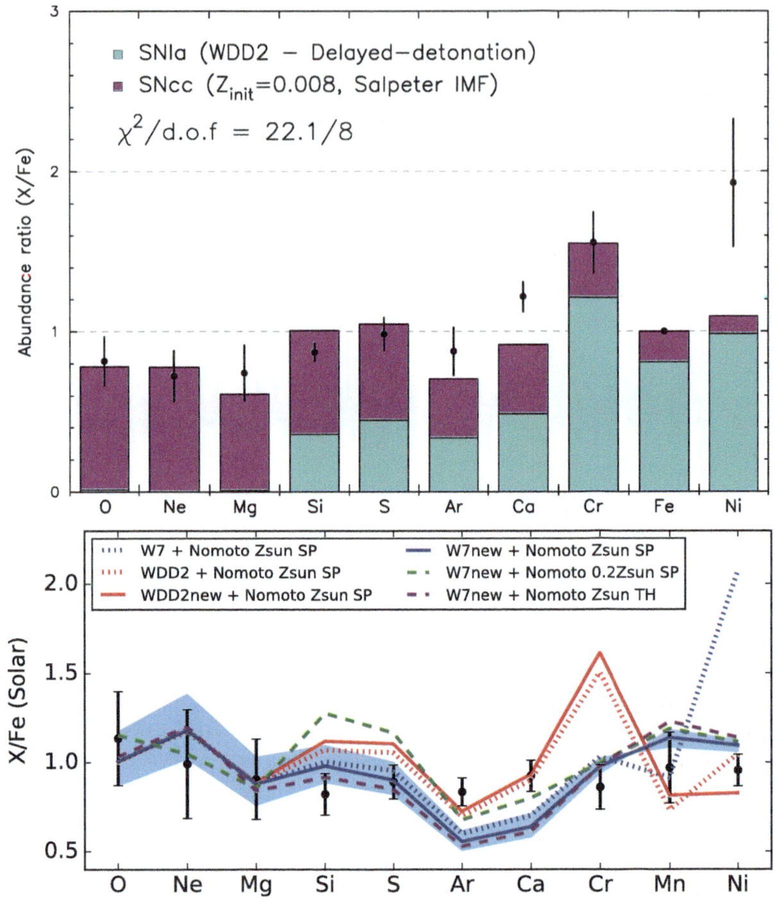

Fig. 8.12 Metal abundances compared to Fe. (Top panel) Average abundance ratios for the CHEERS sample, taken from [69]. The O/Fe and Ne/Fe were measured using RGS data, while the others use lower resolution CCD data. They are compared to a best-fitting model of made up of contributions from a classical SNIa model (WDD2) and a SNcc model. (Bottom panel) The ratios relative to Fe from the *Hitomi* spectrum of the Perseus cluster, or using RGS for the lighter elements (O, Ne, Mg), taken from [102]. The obtained ratios are compared to various combinations of SNIa and SNcc models

of more clusters with high spectral resolution. In addition, we would like to be able to spatially resolve the cluster, to study how metals are transported through the intracluster medium, which is an indirect measurement of motions [90]. Using CCD-quality spectra maps of metallicity in clusters have been made (e.g. [101]), showing how they are transported in the cluster, sloshing with the surrounding hot gas, but having a patchy appearance.

8.7 Beyond CIE Plasma Models

We have already discussed how resonant scattering can affect the spectra of clusters. However, there are other phenomena which may be important in understanding these objects, which can only be addressed by high-resolution spectroscopy.

The non-baryonic process could contribute to the X-ray spectrum in clusters, leading to extra emission lines or modified spectra. One interesting candidate is an emission line at ~ 3.5 keV found in stacked samples of clusters, or in some individual objects, including the Perseus cluster [11, 12, 14], using low spectral resolution CCD data. A possible origin for this line is that could originate from the radiative decay of a sterile neutrino. This is particularly interesting, as such a particle is a candidate for dark matter [10]. However, such claims are controversial as some other analyses did not find similar lines (e.g. [107]). With lower spectral resolution CCD data there are a number of potential systematics which could artificially create features which could be interpreted as lines. Instrumental systematics uncertainties, such as the calibration of the effective area, have been suggested, although this is made less likely by the stacked objects lying at different redshifts. Furthermore, the spectral model of the underlying cluster could introduce such features if there are lines missing or at the wrong strengths, leading to residuals at this energy.

To confirm the presence of the 3.5 keV line requires high spectral resolution data. The line was not found at the expected strength in the *Hitomi* spectrum of Perseus [43]. However, the flux of the line previously found in Perseus was much stronger than the average stacked signal (assuming an origin from dark matter), and so these data do not rule out a line at this level. Further deep *Hitomi*-quality spectra will undoubtedly help, although the line itself may be broad if its origin is from dark matter.

Another effect which may be generated by a different dark matter candidate, axion-like particles (ALPs), is an energy-dependent modulation of the spectra as photons and ALPs are interconverted [61, 119]. The strength of this interaction varies with the magnetic field in the cluster as the photon travels through the cluster. To look for the signal the spectrum of the cluster central AGN can be examined to look for variations from a power-law model. CCD data analyses can be affected by various systematics (e.g. pileup, due to too many photons). Deep *Chandra* grating observations of Perseus were recently made [91], placing very stringent limits on the strength of any possible ALP-photon coupling. Future measurements, however, require excellent calibration of any used instrument to make further improvements.

An effect which may be present in the spectra of clusters is charge exchange (CX; Chap. 10). This phenomenon occurs of a neutral atom interacts with a charged ion, where the ion then recombines into a highly excited state. One region this may occur is in the filamentary systems seen in cool core clusters like Perseus [29]. In this model the hot gas penetrates the cooler filamentary gas through reconnection diffusion, producing soft X-ray emission. Analysis of low resolution *Chandra* spectra indicates CX could contribute to the X-ray flux in Perseus, but the RGS spectra indicate that is not the case in Centaurus [112].

CX has also been proposed to be the origin for the 3.5 keV emission line [37], where it is due to a S XVI line. A search for CX O VIII emission in stacked cluster RGS data found marginal evidence for this line [38].

The assumption of a CIE equilibrium in the ICM of a galaxy cluster is also likely to be incorrect in some locations, for example in the outskirts of a galaxy cluster or in shocked regions [31]. The level of non-equilibrium is important as it can bias important cluster observables, such as surface brightness, and physical quantities, such as hydrostatic mass and temperature (e.g. [6]). Furthermore, it affects the spectrum in detail, allowing it to be detected from the anomalous line ratios in high-resolution spectra. One which is important in clusters is the He-like O VII triplet to H-like O VIII doublet line ratio [118]. Such a measurement could be difficult due to the surrounding complex warm-hot plasma and the requirements of future instrumentation.

8.8 Future Missions

There are many ways that high-resolution spectroscopy can aid our understanding of galaxy clusters. However, we are currently limited by instrumentation. Our primary workhorse is the RGS grating detectors onboard *XMM-Newton*. These have produced amazing results, but to progress further, we need higher spectral resolution and more effective area. In addition, the Fe-K lines cannot be examined using RGS, and these are particularly useful for velocity studies.

Hitomi has shown the power of microcalorimetry in clusters. These non-dispersive instruments are excellent for extended objects, because of the lack of spatial broadening of the spectra. In addition, they have good spectral resolution over their entire range. *Hitomi* studied only one cluster, Perseus, but fortunately, the replacement mission *XRISM* is due to be shortly launched (see Chap. 5). This will have extremely similar capabilities to *Hitomi*, allowing a much greater and more representative range of clusters and groups to be studied. One particularly interesting target is the Coma cluster, where turbulence inside this merging cluster could be studied in detail [123].

Although *XRISM* will make great advances, further future progress will also require a larger effective area and better spatial resolution. Spatial resolution is required to study in detail the different regions inside a cluster. In the core, in particular, we would like to study in detail the individual regions surrounding the central AGN to understand feedback. One mission that is hoped to provide these capabilities is the proposed *Athena* observatory [75]. The currently-designed X-IFU detector on *Athena* would provide ~ 5 arcsec spatial resolution combined with a large effective area. Figure 8.13 shows simulations of AGN feedback in two galaxy clusters as observed by an X-IFU-like instrument. It shows that the shape of the Fe-K lines can be examined in detail in different locations around the cluster, allowing the velocity of the material surrounding the inflated bubble to be directly measured. In addition, *Athena* would be able to measure turbulence and motions generated in the surround-

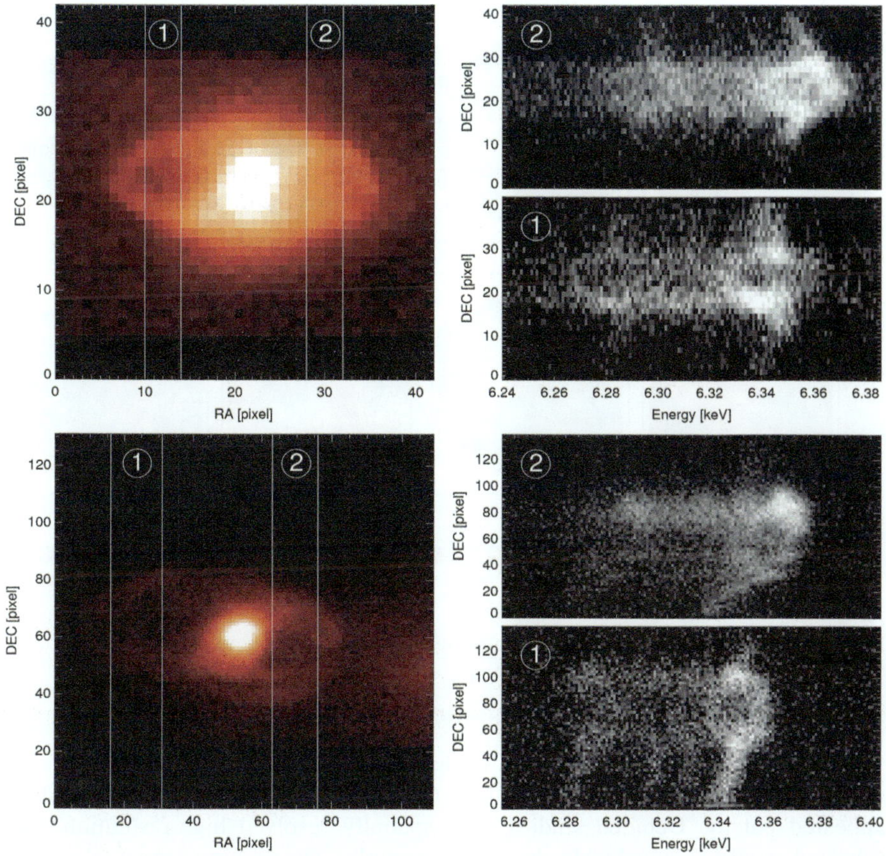

Fig. 8.13 Simulations of AGN feedback of a Cygnus A-like cluster (top) and Hydra A (bottom) for the proposed *Athena* X-IFU microcalorimeter array detector (taken from [18]) based on hydrodynamical simulations [40]. The left panels show images on the detector and 'virtual slit slices' labelled 1 and 2 across the cavities in the clusters. The right panels show the spectra around the 6.7 keV Fe XXV Kα along the virtual slits, for the 250 ks simulated observation. Using these spectra the velocity of the inflated bubble away and towards us can be inferred, along with its age and the jet power. In these simulations, the cavities have ages of 21 and 170 Myr, respectively

ing region. Such measurements would be extremely powerful in understanding the feedback process.

Athena will also allow detailed maps of the motions in clusters to be made, including the velocity and velocity dispersion. Figure 8.14 shows maps of a Coma-like galaxy cluster, where input maps are simulated and recovered [92]. In these simulated data, the 2D power spectrum was obtained, which is an important measurement to with which turbulence can be studied. The higher signal-to-noise ratio *Athena* would produce also has a big advantage in distinguishing studying multiple velocity components along the line as sight, as occurs in a major merger [8].

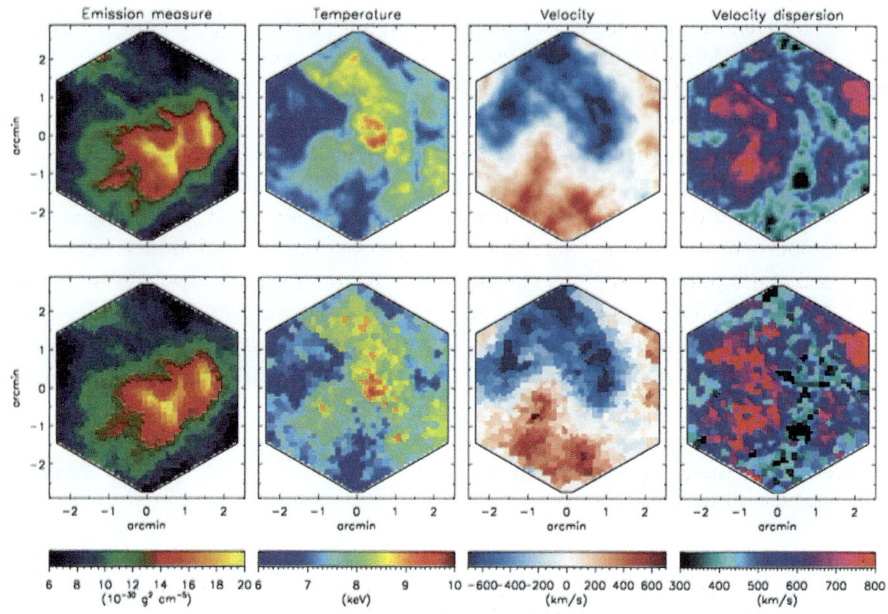

Fig. 8.14 Maps of input properties (top) and recovered values (bottom) for a simulated X-IFU observation (taken from [92])

Metallicity maps made with CCD detectors have shown that metals are not smoothly distributed through the cluster (e.g. [101]), but they sloshed around and appeared patchy. Detailed studies with spatially resolved high-resolution spectroscopy would be invaluable due to the complex temperature structure of these objects. *Athena* X-IFU should have the capability of mapping temperature, metallicity and density out to a large radius in a galaxy cluster, studying chemical evolution in clusters and metal transport through the ICM [19].

References

1. J. Ahoranta, A. Finoguenov, C. Pinto, J. Sanders, J. Kaastra, J. de Plaa, A. Fabian, Observations of asymmetric velocity fields and gas cooling in the NGC 4636 galaxy group X-ray halo. A&A **592**, A145 (2016)
2. S.W. Allen, A.E. Evrard, A.B. Mantz, Cosmological parameters from observations of galaxy clusters. ARA&A **49**(1), 409–470 (2011)
3. E. Anders, N. Grevesse, Abundances of the elements - meteoritic and solar. Geochimica Cosmochimica Acta **53**, 197–214 (1989)
4. K.A. Arnaud, XSPEC: the first ten years, in *ASP Conf. Ser. 101: Astronomical Data Analysis Software and Systems V*, ed. by G.H. Jacoby, J. Barnes (1996), p. 17
5. Y. Ascasibar, M. Markevitch, The origin of cold fronts in the cores of relaxed galaxy clusters. ApJ **650**, 102–127 (2006)

6. C. Avestruz, D. Nagai, E.T. Lau, K. Nelson, Non-equilibrium electrons in the outskirts of galaxy clusters. ApJ **808**(2), 176 (2015)
7. C.J. Bambic, C. Pinto, A.C. Fabian, J. Sanders, C.S. Reynolds, Limits on turbulent propagation of energy in cool-core clusters of galaxies. MNRAS **478**(1), L44–L48 (2018)
8. V. Biffi, J.A. ZuHone, T. Mroczkowski, E. Bulbul, W. Forman, The velocity structure of the intracluster medium during a major merger: simulated microcalorimeter observations. A&A **663**, A76 (2022)
9. M.A. Bourne, D. Sijacki, AGN jet feedback on a moving mesh: cocoon inflation, gas flows and turbulence. MNRAS **472**(4), 4707–4735 (2017)
10. A. Boyarsky, M. Drewes, T. Lasserre, S. Mertens, O. Ruchayskiy, Sterile neutrino Dark Matter. Prog. Part. Nucl. Phys. **104**, 1–45 (2019)
11. A. Boyarsky, O. Ruchayskiy, D. Iakubovskyi, J. Franse, Unidentified line in X-ray spectra of the andromeda galaxy and Perseus galaxy cluster. Phys. Rev. Lett. **113**(25), 251301 (2014)
12. E. Bulbul, M. Markevitch, A. Foster, R.K. Smith, M. Loewenstein, S.W. Randall, Detection of an unidentified emission line in the stacked X-ray spectrum of galaxy clusters. ApJ **789**(1), 13 (2014)
13. G.E. Bulbul, R.K. Smith, A. Foster, J. Cottam, M. Loewenstein, R. Mushotzky, R. Shafer, High-resolution XMM-newton spectroscopy of the cooling flow cluster A3112. ApJ **747**, 32 (2012)
14. N. Cappelluti, E. Bulbul, A. Foster, P. Natarajan, M.C. Urry, M.W. Bautz, F. Civano, E. Miller, R.K. Smith, Searching for the 3.5 keV Line in the deep fields with Chandra: the 10 Ms Observations. ApJ **854**(2), 179 (2018)
15. A.G. Cavaliere, H. Gursky, W.H. Tucker, Extragalactic X-ray sources and associations of galaxies. Nature **231**(5303), 437–438 (1971)
16. E. Churazov, M. Brüggen, C.R. Kaiser, H. Böhringer, W. Forman, Evolution of Buoyant bubbles in M87. ApJ **554**, 261–273 (2001)
17. C.S. Crawford, A.C. Fabian, Emission-line nebulae around central cluster galaxies in cooling flows. MNRAS **259**, 265–280 (1992)
18. J.H. Croston, J.S. Sanders, S. Heinz, M.J. Hardcastle, I. Zhuravleva, L. Bîrzan, R.G. Bower, M. Brüggen, E. Churazov, A.C. Edge, S. Ettori, A.C. Fabian, A. Finoguenov, J. Kaastra, M. Gaspari, M. Gitti, P. E.J. Nulsen, B.R. McNamara, E. Pointecouteau, T.J. Ponman, G.W. Pratt, D.A. Rafferty, T.H. Reiprich, D. Sijacki, D.M. Worrall, R.P. Kraft, I. McCarthy, M. Wise. The hot and energetic universe: AGN feedback in galaxy clusters and groups (2013). arXiv:1306.2323
19. E. Cucchetti, E. Pointecouteau, P. Peille, N. Clerc, E. Rasia, V. Biffi, S. Borgani, L. Tornatore, K. Dolag, M. Roncarelli, M. Gaspari, S. Ettori, E. Bulbul, T. Dauser, J. Wilms, F. Pajot, D. Barret, Athena X-IFU synthetic observations of galaxy clusters to probe the chemical enrichment of the Universe. A&A **620**, A173 (2018)
20. S. De Grandi, S. Molendi, Metallicity gradients in X-ray clusters of galaxies. ApJ **551**(1), 153–159 (2001)
21. J. de Plaa, J.S. Kaastra, N. Werner, C. Pinto, P. Kosec, Y.Y. Zhang, F. Mernier, L. Lovisari, H. Akamatsu, G. Schellenberger, F. Hofmann, T.H. Reiprich, A. Finoguenov, J. Ahoranta, J.S. Sanders, A.C. Fabian, O. Pols, A. Simionescu, J. Vink, H. Böhringer, CHEERS: the chemical evolution RGS sample. A&A **607**, A98 (2017)
22. J. de Plaa, N. Werner, A.M. Bykov, J.S. Kaastra, M. Méndez, J. Vink, J.A.M. Bleeker, M. Bonamente, J.R. Peterson, Chemical evolution in Sérsic 159–03 observed with XMM-Newton. A&A **452**(2), 397–412 (2006)
23. J. de Plaa, I. Zhuravleva, N. Werner, J.S. Kaastra, E. Churazov, R.K. Smith, A.J.J. Raassen, Y.G. Grange, Estimating turbulent velocities in the elliptical galaxies NGC 5044 and NGC 5813. A&A **539**, A34 (2012)
24. A.C. Fabian, Cooling flows in clusters of galaxies. ARA&A **32**, 277–318 (1994)
25. A.C. Fabian, Observational evidence of active galactic nuclei feedback. ARA&A **50**, 455–489 (2012)

26. A.C. Fabian, S.W. Allen, C.S. Crawford, R.M. Johnstone, R.G. Morris, J.S. Sanders, R.W. Schmidt, The missing soft X-ray luminosity in cluster cooling flows. MNRAS **332**(3), L50–L54 (2002)
27. A.C. Fabian, G.J. Ferland, J.S. Sanders, B.R. McNamara, C. Pinto, S.A. Walker, Hidden cooling flows in clusters of galaxies. MNRAS **515**(3), 3336–3345 (2022)
28. A.C. Fabian, J.E. Pringle, On the spatial distribution of heavy elements in X-ray emitting clusters of galaxies. MNRAS **181**, 5P-7P (1977)
29. A.C. Fabian, J.S. Sanders, R.J.R. Williams, A. Lazarian, G.J. Ferland, R.M. Johnstone, The energy source of the filaments around the giant galaxy NGC 1275. MNRAS **417**, 172–177 (2011)
30. A.C. Fabian, S.A. Walker, H.R. Russell, C. Pinto, J.S. Sanders, C.S. Reynolds, Do sound waves transport the AGN energy in the Perseus cluster? MNRAS **464**, L1–L5 (2017)
31. D.C. Fox, A. Loeb, Do the electrons and ions in x-ray clusters share the same temperature? ApJ **491**(2), 459–466 (1997)
32. M. Gaspari, M. Ruszkowski, S.P. Oh, Chaotic cold accretion on to black holes. MNRAS **432**(4), 3401–3422 (2013)
33. E. Gatuzz, J.S. Sanders, R. Canning, K. Dennerl, A.C. Fabian, C. Pinto, H. Russell, T. Tamura, S.A. Walker, J. ZuHone, The velocity structure of the intracluster medium of the Centaurus cluster. MNRAS **513**(2), 1932–1946 (2022)
34. E. Gatuzz, J.S. Sanders, K. Dennerl, C. Pinto, A.C. Fabian, T. Tamura, S.A. Walker, J. ZuHone, Measuring sloshing, merging, and feedback velocities in the Virgo cluster. MNRAS **511**(3), 4511–4527 (2022)
35. V. Ghirardini, Y.E. Bahar, E. Bulbul, A. Liu, N. Clerc, F. Pacaud, J. Comparat, T. Liu, M.E. Ramos-Ceja, D. Hoang, J. Ider-Chitham, M. Klein, A. Merloni, K. Nandra, N. Ota, P. Predehl, T.H. Reiprich, J. Sanders, T. Schrabback, The eROSITA Final Equatorial-Depth Survey (eFEDS). Characterization of morphological properties of galaxy groups and clusters. A&A **661A**, 12 (2022)
36. M.R. Gilfanov, R.A. Syunyaev, E.M. Churazov, Radial brightness profiles of resonance X-ray lines in galaxy clusters. Sov. Astron. Lett. **13**, 3 (1987)
37. L. Gu, J. Kaastra, A.J.J. Raassen, P.D. Mullen, R.S. Cumbee, D. Lyons, P.C. Stancil, A novel scenario for the possible X-ray line feature at $\sim$ 3.5 keV. Charge exchange with bare sulfur ions. A&A **584**, L11 (2015)
38. L. Gu, J. Mao, J. de Plaa, A.J.J. Raassen, C. Shah, J.S. Kaastra, Charge exchange in galaxy clusters. A&A **611**, A26 (2018)
39. N.A. Hatch, C.S. Crawford, A.C. Fabian, Ionized nebulae surrounding brightest cluster galaxies. MNRAS **380**, 33–43 (2007)
40. S. Heinz, M. Brüggen, B. Morsony, Prospects of high-resolution X-ray spectroscopy for active galactic nucleus feedback in galaxy clusters. ApJ **708**, 462–468 (2010)
41. F. Herwig, Evolution of asymptotic giant branch stars. ARA&A **43**(1), 435–479 (2005)
42. Hitomi Collaboration, The quiescent intracluster medium in the core of the Perseus cluster. Nature **535**, 117–121 (2016)
43. Hitomi Collaboration, Hitomi constraints on the 3.5 keV Line in the Perseus galaxy Cluster. ApJ **837**(1), L15 (2017)
44. Hitomi Collaboration, Atmospheric gas dynamics in the Perseus cluster observed with Hitomi. PASJ **70**, 9 (2018)
45. Hitomi Collaboration, Atomic data and spectral modeling constraints from high-resolution X-ray observations of the Perseus cluster with Hitomi. PASJ **70**, 12 (2018)
46. Hitomi Collaboration, Measurements of resonant scattering in the Perseus cluster core with Hitomi SXS. PASJ **70**, 10 (2018)
47. Hitomi Collaboration, Temperature structure in the Perseus cluster core observed with Hitomi. PASJ **70**, 11 (2018)
48. J. Hlavacek-Larrondo, A.C. Fabian, A.C. Edge, H. Ebeling, J.S. Sanders, M.T. Hogan, G.B. Taylor, Extreme AGN feedback in the massive cluster survey: a detailed study of X-ray cavities at z>0.3. MNRAS **421**(2), 1360–1384 (2012)

49. D.S. Hudson, R. Mittal, T.H. Reiprich, P.E.J. Nulsen, H. Andernach, C.L. Sarazin, What is a cool-core cluster? a detailed analysis of the cores of the X-ray flux-limited HIFLUGCS cluster sample. A&A **513**, A37 (2010)
50. R.M. Johnstone, A.C. Fabian, P.E.J. Nulsen, The optical spectra of central galaxies in southern clusters: evidence for star formation. MNRAS **224**, 75–91 (1987)
51. J.S. Kaastra, C. Ferrigno, T. Tamura, F.B.S. Paerels, J.R. Peterson, J.P.D. Mittaz, XMM-Newton observations of the cluster of galaxies Sérsic 159–03. A&A **365**, L99–L103 (2001)
52. J.S. Kaastra, R. Mewe, H. Nieuwenhuijzen. SPEX: a new code for spectral analysis of X & UV spectra, in *UV and X-ray Spectroscopy of Astrophysical and Laboratory Plasmas* (1996), pp. 411–414
53. J.S. Kaastra, F.B.S. Paerels, F. Durret, S. Schindler, P. Richter, Thermal radiation processes. Space Sci. Rev. **134**(1–4), 155–190 (2008)
54. F. Lach, F.K. Röpke, I.R. Seitenzahl, B. Coté, S. Gronow, A.J. Ruiter, Nucleosynthesis imprints from different Type Ia supernova explosion scenarios and implications for galactic chemical evolution. A&A **644**, A118 (2020)
55. E.T. Lau, A.V. Kravtsov, D. Nagai, Residual gas motions in the intracluster medium and bias in hydrostatic measurements of mass profiles of clusters. ApJ **705**, 1129–1138 (2009)
56. D.A. Liedahl, A.L. Osterheld, W.H. Goldstein, New calculations of Fe L-shell X-ray spectra in high-temperature plasmas. ApJ **438**, L115–L118 (1995)
57. H. Liu, A.C. Fabian, C. Pinto, H.R. Russell, J.S. Sanders, B.R. McNamara, Suppressed cooling and turbulent heating in the core of X-ray luminous clusters RXCJ1504.1-0248 and Abell 1664. MNRAS **505**(2), 1589–1599 (2021)
58. H. Liu, C. Pinto, A.C. Fabian, H.R. Russell, J.S. Sanders, Searching for cool and cooling X-ray emitting gas in 45 galaxy clusters and groups. MNRAS **485**(2), 1757–1774 (2019)
59. W. Liu, M. Chiao, M.R. Collier, T. Cravens, M. Galeazzi, D. Koutroumpa, K.D. Kuntz, R. Lallement, S.T. Lepri, D. McCammon, K. Morgan, F.S. Porter, S.L. Snowden, N.E. Thomas, Y. Uprety, E. Ursino, B.M. Walsh, The structure of the local hot bubble. ApJ **834**(1), 33 (2017)
60. J. Mao, J. de Plaa, J.S. Kaastra, C. Pinto, L. Gu, F. Mernier, H.-L. Yan, Y.-Y. Zhang, H. Akamatsu, Nitrogen abundance in the X-ray halos of clusters and groups of galaxies. A&A **621**, A9 (2019)
61. M.C.D. Marsh, H.R. Russell, A.C. Fabian, B.R. McNamara, P. Nulsen, C.S. Reynolds, A new bound on axion-like particles. J. Cosmol. Astropart. Phys. **2017**(12), 036 (2017)
62. W.G. Mathews, D.A. Buote, F. Brighenti, Spatial diffusion of X-ray emission lines in the M87 cooling flow. Evidence for absorption. ApJ **550**(1), L31–L34 (2001)
63. M. McDonald, M. Bayliss, B.A. Benson, R.J. Foley, J. Ruel, P. Sullivan, S. Veilleux, K.A. Aird, M.L.N. Ashby, M. Bautz, G. Bazin, L.E. Bleem, M. Brodwin, J.E. Carlstrom, C.L. Chang, H.M. Cho, A. Clocchiatti, T.M. Crawford, A.T. Crites, T. de Haan, S. Desai, M.A. Dobbs, J.P. Dudley, E. Egami, W.R. Forman, G.P. Garmire, E.M. George, M.D. Gladders, A.H. Gonzalez, N.W. Halverson, N.L. Harrington, F.W. High, G.P. Holder, W.L. Holzapfel, S. Hoover, J.D. Hrubes, C. Jones, M. Joy, R. Keisler, L. Knox, A.T. Lee, E.M. Leitch, J. Liu, M. Lueker, D. Luong-van, A. Mantz, D.P. Marrone, J.J. McMahon, J. Mehl, S.S. Meyer, E.D. Miller, L. Mocanu, J.J. Mohr, T.E. Montroy, S.S. Murray, T. Natoli, S. Padin, T. Plagge, C. Pryke, T.D. Rawle, C.L. Reichardt, A. Rest, M. Rex, J.E. Ruhl, B.R. Saliwanchik, A. Saro, J.T. Sayre, K.K. Schaffer, L. Shaw, E. Shirokoff, R. Simcoe, J. Song, H.G. Spieler, B. Stalder, Z. Staniszewski, A.A. Stark, K. Story, C.W. Stubbs, R. Šuhada, A. van Engelen, K. Vanderlinde, J.D. Vieira, A. Vikhlinin, R. Williamson, O. Zahn, A. Zenteno, A massive, cooling-flow-induced starburst in the core of a luminous cluster of galaxies. Nature **488**, 349–352 (2012)
64. M. McDonald, B.A. Benson, A. Vikhlinin, B. Stalder, L.E. Bleem, T. de Haan, H.W. Lin, K.A. Aird, M.L.N. Ashby, M.W. Bautz, M. Bayliss, S. Bocquet, M. Brodwin, J.E. Carlstrom, C.L. Chang, H.M. Cho, A. Clocchiatti, T.M. Crawford, A.T. Crites, S. Desai, M.A. Dobbs, J.P. Dudley, R.J. Foley, W.R. Forman, E.M. George, D. Gettings, M. D. Gladders, A. H. Gonzalez, N. W. Halverson, F. W. High, G. P. Holder, W. L. Holzapfel, S. Hoover, J.D. Hrubes, C. Jones, M. Joy, R. Keisler, L. Knox, A.T. Lee, E.M. Leitch, J. Liu, M. Lueker, D. Luong-Van, A. Mantz, D.P. Marrone, J.J. McMahon, J. Mehl, S.S. Meyer, E.D. Miller, L. Mocanu, J.J. Mohr, T.E.

Montroy, S.S. Murray, D. Nurgaliev, S. Padin, T. Plagge, C. Pryke, C.L. Reichardt, A. Rest, J. Ruel, J.E. Ruhl, B.R. Saliwanchik, A. Saro, J.T. Sayre, K.K. Schaffer, E. Shirokoff, J. Song, R. Šuhada, H.G. Spieler, S.A. Stanford, Z. Staniszewski, A.A. Stark, K. Story, A. van Engelen, K. Vanderlinde, J.D. Vieira, R. Williamson, O. Zahn, A. Zenteno, The growth of cool cores and evolution of cooling properties in a sample of 83 galaxy clusters at $0.3 < z < 1.2$ selected from the SPT-SZ survey. ApJ **774**(1), 23 (2013)
65. M. McDonald, M. Gaspari, B.R. McNamara, G.R. Tremblay, Revisiting the cooling flow problem in galaxies, groups, and clusters of galaxies. ApJ **858**(1), 45 (2018)
66. B.R. McNamara, H.R. Russell, P.E.J. Nulsen, M.T. Hogan, A.C. Fabian, F. Pulido, A.C. Edge, A mechanism for stimulating AGN feedback by lifting gas in massive galaxies. ApJ **830**(2), 79 (2016)
67. F. Mernier, V. Biffi, H. Yamaguchi, P. Medvedev, A. Simionescu, S. Ettori, N. Werner, J.S. Kaastra, J. de Plaa, L. Gu, Enrichment of the hot intracluster medium: observations. Space Sci. Rev. **214**(8), 129 (2018)
68. F. Mernier, J. de Plaa, J.S. Kaastra, Y.Y. Zhang, H. Akamatsu, L. Gu, P. Kosec, J. Mao, C. Pinto, T.H. Reiprich, J.S. Sanders, A. Simionescu, N. Werner, Radial metal abundance profiles in the intra-cluster medium of cool-core galaxy clusters, groups, and ellipticals. A&A **603**, A80 (2017)
69. F. Mernier, J. de Plaa, C. Pinto, J.S. Kaastra, P. Kosec, Y.Y. Zhang, J. Mao, N. Werner, O.R. Pols, J. Vink, Origin of central abundances in the hot intra-cluster medium. II. Chemical enrichment and supernova yield models. A&A **595**, A126 (2016)
70. R. Mewe, E.H.B.M. Gronenschild, G.H.J. van den Oord, Calculated X-radiation from optically thin plasmas - part five. A&AS **62**, 197 (1985)
71. R. Mewe, J.R. Lemen, G.H.J. van den Oord, Calculated X-radiation from optically thin plasmas. VI - Improved calculations for continuum emission and approximation formulae for nonrelativistic average Gaunt actors. A&AS **65**, 511–536 (1986)
72. R.J. Mitchell, J.L. Culhane, P.J.N. Davison, J.C. Ives, Ariel 5 observations of the X-ray spectrum of the Perseus cluster. MNRAS **175**, 29P-34P (1976)
73. R.F. Mushotzky, A.E. Szymkowiak, Einstein observatory solid state detector observations of cooling flows in clusters of galaxies, in *Cooling Flows in Clusters and Galaxies*, ed. by A.C. Fabian, *NATO Advanced Study Institute (ASI) Series C*, vol. 229 (1988), p. 53
74. D. Nagai, A. Vikhlinin, A.V. Kravtsov, Testing X-Ray measurements of galaxy clusters with cosmological simulations. ApJ **655**(1), 98–108 (2007)
75. K. Nandra, D. Barret, X. Barcons, A. Fabian, J.-W. den Herder, L. Piro, M. Watson, C. Adami, J. Aird, J.M. Afonso et al., The Hot and Energetic Universe: A White Paper presenting the science theme motivating the Athena+ mission (2013)
76. K. Nomoto, C. Kobayashi, N. Tominaga, Nucleosynthesis in stars and the chemical enrichment of galaxies. ARA&A **51**(1), 457–509 (2013)
77. P.E.J. Nulsen, R.M. Johnstone, A.C. Fabian, Star formation in cooling flows. Publ. Astron. Soc. Australia **7**(2), 132–135 (1987)
78. A. Ogorzalek, I. Zhuravleva, S.W. Allen, C. Pinto, N. Werner, A.B. Mantz, R.E.A. Canning, A.C. Fabian, J.S. Kaastra, J. de Plaa, Improved measurements of turbulence in the hot gaseous atmospheres of nearby giant elliptical galaxies. MNRAS **472**, 1659–1676 (2017)
79. N. Ota, H. Yoshida, Search for gas bulk motions in eight nearby clusters of galaxies with Suzaku. PASJ **68**, S19 (2016)
80. E.K. Panagoulia, A.C. Fabian, J.S. Sanders, Searching for the missing iron mass in the core of the Centaurus cluster. MNRAS **433**, 3290–3296 (2013)
81. F. Peng, D. Nagai, Effect of helium sedimentation on X-Ray measurements of galaxy clusters. ApJ **693**(1), 839–846 (2009)
82. J.R. Peterson, S.M. Kahn, F.B.S. Paerels, J.S. Kaastra, T. Tamura, J.A.M. Bleeker, C. Ferrigno, J.G. Jernigan, High-resolution X-Ray spectroscopic constraints on cooling-flow models for clusters of galaxies. ApJ **590**, 207–224 (2003)
83. J.R. Peterson, P.J. Marshall, K. Andersson, Smoothed particle inference: a kilo-parametric method for X-Ray galaxy cluster modeling. ApJ **655**, 109–127 (2007)

84. J.R. Peterson, F.B.S. Paerels, J.S. Kaastra, M. Arnaud, T.H. Reiprich, A.C. Fabian, R.F. Mushotzky, J.G. Jernigan, I. Sakelliou, X-Ray imaging-spectroscopy of Abell 1835. A&A **365**, L104–L109 (2001)
85. C. Pinto, A.C. Fabian, A. Ogorzalek, I. Zhuravleva, N. Werner, J. Sanders, Y.Y. Zhang, L. Gu, J. de Plaa, J. Ahoranta, A. Finoguenov, R. Johnstone, R.E.A. Canning, Insights into the location and dynamics of the coolest X-ray emitting gas in clusters of galaxies. MNRAS **461**(2), 2077–2084 (2016)
86. C. Pinto, A.C. Fabian, N. Werner, P. Kosec, J. Ahoranta, J. de Plaa, J.S. Kaastra, J.S. Sanders, Y.-Y. Zhang, A. Finoguenov, Discovery of O VII line emitting gas in elliptical galaxies. A&A **572**, L8 (2014)
87. C. Pinto, J.S. Kaastra, E. Costantini, C. de Vries, Interstellar medium composition through X-ray spectroscopy of low-mass X-ray binaries. A&A **551**, A25 (2013)
88. C. Pinto, J.S. Sanders, N. Werner, J. de Plaa, A.C. Fabian, Y.-Y. Zhang, J.S. Kaastra, A. Finoguenov, J. Ahoranta, Chemical enrichment RGS cluster sample (CHEERS): constraints on turbulence. A&A **575**, A38 (2015)
89. D.A. Rafferty, B.R. McNamara, P.E.J. Nulsen, M.W. Wise, The feedback-regulated growth of black holes and bulges through gas accretion and starbursts in cluster central dominant galaxies. ApJ **652**, 216–231 (2006)
90. P. Rebusco, E. Churazov, H. Böhringer, W. Forman, Impact of stochastic gas motions on galaxy cluster abundance profiles. MNRAS **359**, 1041–1048 (2005)
91. C.S. Reynolds, M.C.D. Marsh, H.R. Russell, A.C. Fabian, R. Smith, F. Tombesi, S. Veilleux, Astrophysical limits on very light axion-like particles from Chandra grating spectroscopy of NGC 1275. ApJ **890**(1), 59 (2020)
92. M. Roncarelli, M. Gaspari, S. Ettori, V. Biffi, F. Brighenti, E. Bulbul, N. Clerc, E. Cucchetti, E. Pointecouteau, E. Rasia, Measuring turbulence and gas motions in galaxy clusters via synthetic Athena X-IFU observations. A&A **618**, A39 (2018)
93. J.S. Sanders, K. Dennerl, H.R. Russell, D. Eckert, C. Pinto, A.C. Fabian, S.A. Walker, T. Tamura, J. ZuHone, F. Hofmann, Measuring bulk flows of the intracluster medium in the Perseus and Coma galaxy clusters using XMM-Newton. A&A **633**, A42 (2020)
94. J.S. Sanders, A.C. Fabian, Spatially resolved X-ray spectroscopy of the core of the Centaurus cluster. MNRAS **331**, 273–283 (2002)
95. J.S. Sanders, A.C. Fabian, Revealing O VII from stacked X-ray grating spectra of clusters, groups and elliptical galaxies. MNRAS **412**(1), L35–L39 (2011)
96. J.S. Sanders, A.C. Fabian, Velocity width measurements of the coolest X-ray emitting material in the cores of clusters, groups and elliptical galaxies. MNRAS **429**, 2727–2738 (2013)
97. J.S. Sanders, A.C. Fabian, S.W. Allen, R.G. Morris, J. Graham, R.M. Johnstone, Cool X-ray emitting gas in the core of the Centaurus cluster of galaxies. MNRAS **385**, 1186–1200 (2008)
98. J.S. Sanders, A.C. Fabian, K.A. Frank, J.R. Peterson, H.R. Russell, Deep high-resolution X-ray spectra from cool-core clusters. MNRAS **402**, 127–144 (2010)
99. J.S. Sanders, A.C. Fabian, R.K. Smith, Constraints on turbulent velocity broadening for a sample of clusters, groups and elliptical galaxies using XMM-Newton. MNRAS **410**(3), 1797–1812 (2011)
100. J.S. Sanders, A.C. Fabian, R.K. Smith, J.R. Peterson, A direct limit on the turbulent velocity of the intracluster medium in the core of Abell 1835 from XMM-Newton. MNRAS **402**, L11–L15 (2010)
101. J.S. Sanders, A.C. Fabian, G.B. Taylor, H.R. Russell, K.M. Blundell, R.E.A. Canning, J. Hlavacek-Larrondo, S.A. Walker, C.K. Grimes, A very deep Chandra view of metals, sloshing and feedback in the Centaurus cluster of galaxies. MNRAS **457**, 82–109 (2016)
102. A. Simionescu, S. Nakashima, H. Yamaguchi, K. Matsushita, F. Mernier, N. Werner, T. Tamura, K. Nomoto, J. de Plaa, S.C. Leung, A. Bamba, E. Bulbul, M.E. Eckart, Y. Ezoe, A.C. Fabian, Y. Fukazawa, L. Gu, Y. Ichinohe, M.N. Ishigaki, J.S. Kaastra, C. Kilbourne, T. Kitayama, M. Leutenegger, M. Loewenstein, Y. Maeda, E.D. Miller, R.F. Mushotzky, H. Noda, C. Pinto, F.S. Porter, S. Safi-Harb, K. Sato, T. Takahashi, S. Ueda, S. Zha, Constraints on the chemical enrichment history of the Perseus Cluster of galaxies from high-resolution X-ray spectroscopy. MNRAS **483**(2), 1701–1721 (2019)

103. R.K. Smith, N.S. Brickhouse, D.A. Liedahl, J.C. Raymond, Collisional plasma models with APEC/APED: emission-line diagnostics of hydrogen-like and helium-like ions. ApJ **556**, L91–L95 (2001)
104. S.L. Snowden, R. Egger, M.J. Freyberg, D. McCammon, P.P. Plucinsky, W.T. Sanders, J.H.M.M. Schmitt, J. Trümper, W. Voges, ROSAT survey diffuse X-ray background maps. II. ApJ **485**(1), 125–135 (1997)
105. T. Takahashi, M. Kokubun, K. Mitsuda, R. Kelley, T. Ohashi, F. Aharonian, H. Akamatsu, F. Akimoto, S. Allen, N. Anabuki et al., The ASTRO-H (Hitomi) X-ray astronomy satellite, in *Space Telescopes and Instrumentation 2016: Ultraviolet to Gamma Ray*, *Proc*, vol. 9905 (2016), p. 99050U
106. T. Tamura, J.A.M. Bleeker, J.S. Kaastra, C. Ferrigno, S. Molendi, XMM-Newton observations of the cluster of galaxies Abell 496. Measurements of the elemental abundances in the intracluster medium. A&A **379**, 107–114 (2001)
107. T. Tamura, R. Iizuka, Y. Maeda, K. Mitsuda, N.Y. Yamasaki, An X-ray spectroscopic search for dark matter in the Perseus cluster with Suzaku. PASJ **67**(2), 23 (2015)
108. T. Tamura, J.S. Kaastra, J.R. Peterson, F.B.S. Paerels, J.P.D. Mittaz, S.P. Trudolyubov, G. Stewart, A.C. Fabian, R.F. Mushotzky, D.H. Lumb, Y. Ikebe, X-ray spectroscopy of the cluster of galaxies Abell 1795 with XMM-Newton. A&A **365**, L87–L92 (2001)
109. T. Tamura, N.Y. Yamasaki, R. Iizuka, Y. Fukazawa, K. Hayashida, S. Ueda, K. Matsushita, K. Sato, K. Nakazawa, N. Ota, M. Takizawa, Gas bulk motion in the Perseus cluster measured with Suzaku. ApJ **782**, 38 (2014)
110. O. Urban, N. Werner, S.W. Allen, A. Simionescu, A. Mantz, A uniform metallicity in the outskirts of massive, nearby galaxy clusters. MNRAS **470**(4), 4583–4599 (2017)
111. F. Vazza, G. Brunetti, C. Gheller, R. Brunino, M. Brüggen, Massive and refined. II. The statistical properties of turbulent motions in massive galaxy clusters with high spatial resolution. A&A **529**, A17+ (2011)
112. S.A. Walker, P. Kosec, A.C. Fabian, J.S. Sanders, X-ray analysis of filaments in galaxy clusters. MNRAS **453**(3), 2480–2489 (2015)
113. N. Werner, A. Simionescu, E.T. Million, S.W. Allen, P.E.J. Nulsen, A. von der Linden, S.M. Hansen, H. Böhringer, E. Churazov, A.C. Fabian, W.R. Forman, C. Jones, J.S. Sanders, G.B. Taylor. Feedback under the microscope-II. Heating, gas uplift and mixing in the nearest cluster core. MNRAS **407**, 2063–2074 (2010)
114. N. Werner, O. Urban, A. Simionescu, S.W. Allen, A uniform metal distribution in the intergalactic medium of the Perseus cluster of galaxies. Nature **502**(7473), 656–658 (2013)
115. N. Werner, I. Zhuravleva, E. Churazov, A. Simionescu, S.W. Allen, W. Forman, C. Jones, J.S. Kaastra, Constraints on turbulent pressure in the X-ray haloes of giant elliptical galaxies from resonant scattering. MNRAS **398**, 23–32 (2009)
116. R. Willingale, R.L.C. Starling, A.P. Beardmore, N.R. Tanvir, P.T. O'Brien, Calibration of X-ray absorption in our galaxy. MNRAS **431**(1), 394–404 (2013)
117. J. Wilms, A. Allen, R. McCray, On the absorption of X-rays in the interstellar medium. ApJ **542**(2), 914–924 (2000)
118. K.-W. Wong, C.L. Sarazin, L. Ji, X-ray signatures of non-equilibrium ionization effects in galaxy cluster accretion shock regions. ApJ **727**(2), 126 (2011)
119. D. Wouters, P. Brun, Constraints on axion-like particles from X-ray observations of the hydra galaxy cluster. ApJ **772**(1), 44 (2013)
120. XRISM Science Team, Science with the X-ray Imaging and Spectroscopy Mission (XRISM) (2020). arXiv:2003.04962
121. H. Xu, S.M. Kahn, J.R. Peterson, E. Behar, F.B.S. Paerels, R.F. Mushotzky, J.G. Jernigan, A.C. Brinkman, K. Makishima, High-resolution observations of the elliptical galaxy NGC 4636 with the reflection grating spectrometer on board XMM-Newton. ApJ **579**, 600–606 (2002)
122. J.A. ZuHone, M.W. Kunz, M. Markevitch, J.M. Stone, V. Biffi, The effect of anisotropic viscosity on cold fronts in galaxy clusters. ApJ **798**, 90 (2015)

123. J.A. ZuHone, M. Markevitch, I. Zhuravleva, Mapping the gas turbulence in the coma cluster: predictions for astro-H. ApJ **817**(2), 110 (2016)
124. J.A. ZuHone, E.D. Miller, E. Bulbul, I. Zhuravleva, What do the Hitomi observations tell us about the turbulent velocities in the Perseus cluster? Probing the velocity field with mock observations. ApJ **853**, 180 (2018)

Chapter 9
Active Galactic Nuclei with High-Resolution X-Ray Spectroscopy

Luigi C. Gallo, Jon M. Miller, and Elisa Costantini

9.1 Introduction

Active galactic nuclei (AGN) are unlike any other class of astronomical object. They cannot be described by a single, dominating process. Instead, AGN radiate energy over the entire electromagnetic spectrum, and are the sites of pair-production, cosmic rays, and gravitational waves. Radiation is created through multiple processes, including blackbody, synchrotron, Comptonisation, bremsstrahlung, and line emission. Gravity attracts material inward toward the black hole, but mass and energy can also be ejected into the host galaxy and beyond. Gas can be viewed in absorption and emission, and it exists in various physical states that can be optically thick and thin, as well as neutral and completely ionised. Ionisation (see Eq. 9.4) can occur through collisional and radiative excitation. Moreover, all these physical processes are subject to extreme gravity and magnetic fields, often invoking special and general relativity and relativistic magnetohydrodynamics (MHD). Researchers must draw from many areas of physics to understand AGN.

L. C. Gallo (✉)
Department of Astronomy and Physics, Saint Mary's University, 923 Robie Street, Halifax NS B3H 3C3, Canada
e-mail: luigi.gallo@smu.ca

J. M. Miller
Department of Astronomy, University of Michigan, 500 Church Street, Ann Arbor, MI 48109, USA
e-mail: jonmm@umich.edu

E. Costantini
SRON Netherlands Institute for Space Research, Niels Bohrweg 4, 2333CA, Leiden, The Netherlands
e-mail: e.costantini@sron.nl

Anton Pannekoek Astronomical Institute, University of Amsterdam, Postbus 94249, NL-1090 GE, Amsterdam, The Netherlands

C. Bambi and J. Jiang (eds.), *High-Resolution X-ray Spectroscopy*, Springer Series in Astrophysics and Cosmology, https://doi.org/10.1007/978-981-99-4409-5_9

Though fascinating in their own right, AGN have far-reaching influence in other fields of astronomy. The AGN system is tiny in comparison to the host galaxy mass ($M_{BH}/M_{gal} \sim 0.001$; e.g. [62, 75]), but its rest mass energy is comparable to the gravitational binding energy of the host galaxy. If a small fraction of the SMBH energy is deposited into the host galaxy or intracluster medium, this AGN feedback will influence how galaxies evolve. AGN feedback (e.g. [16, 58, 117]) will be important if the kinetic luminosity

$$L_{KE} = \frac{1}{2}\dot{M}_{out} v_r^2 \tag{9.1}$$

of the wind deposits into the host galaxy approximately 0.5–5% of the AGN bolometric luminosity (L_{bol}) [51, 97, 119, 194, 199]. In the kinetic luminosity expression (Eq. 9.1), v_r is the radial velocity of the wind and

$$\dot{M}_{out} = \mu N_H m_p \Omega r f v_r \tag{9.2}$$

is the mass outflow rate in terms of the total column density (N_H), solid angle subtended by the outflow (Ω), distance from the black hole (r), proton mass (m_p), the volume filling factor (f), v_r, and the mean atomic weight correction (μ), which is $\sim$1.23 for cosmic abundances [24].

A simple illustration of the AGN region is depicted in Fig. 9.1. The so-called central engine of the AGN is defined by the accretion disc that funnels material toward the supermassive ($M_{BH} \simeq 10^{6-9}\, M_{\odot}$) black hole (SMBH). The accretion rate, which is often parameterized by the ratio of the bolometric luminosity over the Eddington luminosity[1] ($\lambda = L_{bol}/L_{Edd}$), will determine the structure of the accretion disc. For moderate values of the Eddington ratio ($\sim$0.01−0.3), the disc can be approximated as a standard Shakura-Sunyaev disc [192] that is optically-thick, geometrically-thin, and radiative efficient.

The X-ray emission from the inner-most region within 10's of gravitational radii ($R_G = GM/c^2$) is dominated by the hot corona. This primary X-ray source can illuminate the inner disc leading to the production of the reflection spectrum that is blurred by relativistic effects close to the black hole (e.g. [10, 186, 187]). The dominant spectral feature here is the relativistic Fe Kα emission line (e.g. [59, 135]). Alternatively, the inner disc region might be blanketed by a warm corona that is optically-thick and conceals the fast inward flow (e.g. [9, 42, 139]).

Winds in the accretion disc are important for transporting angular momentum outward so that material can flow inward (e.g. [7, 22, 156]). Magnetic fields (orange curves in Fig. 9.1) that are capable of launching material out of the system (Sect. 9.4.2.3) will thread the accretion disc over large distances. The material in the MHD-driven wind closest to the black hole and moving fastest will also be the most highly ionised since it is closest to the primary X-ray source (i.e. the corona).

[1] The Eddington luminosity, $L_{Edd} = 1.26 \times 10^{38} M_{BH}/M_{\odot}$erg s^{-1}, is the maximum luminosity a system can have such that the gravitational infall of ionised hydrogen gas is exactly balanced by the outward radiation pressure.

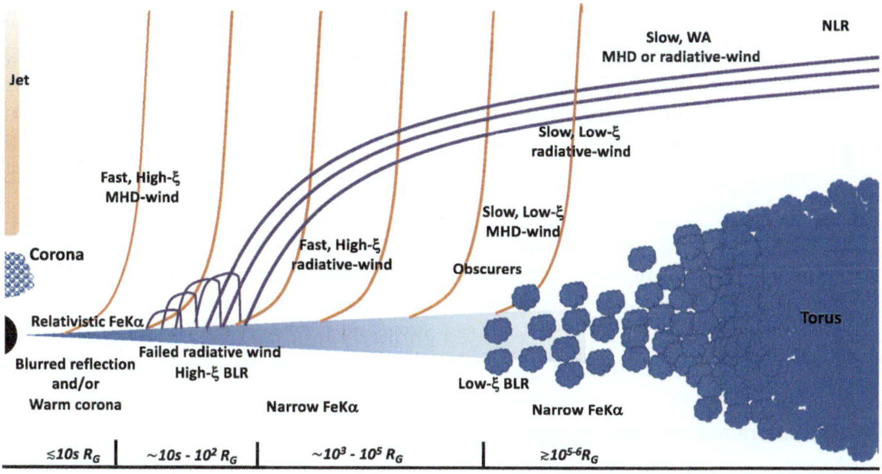

Fig. 9.1 A simple illustration of the AGN region depicting the scales and main components responsible for X-ray emission. The standard accretion disc extends down to the innermost stable circular orbit (ISCO) between $\sim$1.2$-$6 R_G depending on the black hole spin. The optically thin, hot corona is the primary X-ray source, and it is located close to the SMBH. A collimated jet whose base may coincide with the corona, may or may not be present. The magnetic field lines (orange curves) are the sites of the MHD-driven winds (Sect. 9.4.2.3), and they thread the accretion disc over a large range. The streamlines from radiation-driven winds (purple curves; Sect. 9.4.2.2) are successfully launched at larger radii where the radiation pressure exceeds the gravitation force in the disc. Within this launching radius, the wind falls back toward the disc, forming a failed wind, which may constitute the inner broad-line region (BLR, Sect. 9.3.2) and/or shielding gas [176]. The disc morphs into the obscuring torus (Sect. 9.3.2) at large distances, and the traditional warm absorber (WA, Sect. 9.2) on similar scales. The narrow-line region (NLR) occupies galactic scales, but is photoionised by the AGN. The observer's line-of-sight will dictate the type of system that is seen and the dominant processes at work. High-resolution spectroscopy will probe the X-ray emission from all these regions

Radiative winds (Sect. 9.4.2.2) will also be launched from a distance corresponding to where the outward velocity from radiation pressure exceeds the escape velocity in the disc (purple streamlines in Fig. 9.1). At distances closer than the launching radius, the radiation-driven wind will fall back onto the disc. This failed wind region may manifest as many AGN components [78, 79] and may form part of the highly ionised broad-line region (BLR; Sect. 9.3.2) in AGN.

On parsec scales, the outer disc morphs into the torus, which is significantly optically thick and neutral. Compton scattering and neutral iron emission are evident here (Sect. 9.3.2). The traditional warm absorber (WA) will occupy scales similar to these outer disc regions (e.g. [133, 134]). The WA is responsible for the "normal" velocity winds (Sect. 9.2) that may be driven by either radiative or MHD effects. The narrow-line region (NLR) occupies galactic scales, but it is still photoionised by the AGN.

There are many intrinsic variations in the AGN phenomenon that likely arise from different accretion rates. However, the observer's line-of-sight also plays a significant factor [4, 214]. The observer's view through the obscuring torus will determine which disc regions are observable and if the AGN is defined as an unobscured Type I (e.g. Seyfert 1) or obscured Type II (e.g. Seyfert 2). Whether the jet is present, its relative dominance over other AGN components, and the observer's perspective will dictate if the AGN is radio-loud (jetted) or radio-quiet (non-jetted). The line-of-sight through the winds to the primary X-ray source might also influence the types of winds that are observed.

In the past few decades, tremendous advances were made regarding the X-ray studies of AGN, particularly relating to broadband spectroscopy and variability. The transmission gratings on *Chandra* (see Chap. 3) and the reflection gratings on *XMM-Newton* (see Chap. 2) provided glimpses into the discovery space opened by high-resolution spectroscopy. AGN grating data were rich in spectral features, delivering information on gas temperatures, densities, dynamics, and origins. AGN X-ray spectra are not just "power laws".

As we enter the era of calorimeter spectrometers with XRISM [205] and Athena [13], a new discovery space will be unveiled. With a resolving power $E/\Delta E \approx 1400$, corresponding to ~5 eV resolution at ~7 keV, and a collecting area about 10-times that of *Chandra* in the Fe Kα band, XRISM will transform AGN science in the coming year. We had but a brief, exciting view of this with *Hitomi* (e.g. [93, 94, 202]).

In this chapter, we will review and explore the areas of AGN research where high-resolution spectroscopy will make a certain impact. In Sect. 9.2, the ionised (warm) absorber science that has benefited greatly from grating spectrometers will be reviewed. Later sections focus on the Fe Kα band, where calorimeters will resolve these data for the first time. In Sect. 9.3, the emphasis will be on the nature and origin of the narrow (and broad) Fe Kα emission lines. In Sect. 9.4, the highly-ionised iron seen in absorption and forming ultrafast outflows is examined.

9.2 Warm Absorbers

The first detection of absorption from ionised gas in an AGN X-ray spectrum [89] opened a new window to study highly ionised nuclear winds. These outflows were subsequently detected and studied by all moderate-resolution CCD cameras: *ASCA* [77, 181], *BeppoSAX* (e.g., [39, 162]), and *ROSAT* (e.g., [121]). From these early measurements, the clearest feature in the spectrum was identified as the photoelectric bound-free transition of O VII at $E = 0.74$ keV. This feature, detected with different optical depths in most pointed observations of bright AGN, indicated gas with column densities $N_H > 10^{21}$ cm^{-2} [77]. A second feature at $E = 0.87$ keV was attributed to the O VIII photoelectric edge, suggesting a more ionized component [181].

A more quantitative characterisation of these WA came with the advent of the high-energy resolution grating spectrometers: *XMM-Newton*-RGS, *Chandra*-HETG and *Chandra*-LETG [19, 106, 114]. Dozens of transitions originating from carbon, nitrogen, oxygen, iron and neon were identified, and the deep absorption edges were no longer the only predictors of ionised absorbers. The feature identified as O VII in earlier studies turned out to be heavily blended with iron transitions from the L-shell to the M-shell, called the iron UTA (unidentified transition array) [19, 189]. Refinements in the atomic database (e.g., [82, 83]) and subsequent studies determined that the iron UTA ions are very sensitive to changes in the ionisation parameter (Fig. 9.2), allowing a robust determination of the state of the gas [19, 201] and a diagnostic for gas changes over time (e.g., [132]).

The column density provides a measure of the quantity of gas intrinsic to the source along our line of sight. This does not provide any information on the geometry of the system since neither the thickness nor the density of the gas is directly measurable. The only constraint is that the thickness (Δr) of the gas cannot be greater than the gas distance from the central source (r) [24]:

$$\Delta r < r. \tag{9.3}$$

A range of column densities spanning more than two orders of magnitude have been reported [134], with the bulk of the gas being in the interval $\log(N_H/\mathrm{cm}^{-2}) \sim 20{-}22$.

In this paper, the ionisation state of the gas is parameterised by [204]:

$$\xi = L_{\mathrm{ion}}/nr^2 \qquad [\mathrm{erg\,cm\,s^{-1}}], \tag{9.4}$$

where ξ is the ionisation parameter, L_{ion} is the ionizing luminosity in the 1–1000 Ryd interval, and n is the gas density. For different values of ξ, distinctive groups of ions will be present in the X-ray spectrum.[2] In Fig. 9.2 (left panel), we show the distribution of ionic column densities for the iron ions as a function of $\log \xi$. From the spectral point of view, this results in Fe absorption features, among others, distributed all over the X-ray spectrum. In Fig. 9.2 (right panel), we show how iron significantly characterises the absorbed spectrum. For instance, at $\log \xi \sim 0.8$, the lower ionisation ions (marked by the iron UTA transitions) are more visible. As $\log \xi$ increases, higher ionisation ions are present, originating from other iron L-shell and K-shell transitions.

The spectral energy distribution (SED) ranging from the ionisation threshold energy of hydrogen (1 Ryd = 13.6 eV) to the end of the canonical X-ray band (1000 Ryd) has a profound impact on the characteristics of the WA. The effects of the UV and soft X-ray spectral shape, as well as the high energy tail of the distribution, have been extensively studied (e.g. [31, 32, 160]). Alongside the distribution of the ionising photons as a function of energy, the metallicity of the WA itself also has a significant influence (e.g. [31, 122]).

A useful visualisation of these influencing factors is given by the so-called stability curve. This describes, in a $\log\xi/T$ *versus* $\log T$ plane, where a WA can exist in equilibrium, for a given SED and metallicity set. The term ξ/T is proportional to the

[2] In this section, all illustrations are produced using the SPEX package [107].

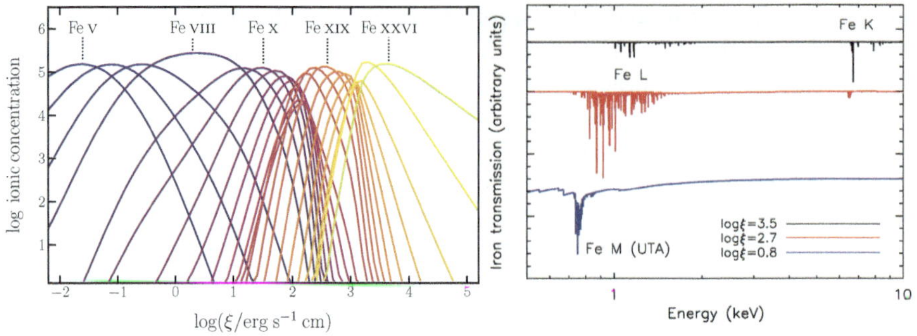

Fig. 9.2 Left panel: The distribution in the concentration of iron ions as a function of logξ. As the ionisation parameters increases, the ion concentration of more ionised Fe is enhanced (courtesy of D. Rogantini). Right panel: The iron transmission spectra for selected values of logξ highlighting the Fe-UTA, L- and K-shell transitions

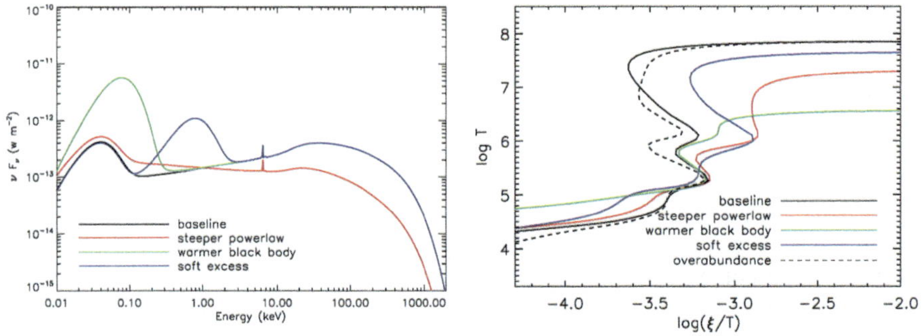

Fig. 9.3 Left panel: Examples of SEDs illuminating the ionized gas (following [31, 32]). Here, the baseline SED is produced by black body radiation with a temperature 0.01 keV, and a Comptonised component acting at soft X-ray energies with a seed temperature commensurate with the disc. The medium energy is characterized by a power law with a photon index $\Gamma = 1.8$ and a reflection component, including the iron emission line, which is cutoff at $E = 300$ keV (black solid curve). Right panel: The thermal stability curves for the SEDs displayed in the left panel (solid curves). The curves have been shifted along the x axis, to highlight the different SED effects [32]. In addition, the effect of a metal overabundance with respect to solar values is displayed with a dashed curve

ionisation pressure parameter, Ξ, which is the ratio of the radiation pressure over the gas thermal pressure (e.g. [188]):

$$\Xi \equiv \frac{P_{\rm rad}}{P_{\rm th}} = \frac{L}{4\pi r^2 c}\frac{1}{nk_B T} \propto \frac{\xi}{T}. \tag{9.5}$$

Here, k_B is the Boltzmann constant and Eq.9.4 is used to simplify the final formulation. Following [31, 32], Fig. 9.3 illustrates the behaviour of the thermal stability curve (right panel) as a function of a few key parameters of the SED (left panel).

The presence of a stable branch in the curve, that is an almost constant pressure zone for a range of temperatures (e.g., [129]), is enhanced as the power law slope is relatively flat ($\Gamma = 1.8$ in this example). Under these conditions, more WA components with different ionisations can co-exist in pressure equilibrium [131]. However, the power law slopes are often significantly steeper (up to $\Gamma \sim 2.8$ [21]). In these cases ($\Gamma = 2.1$, red solid curve in Fig. 9.3), the pressure equilibrium zone is disrupted.

The effect of a higher temperature seed black body, or of an enhanced soft excess (green and blue solid curve, respectively in Fig. 9.3), impact the curve at lower values of ξ/T, where warm absorbers exist at higher temperatures than in the baseline case, due to heating by iron ions [32].

Element abundances in the WA also have an impact on the stability curve. To emphasize this effect, in Fig. 9.3 (right panel) an overabundance by a factor of five with respect to solar values is displayed (dashed line). The effect is to create a larger zone of pressure stability.

The shape of the SED also influences the type of ions that appear in the X-ray absorption spectrum [160]. A SED typical of a Seyfert 1 galaxy shows a number of absorption lines and photoelectric edges from C, N, and O. Instead, a steeper energy spectrum, for example, characteristic of narrow-line Seyfert 1 (e.g. [26, 27, 71]), stimulates iron transitions at different ionisation stages, producing the typical UTA and L- transition arrays at soft energies [160]. For higher ionisation parameters, narrow-line Seyfert 1s display more pronounced iron absorption in the 6–7 keV range (e.g. [70]).

The spectra obtained in the X-rays with high-energy-resolution clearly show that absorption lines displayed a significant blueshift in velocity with respect to the redshift of the AGN host [106, 114]. The blueshift corresponds to WA with velocities in the range $10^{2-3}\,\mathrm{km\,s^{-1}}$ (e.g. [134]). This parameter added the important information that other than an absorbing gas along the line of sight, these components were ejected from the nuclear region towards the host environment. In the same object, different WA components do not necessarily share the same outflow velocity. The distribution of outflow velocities for WA was found to be weakly correlated with the ionisation parameter [134].

Some high-quality spectra of Seyfert 1 galaxies are known to host multiple WA components (e.g. [46, 104]), that differ in ionisation, column density, and outflow velocity. How these components are organized in the AGN system is one of the open questions in this field. As seen above, WA components are only sometimes found to be in pressure equilibrium, where a diffuse, more ionized absorber contains a colder, possibly clumpy, component. Pressure equilibrium would ensure a long-lived gas outflowing structure. In the same object, some WA components are found in equilibrium with each other, in the constant pressure branch of the curve, while others sit in the low ionisation branch (e.g. [46]).

An intuitive picture (Fig. 9.1) sees the outflows along our line of sight as part of a continuous stream, with the more ionized components located closer to the ionising source (Eq. 9.4). Using Eq. 9.3 as an upper limit for the radius and assuming that the outflow velocity v_{out} is larger than or equal to the gas escape velocity, a rough range for the location of the absorbers can be found, using this simple geometry [24, 133]. The lower limit on the radius is then given by $r > 2GM_{BH}/v_{out}^2$. With these estimates, the WA location encompasses a radius range of more than two orders of magnitude, with lower ionisation components located further away [133]. Their location seems to be roughly between the broad line region and the molecular torus [24].

The WA components are generally modelled as discrete components, with well-defined parameters. The absorption measure distribution (AMD, [18, 95, 116]), describes instead the absorption spectra in terms of a continuous distribution of column densities per unit of $\log\xi$ ($dN_H/d(\log\xi)$) as a function of $\log\xi$. The integral over $\log\xi$ would result in the total column density of the WA components. A linear fit of the AMD distribution may be only a simple parameterization, but provides a slope a that may be used as a diagnostic to be compared with theoretical models (see below and [18]). The value of a has been reported to range between 0–0.4 [18].

Absorption by ionized gas has been mostly associated with radio-quiet objects. The sparse detection in radio-loud, non-blazar objects (e.g. [47, 180, 211]), suggests that radio loudness could somehow interfere with the detection of WA, for example, if the object was observed at an unfavourable angle. The X-ray radiation of radio-loud objects being more intense ([86] and references therein), was thought to fully ionise the surrounding medium. A systematic study of radio-loud objects in X-rays showed a robust anti-correlation between the power of the radio emission and the column density of the warm absorbers [146], independent of the inclination angle and X-ray luminosity. This anti-correlation pointed to a bi-modality between the radio activity and the disk ultimately originating the WA, providing further clues on the origin of the winds.

9.2.1 Obscurers

A category of ionized gas that has been relatively overlooked due to their unpredictable occurrence, has been the so-called "obscurers" (e.g., [104, 112, 138, 147]). This obscuring gas can be formed by one or more high column density ($N_H \sim 10^{22-23}\,\mathrm{cm}^{-2}$) gas mass, that temporarily obliterates the X-ray soft energy spectrum. This phenomenon became well known from long-term *RXTE* and *Swift* light curves, as seemingly normal Seyfert 1 sources, would undergo periods of very hard spectra (high hardness ratios), due to the suppression of the soft X-rays [140]. These sources were sometimes caught in this state, for example by *XMM-Newton*-pn (e.g. NGC 3516 [212], Mrk766 [183], 1H0419-577 [50]), but without any UV spectroscopic coverage. The limited energy coverage raises different interpretations on whether the hardness ratio is due to absorption or to intrinsic changes in the source (e.g. [100]).

The fortuitous simultaneous observations of NGC 5548 in this spectral state with *XMM-Newton*, *HST*-COS, *NuSTAR* and *Integral* [104], allowed for the first time a comprehensive study of this phenomenon. The *HST*-COS spectrum showed, for every major transition (from C II up to C IV, N V, and Ly α [104]), the presence of deep outflows, with velocities in the range 1000–5000 km s^{-1}. The ionisation parameter of this high-column density gas component is relatively low with logξ $\sim$–(2–1), possibly pointing at a dense gas mass. This mass is also not completely covering our line-of-sight ($C_V \sim$ 0.5–1), changing in both covering factor and column density on a time scale of days–months [49], hinting at a patchy nature for the obscurer.

The degree of absorption correlated with obscuration in the soft X-rays, unequivocally linked the two absorbing agents as the same gas. The location of the obscurer has been estimated to be between the UV-emitting broad line region and the WA. The obscuring gas indeed absorbs the blue side of the broad emission lines in the UV (see Fig. 9.1 in [104]). At the same time, the WA components have been found to be photoionised by the central source SED, but significantly modified in shape by the obscurer. This meant that the WA components must have been located at larger distances with respect to the obscuring gas. Instabilities and eruptions in the accretion disks have been invoked to explain the occasional rise of this high-column density gas. However, it is still uncertain why some obscuration events last for almost 10 years (e.g., NGC 5548 [148]) and others, occurring in objects of seemingly similar evolutionary state, last only weeks (e.g., NGC 3783 [147]). The decline of these disk-wind obscurers is not directly connected to the SED changes, nor to the frequency of their appearance [105, 148].

Dedicated campaigns, covering the UV and the X-ray band, revealed that many Seyfert 1 galaxies may undergo periods of obscuration during their active life. The study of obscurers also brought to light a complex interplay between the illumination and the covering factors of the UV and X-ray obscurer [45, 145, 148]. The presence of the obscurer may also describe the X-ray shielding invoked to explain the survival of UV absorbers [45, 176]. In addition to the cold patchy components, the occurrence of obscurers has also been associated with an additional component of very high-ionisation gas, with velocity consistent with the cold component, indicating inhomogeneity in the medium, where lower density hems become more ionised [147].

9.2.2 *The Importance of WA and the Density Determination*

As seen above, WA and outflows in general are promising conveyors of feedback into the host galaxy, with important implications for galaxy evolution and formation. In any model predicting the launching mechanism and the impact of outflows on the surrounding medium, it is fundamental to know at what distance the outflow is launched. The rough estimates reported above indicate a range of distances that span orders of magnitudes. This uncertainty is then reflected in Eq. 9.2. On the other hand, the distance r in Eq. 9.4 can be calculated from observable parameters only if the density n is known.

A method that has been successfully used in the UV band is density evaluation through density-sensitive absorption lines. These absorption lines are the result of the electron population of a so-called metastable level, above the ground level. The population of this unstable level may be due to both an excess of optical photons in the SED or to the gas with a density above a given threshold, which is different depending on the ion. The column density of the metastable level line is compared to the ground transition and to theoretical curves to find the best fit for the gas density (e.g. [123]). Metastable levels of C III, N III have been regularly used (e.g. [68]) as well as a number of other UV (Si II, S III, P III, Fe III [5]), and optical transitions (e.g. Fe II [6]). Sometimes, the metastable line happens to be part of a WA component that is visible both in the UV and the X-rays, providing a density estimate also for the X-ray absorber [35, 48, 56]. In the X-ray band, only one detection of O V has been reported for an AGN so far [108], leading to a lower limit for the gas density.

Methods that may be easier to apply, make use of the response of the gas as a function of the ionizing luminosity (Eq. 9.4). Every ion in a gas will have its own reaction time, depending on the density of the gas [128, 161, 185]. In particular, the time taken for the gas to react is inversely proportional to the gas density. The time dependence of the ionic concentration of a certain element X, n_{X^i} can be written as:

$$\frac{dn_{X^i}}{dt} = -n_e n_{X^i}\alpha_{\mathrm{rec},X^{i-1}} - n_{X^i} I_{X^i} + n_e n_{X^{i+1}}\alpha_{\mathrm{rec},X^i} + n_{X^{i-1}} I_{X^{i-1}}. \quad (9.6)$$

Here, n_e is the electron density, while I_{X^i} and $\alpha_{\mathrm{rec},X^i}$ are the ionisation and recombination rates between state i and $i+1$.

In Fig. 9.4, the behaviour of Fe XX as a function of time is illustrated for a range of gas densities. The higher the column density the more the gas response approaches an instantaneous change as a function of the flux variation, leaving the gas in equilibrium conditions at every time. If, on the contrary, the gas density is very low, the signal is diluted and no variation in the gas is observed. For a range of densities ($\log(N_H/\mathrm{cm}^{-2}) \sim 5$–7), Fe XX reaction time is significantly delayed.

In principle, Eq. 9.6 can be used for every source whose flux significantly changes in time. In practice, this method, that relies on time-resolved spectroscopy, is often limited by the signal-to-noise per time bin of the spectrum. A slow variation may lead to subtle variations in the WA that are difficult to detect [103], while strong and sudden variations require that the WA is analysed on few-kilosecond time bins, therefore reducing the quality of the spectrum [132]. Several estimates of the density, and therefore on the distance, of the WA have been reported. Only a few experiments were successful in putting limits on the distance of the WA. For example, [131] derived an upper limit of $r < 6$ pc for a gas component in NGC 3783, consistent with the limit ($r < 25$ pc) derived for the same gas component from the analysis of the UV data [69].

In Mrk 509, the general lack of variability in the WA set the lower limits to relatively far distances for the different components (ranging from about 5 pc up to 70 pc, with some components at kpc scales [103]). A similar range, spanning from parsecs to tens of parsecs, has been found for the WA components of NGC 985 [54]. Smaller black hole systems, like in the narrow-line Seyfert 1 galaxies, show instead the desired large variation in flux on time scales of few kiloseconds. An intensively studied object of this class is NGC 4051, where the WA components were found to be closer to the black hole, around 1 light day distance (e.g., [132, 200]).

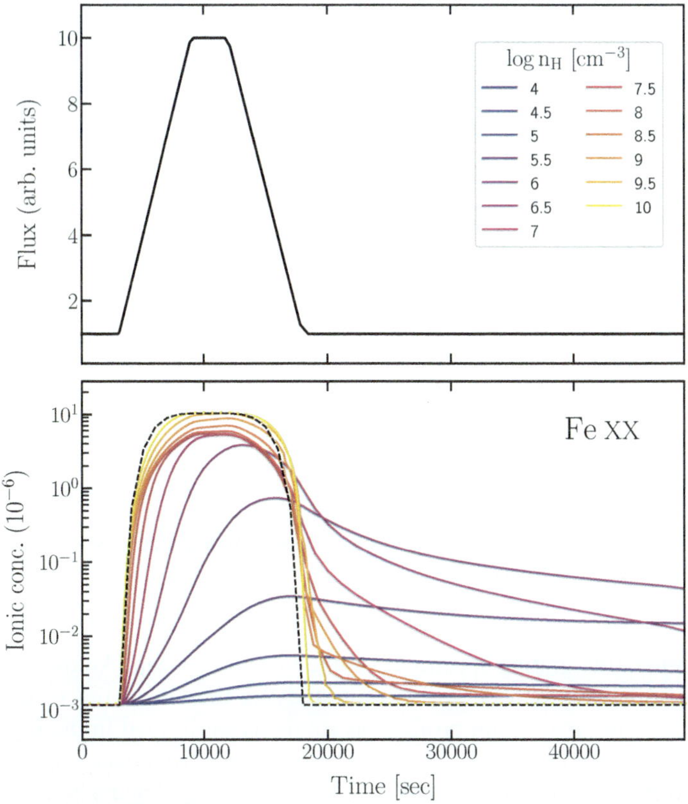

Fig. 9.4 Upper panel: An example of a flaring light curve. Lower panel: The ionic concentration of Fe XX response to the time evolution of the flux change. The equilibrium condition (i.e. instantaneous response, following Eq. 9.4) is marked with a dashed line. Depending on the density, the gas will respond with a different delay, as highlighted by the colour-coded solid lines. Adapted from [185]

9.2.3 Future Outlook on Warm Absorbers

The design of future high-resolution instruments would certainly reward the study of ionised absorbers. High-sensitivity, high-resolution ($\Delta E \leq 5$ eV) calorimeters (XRISM, [205] and Athena-XIFU [13]) will bring significant advancement, for example, in the study of variable WA and the determination of the gas density. In the future, time-resolved spectra will allow us to follow the evolution of multi-components in WA as a function of time [185]. At the same time, different applications of Eq. 9.6 will be possible, for instance combining timing and spectroscopy [195]. The coherence between two signals, in this case between the continuum and the warm absorber, as a function of energy, bears the signature of the time delay of the recombining gas and therefore of the gas density [101]. This can be detected and studied provided high-sensitivity and resolution data (Fig. 9.5).

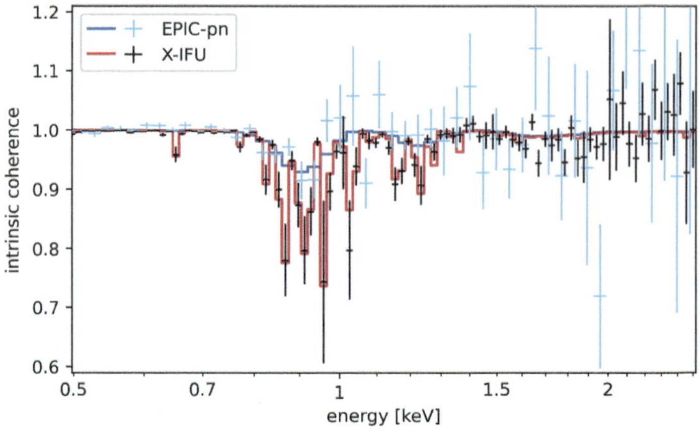

Fig. 9.5 Athena-XIFU simulation (red data points and solid line) of the coherence as a function of energy for a typical narrow-line Seyfert 1 source hosting a WA. This is compared to the capabilities of current high-sensitivity instruments (e.g., *XMM-Newton* EPIC-pn, blue data points and solid line). The shape of the coherence spectrum as a function of frequency can be modelled to solve for the gas density. Modified from [101]

New generation grating spectra (as in the mission concept Arcus [198]), operating in the soft-X-ray energy, will permit studying the kinematics of the absorbers (Fig. 9.6) and line absorption profiles, as routinely performed in the UV band. The envisaged resolution ($\lambda/\Delta\lambda > 2500$) and a large effective area compared to *XMM-Newton*-RGS, will allow us to perform different density diagnostic tests, including the detection of metastable levels [144] and time-resolved spectroscopy [185] on a large number of sources (Fig. 9.7).

9.3 Fe K Emission Lines in AGN

9.3.1 The Atomic Physics of Fe K Emission Lines

Narrow emission lines from the Fe K shell are the most prominent atomic feature in the X-ray spectra of AGN. Typically, a single emission line is seen in addition to a local continuum that can be approximated with a power law. These lines are produced when ionizing X-ray radiation illuminates relatively cold, dense gas. In this sense, narrow Fe K lines trace the interaction of the central engine with the accretion flow on all scales, and can serve to test models for its (sub-)structure and evolution. Indeed, hard X-ray emission from the central engine can even excite narrow Fe K lines on scales that are better associated with the larger host galaxy than the accretion flow.

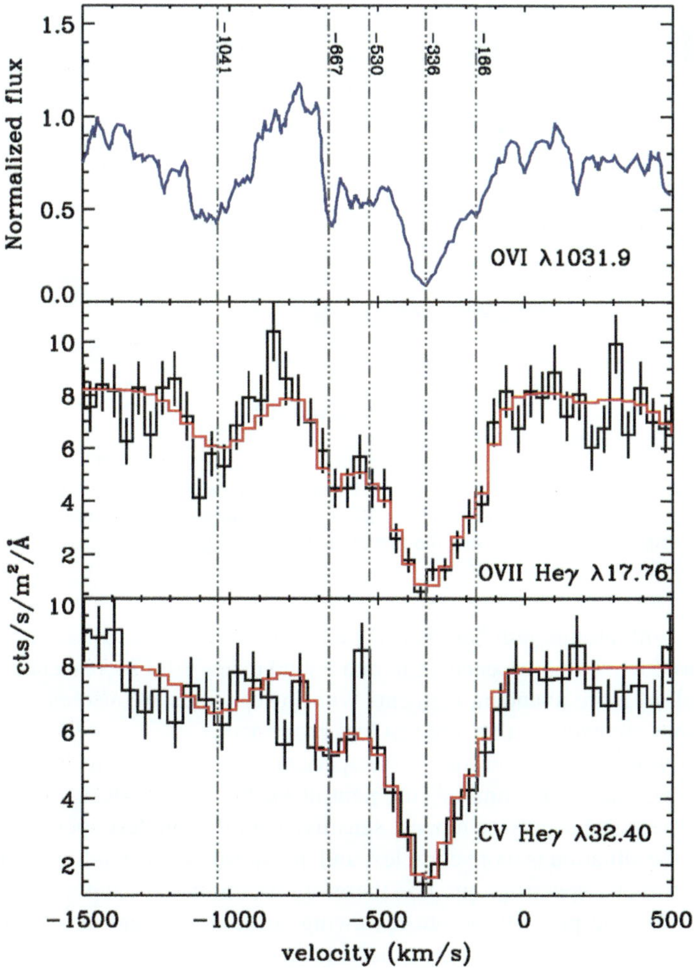

Fig. 9.6 Arcus simulation of the non-saturated γ lines of the He-like series for O VII and C V, respectively (middle and lower panels), compared to the *HST*-STIS velocity resolved profile of one of the doublet components of O VI (λ1031.9Å) (upper panel [41])

The "X-ray reflection nebulae" in the center of the Milky Way, for instance, likely indicate that Sgr A* was much more luminous *few* $\times$ 10^4 years ago [124] (also see [173]).

It is important to clarify that Fe K-shell lines are only prominent within the spectra of AGN owing to a combination of three key factors: (1) a relatively high fluorescence yield, (2) a relatively high abundance, and (3) the Fe K lines fall within a part of the spectrum that is otherwise relatively simple. The fluorescence yield of an atomic shell is simply the probability that a vacancy results in a radiative transition, rather than the ejection of an Auger electron. The K-shell fluorescence yield is positively

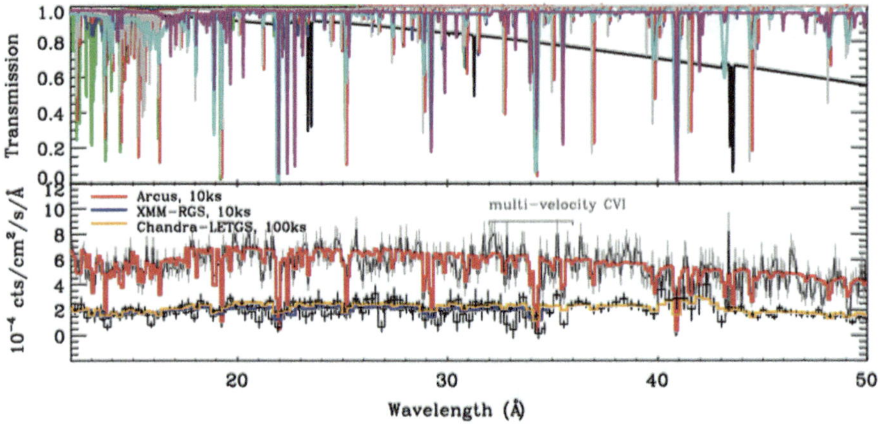

Fig. 9.7 A 10 ks simulation using a next generation soft-X-ray grating ($\lambda/\Delta\lambda > 2500$). The simulation includes a six-component WA [104] and Galactic cold absorption (upper panel). The grating simulation is compared with *XMM-Newton*-RGS and *Chandra*-LETGS (lower panel). The data have been rebinned and shifted along the vertical axis for clarity

correlated with atomic number [12]. By a factor of ~5, Fe is the element with the highest product of fluorescence yield and abundance [76]. Fluorescence lines from the K-shell of more abundant elements with lower yields are also excited when Fe K-shell lines are excited; they are just less prominent.

A narrow Fe K emission line in the spectrum of a given AGN is often referred to as "*the* neutral Fe K" line or "the neutral Fe Kα line." These colloquial terms are convenient, especially at modest sensitivity and/or modest spectral resolution. However, the situation is more complex, and this may become readily apparent in the era of calorimeter spectroscopy. It is therefore worth undertaking a quick overview of the key atomic physics, before reviewing some key recent developments with *Chandra*, *XMM-Newton*, and other telescopes.

The neutral Fe Kα fluorescence line arises from a 2p-1s electron transition, and the spin-orbit interaction therefore creates two lines with a small energy difference: Kα_1 at 6.404 keV and Kα_2 and 6.391 keV, with a 2:1 branching ratio [12]. This difference, just $\Delta E = 13$ eV, exceeds the $\Delta E = 5$ eV energy resolution that was achieved with the calorimeter aboard *Hitomi* [202]; the "Resolve" calorimeter that will fly aboard XRISM in 2023 is expected to have the same resolution [205]. The X-ray Integral Field Unit (XIFU) spectrometer expected to fly aboard Athena in the 2030s is expected to have a resolution of just $\Delta E = 2.5$ eV.

The upper panel of Fig. 9.8 shows a model spectrum constructed using MYTORUS [155] owing to its physical self-consistency, and because it has a native resolution of just 2 eV and is therefore suited to even calorimeter spectra [155, 223]. It depicts an Fe K emission spectrum from neutral gas, broadly characteristic of the spectra observed in Seyfert 1 AGN. The model was constructed according to the prescriptions in [111] assuming an obscuring column density of $N_H = 1 \times 10^{22}$ cm^{-2} and an

The more detailed nature of this survey permitted a more nuanced and very important finding: *"There is no universal location of the Fe Kα line-emitting region relative to the optical broad line region (BLR). In general, a given source may have contributions to the Fe Kα line flux from parsec-scale distances from the putative black hole, down to matter a factor of ∼2 closer to the black hole than the BLR"* [193].

Andonie et al. [3] have undertaken the most recent and expansive examination of narrow Fe Kα emission line regions. Their analysis included 38 bright AGN in the *Neil Gehrels Swift Observatory* Burst Alert Telescope (BAT) Spectroscopic Survey. Utilizing *Chandra* images and spectra, *XMM-Newton* spectra, and variability studies, independent estimates of the Fe Kα production radius ($R_{Fe\ K}$) were obtained and compared to plausible estimates of the dust sublimation radius ($R_{subl.}$) in each source. In the cases where data permitted measurements of the line FWHM, the $R_{Fe\ K} < R_{subl.}$ in 90% of the sources (21/24 AGN). Similarly, in the cases where significant line variability was detected, $R_{Fe\ K} < R_{subl.}$ in 83% of the sources. Andonie et al. [3] conclude that Fe Kα lines in unobscured AGN typically originate in the outer part of the BLR, or the outer disk, but carefully note that *"the large diversity of continuum and narrow Fe Kα variability properties are not easily accommodated by a universal scenario."*

The left hand panel in Fig. 9.9 shows the Fe Kα emission radius versus an estimate of the dust sublimation radius. The Fe Kα emission radius was calculated based on the velocity width of the line, and the dust sublimation radius was estimated using an expression derived by [159], and assuming that graphite sublimates at

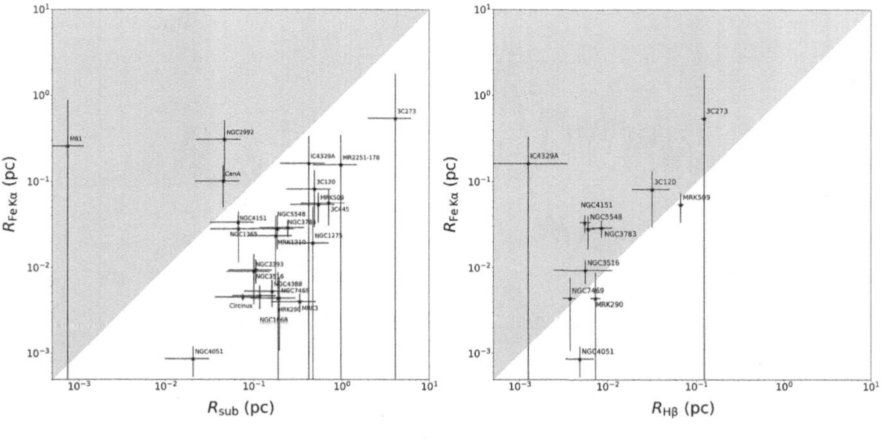

Fig. 9.9 These panels summarize some of the key results obtained in the survey of Seyfert Fe Kα lines undertaken by [3]. At left, the Fe Kα line production radius (measured via line widths) is compared to an estimate of the dust sublimation radius—a proxy for the innermost extent of the dusty torus—in each AGN. The Fe Kα line production regions are found to occur at smaller radii, strongly suggesting that the bulk of the emission seen in unobscured AGN is produced within the BLR or in even more compact regions. At right, the Fe Kα line production radius in each AGN is compared to the Hβ production radius, the characteristic radius of the optical BLR. The data do not appear to permit a systematic statement, but it is fair to conclude that the production radii are broadly comparable. (This figure is reproduced from [3], with permission)

$T = 1500$ K. The figure illustrates that the Fe Kα line production radius is systematically smaller than the dust sublimation radius within the sample. If we take the dust sublimation radius as indicative of the innermost extent of the torus, this finding is at least qualitatively consistent with the dust reverberation results of [153]. The right hand panel in Fig. 9.9 depicts the Fe Kα emission line radius versus the Hβ production radius. There is no clear trend within the data; only a few AGN clearly lie above or below the line that marks an equivalent production radius, and many Fe Kα line production radii carry relatively large uncertainties.

Figure 9.10 shows a comparison of summed *Chandra*-HEG spectra from the Seyfert 1 NGC 4151 in its high and low flux states, and the summed HEG spectrum of the Compton-thick Seyfert 2 NGC 1068. The opposing first-order spectra were added, the spectra have been shifted in energy to their respective rest frames, and the flux of NGC 4151 has been adjusted to the distance of NGC 1068. No spectral fits were made. Recent estimates suggest that the mass of the black hole in NGC 1068 is $M = 1.5 \times 10^7\ M_{\odot}$ [154]; optical reverberation mapping gives a formally equivalent black hole mass in NGC 4151: $M = 1.66^{+0.48}_{-0.34} \times 10^7\ M_{\odot}$ [20]. For these mass estimates, NGC 1068 is likely accreting at an Eddington fraction at or below $\lambda \leq 0.2$ [23], and NGC 4151 at a rate of $\lambda = 0.03{-}0.04$ [151]. Given their similar

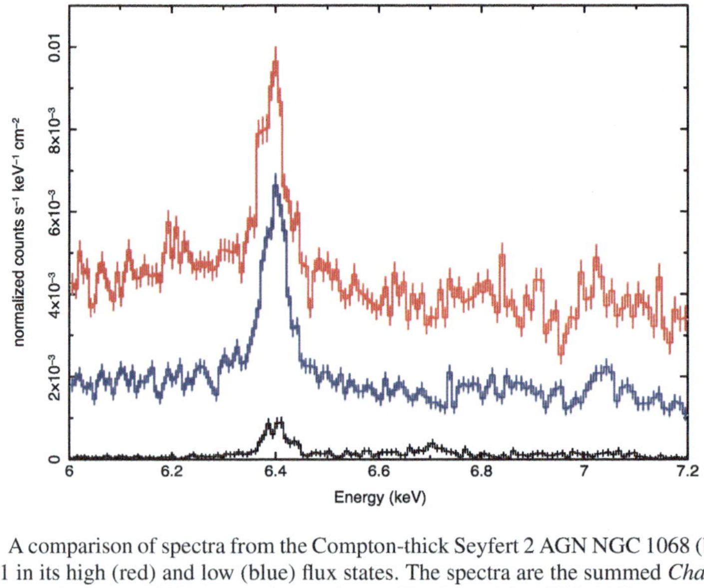

Fig. 9.10 A comparison of spectra from the Compton-thick Seyfert 2 AGN NGC 1068 (black), and NGC 4151 in its high (red) and low (blue) flux states. The spectra are the summed *Chandra*-HEG spectra from each source, shifted to their host frame. The flux of the NGC 4151 spectra have been shifted to the smaller distance to NGC 1068. The inferred black hole masses and Eddington fractions in these AGN are broadly similar (see the text for details), enabling a consistent comparison. If the covering factor of cold, dense gas were the only factor that determined line strength, the narrow Fe Kα line in NGC 1068 should be much stronger. However, factors including the distance of the gas to the responding medium also matter, and this is likely the reason that the line is stronger in NGC 4151. This figure illustrates that different viewing angles and obscuration properties can reveal different regions, not necessarily different aspects of the same emission region

masses, the fact that NGC 1068 is likely surrounded by gas with a higher filling factor and likely accreting at a higher rate than NGC 4151, it should show a stronger neutral Fe Kα emission line in this comparison. However, the opposite is the case. This can be explained if the gas that is emitting the Fe Kα line *within our line of sight* in NGC 4151 is closer to the central engine than the emitting gas *within our line of sight* is in NGC 1068. Without any spectral modeling, this example illustrates that different views of the central engine actually reveal different regions, and that faulty conclusions may be drawn if a single emitting geometry is invoked.

9.3.3 The Approaching Calorimeter Era

A number of recent results highlight the potential of X-ray calorimeter spectroscopy to reveal the inner accretion flow onto massive black holes. Here, we highlight four key results that likely provide an early glimpse of the science that will be enabled by the sharper resolutions and larger effective areas afforded by XRISM and Athena.

In a major departure from phenomenological Gaussian modeling, and simple comparisons of Fe Kα and Hβ line widths, [37] examined the spectrum of the Seyfert 1 Mrk 279 using the Local Optimally emitting Cloud model (LOC; [8]). This model does not assume a geometry for the BLR; rather, it simply assumes that each line is the result of contributions from multiple regions that follow a power law distribution in gas density and radius. Mrk 279 is particularly interesting among Seyferts, in that its "narrow" Fe Kα line appears to be composed of an unresolved core and a $FWHM = 14{,}000$ km s^{-1} component.

By first establishing the parameters of a LOC model using UV and soft X-ray lines and then extending the model to the Fe K band, [36, 37] show that the BLR model can account for only 3–17% of the Fe Kα broad line flux. A subsequent study using the same technique, based on simultaneous optical, UV, and X-ray observations of Mrk 509, [38] established that the contribution of the BLR to the Fe Kα line could be up to 30%. The clear implication is that part of the Fe Kα line in these two objects originates at smaller radii than the optical, UV, and even soft X-ray BLR, while the unresolved line core may be produced in the torus at larger radii. If these cases are not unique, but rather just an AGN that offers a fortuitous view of its inner disk, BLR, and torus, moderately deep observations of Seyfert-1 AGN with XRISM may readily decompose seemingly monolithic lines into such components.

NGC 4151 is the brightest Seyfert AGN in the 4–10 keV band, with the strongest narrow Fe Kα line observed in any Seyfert 1 (e.g., [193]). As such, it can be expected to deliver the most sensitive spectra and important hints of the potential of calorimeter spectroscopy with XRISM and Athena. These expectations led [151] to examine *Chandra*-HEG spectra of NGC 4151. The total summed spectrum (631 ks of exposure), and spectra summed from low (337 ks) and high (294 ks) flux states were examined. In each case, the line profile is found to be asymmetric and red-skewed, suggestive of weak relativistic Doppler shifts (see Fig. 9.11). Fits with a number of independent models imply line production radii in the $R = 500-1000\ \mathrm{GM/c^2}$ range.

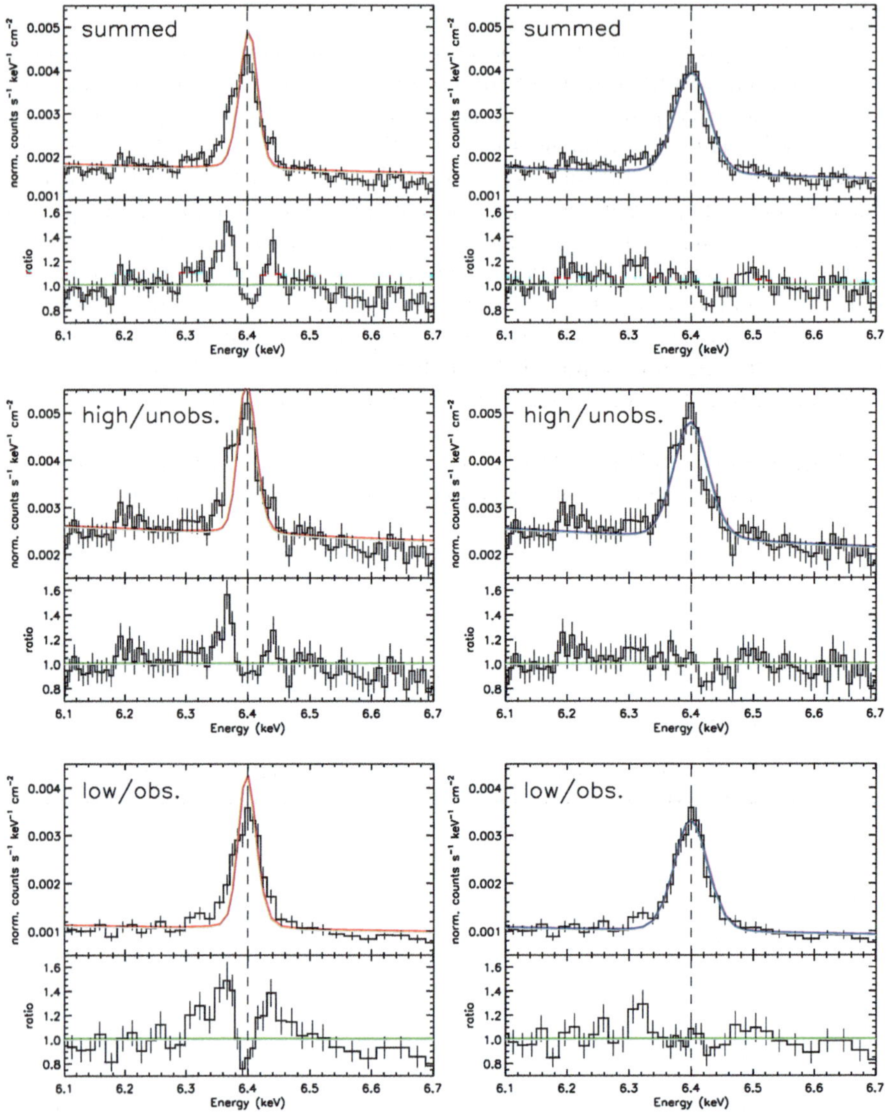

Fig. 9.11 Fits to the "narrow" Fe Kα line in summed *Chandra*-HEG spectra of NGC 4151 [151]. The spectra were shifted to the rest frame and binned for sensitivity and visual clarity. The left hand panels depict the results of fits with Gaussian profiles that only have instrumental broadening; while the right hand panels depict fits with Gaussian profiles with a variable width. The line is clearly asymmetrical, with excess flux to the red; a variety of models that include weak relativistic Doppler shifts and gravitational red-shifts suggest an inner production region of $R = 500{-}1000$ GM/c^2, considerably smaller than the optical BLR. (This figure is reproduced from [151], with permission)

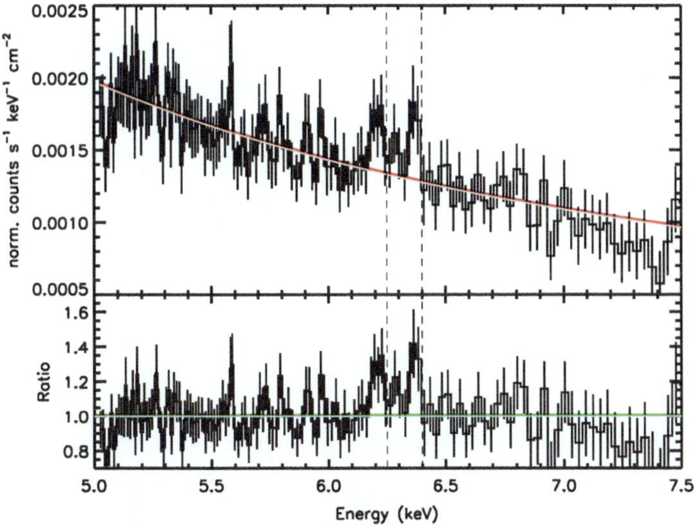

Fig. 9.12 The difference spectrum obtained when the summed *Chandra*-HEG exposure in the low flux state is subtracted from the summed exposure in the high flux state [151]. The spectrum was shifted to its rest frame, binned, and fit with a simple power-law. The lingering Fe Kα emission is red-shifted from the expected value of $E = 6.40$ keV, and equally strong emission is red-shifted from the Compton shoulder at $E = 6.25$ keV, though the narrow core and shoulder are not expected to be comparably strong. The residual structure is suggestive of a warp or a ring in the inner disk extending between $R = 50-130$ GM/c^2, potentially enhanced with a higher mass accretion rate and/or stronger irradiation. (This figure is reproduced from [151], with permission)

This again implies that the bulk of the "narrow" Fe K line flux originates at smaller radii than the optical BLR.

NGC 4151 may also provide early hints that the region between the innermost disk and optical BLR is structured. Miller et al. [151] find that the high-low flux difference spectrum reveals a line profile with two peaks, red-shifted from the expected narrow line core and Compton shoulder; fits to this profile require a narrow ring of emission between $R = 70-120$ GM/c^2 (see Fig. 9.12). Independently, within the spectra typified by a high continuum flux, the Fe Kα line flux appears to vary on time scales of $\Delta t \simeq 2 \times 10^4$ s, implying $R \simeq 50-130$ GM/c^2. These findings are at least qualitatively consistent with a warp or ring-like structure, similar to the features seen in numerical simulation of accretion flows when the angular momenta of the black hole and accretion flow are misaligned (e.g., [163]).

Zoghbi et al. [226] reported the first detection of reverberation in the narrow Fe Kα line in any AGN. Again owing to its flux, this detection was made in NGC 4151, using data from *XMM-Newton* and *Suzaku*. The flux sensitivity of these data outweighed their modest resolution relative to the *Chandra*-HEG. Using the JAVELIN code that is widely implemented to measure lags from the optical BLR [227], Zoghbi et al. [226] measure a delay of $\tau = 3.3^{+1.8}_{-0.7}$ days (see Fig. 9.13). For plausible black

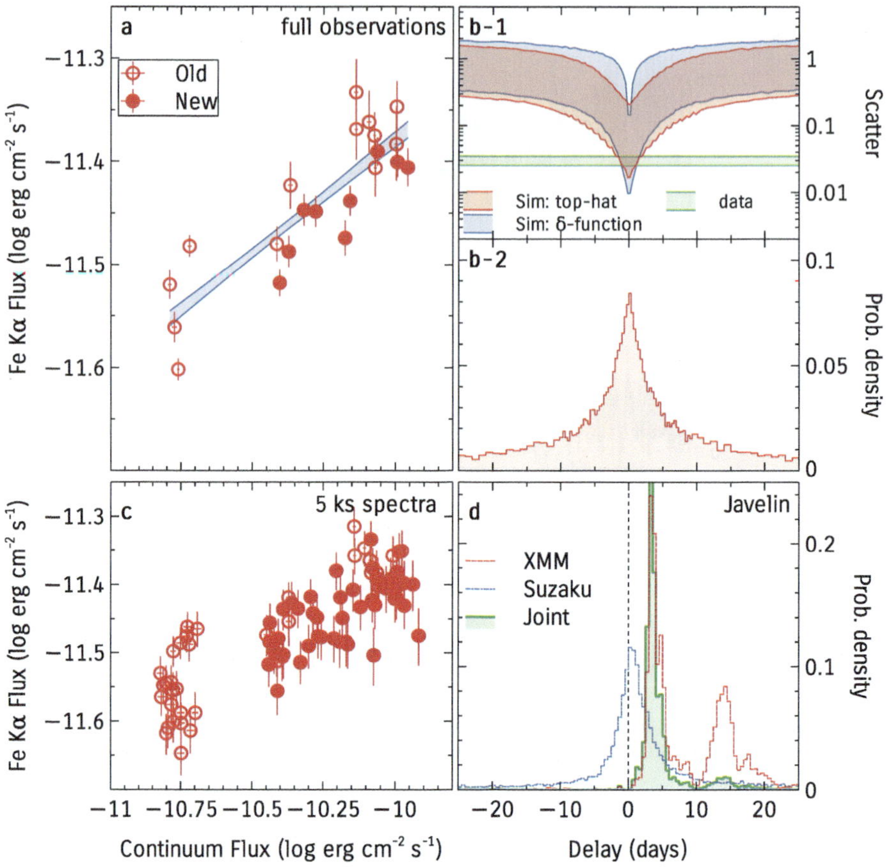

Fig. 9.13 Evidence of a reverberation lag in the narrow Fe Kα line in NGC 4151 [226], based on numerous observations with *XMM-Newton* and *Suzaku*. Panel **a** shows a strong correlation between the Fe Kα flux and continuum flux measured across full exposures. Panel **b1** shows the scatter in the line-continuum relation in different simulations that sample different delay times. Panel **b2** shows the probability density of the different delay times. Panel **c** shows the line-continuum relation, taking 5 ks segments instead of full observations. Finally, panel **d** shows the probability density for different values of the lag using the 5 ks segments. A lag of $\tau = 3.3^{+1.8}_{-0.7}$ days is measured, yielding a black hole mass of $M = 1.8^{+2.2}_{-1.1} \times 10^7\ M_\odot$ assuming the line widths measured by [151]. This is fully consistent with the most recent optical reverberation mass, $M_{BH} = 1.66^{+0.48}_{-0.34} \times 10^7$ M$_\odot$ [20]

hole masses, this light travel time is broadly consistent with the radii measured via fits to the Fe Kα line shape observed with *Chandra* [151].

Using the standard equation to derive the black hole mass, $M_{BH} = fc\tau v^2/G$ (where f is a geometric factor derived via comparisons to direct primary masses, and v is the width of the line), [226] infer a mass of $M_{BH} = 1.8^{+2.2}_{-1.1} \times 10^7\ M_\odot$, assuming a standard value for the geometric factor ($f = 4.13 \pm 1.05$; [81]) and the line width

9.4.4 Progress and Caveats

High-resolution X-rays spectroscopy will carry the massive potential to enhance our understanding of ultrafast outflows in AGN. With XRISM, Arcus, and Athena we will truly begin probing the origins, physics, and mechanisms at work in winds; and finally, test alternative models. Not only will this improve our understanding of UFOs themselves, but better our knowledge of AGN physics on the black hole scales and AGN feedback on galactic scales.

High-resolution spectroscopy will expose our misunderstandings. The current wind models are incredibly sophisticated, but are incomplete as they are based on simplifying assumptions, limited understanding, and finite computational power. Line profiles will depend on a number of factors, for example, the slope of the AMD. Even with high resolution, in some situations the different launching mechanisms might produce similar profiles [64]. The reader is advised to understand the assumptions and limitations of each model before employing them in research.

The physical situation will undoubtedly be much more complex. There is no reason that hybrid winds, which are combinations of thermal, magnetic and radiative, cannot co-exist in a given AGN [57]. Likewise, the blurred reflection will also complicate modelling [170], and both the intrinsic AGN and wind can be simultaneously varying on different time scales.

9.5 Conclusion

The AGN community has benefited immensely from *XMM-Newton*, *Chandra*, and other great X-ray observatories over the past 20 years. Many of these instruments will be producing excellent science for years to come. It is because of the tremendous success we have had with these instruments that the community can move forward with an eye on high resolution spectroscopy.

The launch of XRISM is imminent. This will be followed by the launch of an Athena-like mission (NewAthena) in the early or mid-2030s, covering the same pass band with even sharper resolution and higher sensitivity. In between, mission concepts such as Arcus may provide particularly high resolution grating spectroscopy below 1 keV, potentially simultaneously with UV spectroscopy, providing pristine line profiles to perform different density diagnostic tests, the detection of metastable levels, and time-resolved spectroscopy.

Assuming that the calorimeter aboard XRISM will operate for ten years, what are some optimistic goals for our understanding of AGN? And what are some equivalent goals for the more advanced missions that will follow XRISM?

The sensitivity of XRISM should clearly determine the demographics of both slow X-ray winds, and UFOs in bright, local Seyferts. We should learn if some sources display particularly line-rich spectra because they afford a fortuitous viewing angle, because their Eddington fraction is just right, or because their wind components have fortuitous ionisation levels. It is particularly important to understand the demographics of wind feedback as a function of Eddington fraction and XRISM will begin examining this for local Seyferts. If XRISM finds that UFOs provide insufficient feedback in the Seyfert phase, this does not preclude a larger role for winds at higher Eddington fractions. Athena will make it possible to extend detailed wind studies to quasars at modest red-shifts, and to local sources with lower Eddington fractions, and thus answer this key question.

In a subset of the brightest Seyferts, XRISM should be able to extend the range in ionisation over which AMDs are constructed by two orders of magnitude, offering an improved understanding of wind driving mechanisms. In the same subset, ionisation time scales should reveal gas densities, and therefore the absorption radius in the given wind zone, providing an independent angle on wind-driving mechanisms.

Since new reverberation studies suggest that the innermost wall of the "torus" is only a factor of a few more distant from the central engine than the optical BLR, few composite narrow Fe Kα emission line profiles might be expected in XRISM spectra. However, observations of bright Seyferts should clearly reveal the narrow Fe Kα line production region and its relationship to the BLR and torus. In a subset of the brightest Seyferts, reverberation mapping may offer an independent view on this problem, as well as independent constraints on black hole masses. If the disk is warped between the ISCO and the BLR owing to a misalignment of the disk with the black hole spin vector, resulting in "extra" reflecting area per unit radius, XRISM may also be able to reveal hints of this structure. In the longer term, the added sensitivity afforded by Athena should be able to test related theories in a much larger set of systems.

The epoch of high-resolution X-ray spectroscopy is upon us. For AGN, this is an exciting era that is full of massive potential for uncovering the ins and outs of black hole accretion.

Acknowledgements The authors would like to thank Margaret Buhariwalla, Susmita Chackravorty, Keigo Fukumura, Adam Gonzalez, Jelle Kaastra, Tim Kallman, Gabriele Matzeu, Missagh Mehdipour, Daniel Proga, John Raymond, Daniele Rogantini for discussion, data, code, and help with figures.

References

1. J.T. Allen, P.C. Hewett, N. Maddox, G.T. Richards, V. Belokurov, MNRAS **410**, 860–884 (2011). https://doi.org/10.1111/j.1365-2966.2010.17489.x. [arXiv:1007.3991]
2. W.N. Alston, A.C. Fabian, E. Kara et al., NatAs **4**, 597–602 (2020). https://doi.org/10.1038/s41550-019-1002-x. [arXiv:2001.06454]
3. C. Andonie, F.E. Bauer, R. Carraro et al., A&A **664**, A46 (2022). https://doi.org/10.1051/0004-6361/202142473. [arXiv:2204.09469]
4. R. Antonucci, ARA&A **31**, 473–521 (1993). https://doi.org/10.1146/annurev.aa.31.090193.002353
5. N. Arav, C. Chamberlain, G.A. Kriss et al., A&A **577**, A37 (2015). https://doi.org/10.1051/0004-6361/20142530210.48550/arXiv.1411.2157. [arXiv:1411.2157]

6. N. Arav, M. Moe, E. Costantini et al., ApJ **681**, 954–964 (2008). https://doi.org/10.1086/588651. [arXiv:0807.0228]
7. S.A. Balbus, J.F. Hawley, ApJ **376**, 214 (1991). https://doi.org/10.1086/170270
8. J. Baldwin, G. Ferland, K. Korista, D. Verner, ApJL **455**, L119 (1995). https://doi.org/10.1086/309827. [arXiv:astro-ph/9510080]
9. D.R. Ballantyne, MNRAS **491**, 3553–3561 (2020). https://doi.org/10.1093/mnras/stz3294. [arXiv:1911.10029]
10. D.R. Ballantyne, R.R. Ross, A.C. Fabian, MNRAS **327**, 10–22 (2001). https://doi.org/10.1046/j.1365-8711.2001.04432.x. [arXiv:astro-ph/0102040]
11. C. Bambi, L.W. Brenneman, T. Dauser et al., SSRv **217**, 65 (2021). https://doi.org/10.1007/s11214-021-00841-8. [arXiv:2011.04792]
12. W. Bambynek, B. Crasemann, R.W. Fink et al., Rev. Mod. Phys. **44**, 716–813 (1972)
13. D. Barret, T. Lam Trong, J.-W. den Herder, et al., SPIE **10699**, 106991G (2018). https://doi.org/10.1117/12.2312409. [arXiv:1807.06092]
14. M.A. Bautista, C. Mendoza, T.R. Kallman, P. Palmeri, A&A **403**, 339–355 (2003). https://doi.org/10.1051/0004-6361:20030367. [arXiv:astro-ph/0207323]
15. M.A. Bautista, C. Mendoza, T.R. Kallman, P. Palmeri, A&A **418**, 1171–1178 (2004). https://doi.org/10.1051/0004-6361:20034198
16. M.C. Begelman, cbhg.symp **374** (2004), [arXiv:astro-ph/0303040]
17. M.C. Begelman, C.F. McKee, G.A. Shields, ApJ **271**, 70–88 (1983). https://doi.org/10.1086/161178
18. E. Behar, ApJ **703**, 1346–1351 (2009). https://doi.org/10.1088/0004-637X/703/2/1346. [arXiv:0908.0539]
19. E. Behar, M. Sako, S.M. Kahn, ApJ **563**, 497–504 (2001). https://doi.org/10.1086/323966. [arXiv:astro-ph/0109314]
20. M.C. Bentz, P.R. Williams, T. Treu, ApJ **934**, 168 (2022). https://doi.org/10.3847/1538-4357/ac7c0a. [arXiv:2206.03513]
21. S. Bianchi, M. Guainazzi, G. Matt, N. Fonseca Bonilla, & G. Ponti, A&A **495**, 421-430 (2009). https://doi.org/10.1051/0004-6361:200810620. [arXiv:0811.1126]
22. R.D. Blandford, D.G. Payne, MNRAS **199**, 883–903 (1982). https://doi.org/10.1093/mnras/199.4.883
23. J. Bland-Hawthorn, S.L. Lumsden, G.M. Voit, G.N. Cecil, J.C. Weisheit, Ap&SS **248**, 177–190 (1997). https://doi.org/10.1023/A:1000525513140. [arXiv:astro-ph/9706211]
24. A.J. Blustin, M.J. Page, S.V. Fuerst, G. Branduardi-Raymont, C.E. Ashton, A&A **431**, 111–125 (2005). https://doi.org/10.1051/0004-6361:20041775. [arXiv:astro-ph/0411297]
25. R. Boissay-Malaquin, A. Danehkar, H.L. Marshall, M.A. Nowak, ApJ **873**, 29 (2019). https://doi.org/10.3847/1538-4357/ab0082. [arXiv:1901.06641]
26. T. Boller, W. N. Brandt, & H. Fink, A&A **305**, 53 (1996). https://doi.org/10.48550/arXiv.astro-ph/9504093. [arXiv:astro-ph/9504093]
27. W.N. Brandt, S. Mathur, M. Elvis, MNRAS **285**, L25–L30 (1997). https://doi.org/10.1093/mnras/285.3.L25l10.48550/arXiv.astro-ph/9703100. [arXiv:astro-ph/9703100]
28. H. Brunner, T. Liu, G. Lamer et al., A&A **661**, A1 (2022). https://doi.org/10.1051/0004-6361/202141266. [arXiv:2106.14517]
29. M. Cappi, AN **327**, 1012 (2006). https://doi.org/10.1002/asna.200610639. [arXiv:astro-ph/0610117]
30. J.I. Castor, D.C. Abbott, R.I. Klein, ApJ **195**, 157–174 (1975). https://doi.org/10.1086/153315
31. S. Chakravorty, A.K. Kembhavi, M. Elvis, G. Ferland, MNRAS **393**, 83–98 (2009). https://doi.org/10.1111/j.1365-2966.2008.14249.x10.48550/arXiv.0811.2404. [arXiv:0811.2404]
32. S. Chakravorty, R. Misra, M. Elvis, A.K. Kembhavi, G. Ferland, MNRAS **422**, 637–651 (2012). https://doi.org/10.1111/j.1365-2966.2012.20641.x10.48550/arXiv.1201.5435. [arXiv:1201.5435]
33. G. Chartas, W.N. Brandt, S.C. Gallagher, G.P. Garmire, ApJ **579**, 169–175 (2002). https://doi.org/10.1086/342744. [arXiv:astro-ph/0207196]
34. J. Contopoulos, R.V.E. Lovelace, ApJ **429**, 139 (1994). https://doi.org/10.1086/174307

35. E. Costantini, SSRv **157**, 265–277 (2010). https://doi.org/10.1007/s11214-010-9706-310.48550/arXiv.1011.2384. [arXiv:1011.2384]
36. E. Costantini, J.S. Kaastra, N. Arav et al., A&A **461**, 121–134 (2007). https://doi.org/10.1051/0004-6361:20065390. [arXiv:astro-ph/0609385]
37. E. Costantini, J.S. Kaastra, K. Korista et al., A&A **512**, A25 (2010). https://doi.org/10.1051/0004-6361/200912555. [arXiv:1001.2712]
38. E. Costantini, G. Kriss, J.S. Kaastra et al., A&A **595**, A106 (2016). https://doi.org/10.1051/0004-6361/201527956. [arXiv:1606.06579]
39. E. Costantini, F. Nicastro, A. Fruscione et al., ApJ **544**, 283–292 (2000). https://doi.org/10.1086/31720010.48550/arXiv.astro-ph/0007158. [arXiv:astro-ph/0007158]
40. D.M. Crenshaw, S.B. Kraemer, ApJ **753**, 75 (2012). https://doi.org/10.1088/0004-637X/753/1/75. [arXiv:1204.6694]
41. D.M. Crenshaw, S.B. Kraemer, J.R. Gabel et al., ApJ **594**, 116–127 (2003). https://doi.org/10.1086/37679210.48550/arXiv.astro-ph/0305154. [arXiv:astro-ph/0305154]
42. B. Czerny, M. Nikołajuk, A. Rózańska et al., A&A **412**, 317–329 (2003). https://doi.org/10.1051/0004-6361:20031441. [arXiv:astro-ph/0309242]
43. R.C. Dannen, D. Proga, T.R. Kallman, T. Waters, ApJ **882**, 99 (2019). https://doi.org/10.3847/1538-4357/ab340b. [arXiv:1812.01773]
44. T. Dauser, J. Wilms, C.S. Reynolds, L.W. Brenneman, MNRAS **409**, 1534–1540 (2010). https://doi.org/10.1111/j.1365-2966.2010.17393.x. [arXiv:1007.4937]
45. M. Dehghanian, G.J. Ferland, B.M. Peterson et al., ApJL **882**, L30 (2019). https://doi.org/10.3847/2041-8213/ab3d4110.48550/arXiv.1908.07686. [arXiv:1908.07686]
46. R.G. Detmers, J.S. Kaastra, K.C. Steenbrugge et al., A&A **534**, A38 (2011). https://doi.org/10.1051/0004-6361/201116899. [arXiv:1107.0658]
47. L. Di Gesu, E. Costantini, A&A **594**, A88 (2016). https://doi.org/10.1051/0004-6361/20162867010.48550/arXiv.1607.01943. [arXiv:1607.01943]
48. L. Di Gesu, E. Costantini, N. Arav et al., A&A **556**, A94 (2013). https://doi.org/10.1051/0004-6361/201321416. [arXiv:1305.6232]
49. L. Di Gesu, E. Costantini, J. Ebrero et al., A&A **579**, A42 (2015). https://doi.org/10.1051/0004-6361/201525934. [arXiv:1505.02562]
50. L. Di Gesu, E. Costantini, E. Piconcelli et al., A&A **563**, A95 (2014). https://doi.org/10.1051/0004-6361/201322916. [arXiv:1401.5614]
51. T. Di Matteo, V. Springel, L. Hernquist, Natur **433**, 604–607 (2005). https://doi.org/10.1038/nature03335. [arXiv:astro-ph/0502199]
52. C. Done, M.A. Sobolewska, M. Gierliński, N.J. Schurch, MNRAS **374**, L15–L19 (2007). https://doi.org/10.1111/j.1745-3933.2006.00255.x. [arXiv:astro-ph/0610078]
53. C. Done, R. Tomaru, T. Takahashi, MNRAS **473**, 838–848 (2018). https://doi.org/10.1093/mnras/stx2400. [arXiv:1612.09377]
54. J. Ebrero, V. Domček, G.A. Kriss, J.S. Kaastra, A&A **653**, A125 (2021). https://doi.org/10.1051/0004-6361/202040045. [arXiv:2107.05676]
55. J. Ebrero, G.A. Kriss, J.S. Kaastra, J.C. Ely, A&A **586**, A72 (2016). https://doi.org/10.1051/0004-6361/20152749510.48550/arXiv.1511.07169. [arXiv:1511.07169]
56. D. Edmonds, B. Borguet, N. Arav et al., ApJ **739**, 7 (2011). https://doi.org/10.1088/0004-637X/739/1/7
57. J.E. Everett, ApJ **631**, 689–706 (2005). https://doi.org/10.1086/432678. [arXiv:astro-ph/0506321]
58. A.C. Fabian, ARA&A **50**, 455–489 (2012). https://doi.org/10.1146/annurev-astro-081811-125521. [arXiv:1204.4114]
59. A.C. Fabian, M.J. Rees, L. Stella, N.E. White, MNRAS **238**, 729–736 (1989). https://doi.org/10.1093/mnras/238.3.729
60. A.C. Fabian, C.S. Reynolds, J. Jiang et al., MNRAS **493**, 2518–2522 (2020). https://doi.org/10.1093/mnras/staa482. [arXiv:2002.06388]
61. A.C. Fabian, A. Zoghbi, R.R. Ross et al., Natur **459**, 540–542 (2009). https://doi.org/10.1038/nature08007

62. L. Ferrarese, D. Merritt, ApJL **539**, L9–L12 (2000). https://doi.org/10.1086/312838. [arXiv:astro-ph/0006053]
63. T.C. Fischer, C. Machuca, M.R. Diniz et al., ApJ **834**, 30 (2017). https://doi.org/10.3847/1538-4357/834/1/30. [arXiv:1609.08927]
64. K. Fukumura, M. Dadina, G. Matzeu et al., ApJ **940**, 6 (2022). https://doi.org/10.3847/1538-4357/ac9388. [arXiv:2205.08894]
65. K. Fukumura, D. Kazanas, I. Contopoulos, E. Behar, ApJ **715**, 636–650 (2010). https://doi.org/10.1088/0004-637X/715/1/636. [arXiv:0910.3001]
66. K. Fukumura, D. Kazanas, I. Contopoulos, E. Behar, ApJL **723**, L228–L232 (2010). https://doi.org/10.1088/2041-8205/723/2/L228. [arXiv:1009.5644]
67. K. Fukumura, F. Tombesi, D. Kazanas et al., ApJ **805**, 17 (2015). https://doi.org/10.1088/0004-637X/805/1/17. [arXiv:1503.04074]
68. J.R. Gabel, D.M. Crenshaw, S.B. Kraemer et al., ApJ **583**, 178–191 (2003). https://doi.org/10.1086/34509610.48550/arXiv.astro-ph/0209484. [arXiv:astro-ph/0209484]
69. J.R. Gabel, S.B. Kraemer, D.M. Crenshaw et al., ApJ **631**, 741–761 (2005). https://doi.org/10.1086/43268210.48550/arXiv.astro-ph/0506323. [arXiv:astro-ph/0506323]
70. L.C. Gallo, MNRAS **368**, 479–486 (2006). https://doi.org/10.1111/j.1365-2966.2006.10137.x10.48550/arXiv.astro-ph/0602145. [arXiv:astro-ph/0602145]
71. L. Gallo, rnls.conf **34** (2018). https://doi.org/10.22323/1.328.0034. [arXiv:1807.09838]
72. L.C. Gallo, A.C. Fabian, MNRAS **418**, L59–L63 (2011). https://doi.org/10.1111/j.1745-3933.2011.01143.x. [arXiv:1108.5060]
73. L.C. Gallo, A.C. Fabian, MNRAS **434**, L66–L69 (2013). https://doi.org/10.1093/mnrasl/slt080. [arXiv:1306.3404]
74. L.C. Gallo, A.G. Gonzalez, S.G.H. Waddell et al., MNRAS **484**, 4287–4297 (2019). https://doi.org/10.1093/mnras/stz274. [arXiv:1901.07899]
75. K. Gebhardt, R. Bender, G. Bower et al., ApJL **539**, L13–L16 (2000). https://doi.org/10.1086/312840. [arXiv:astro-ph/0006289]
76. I.M. George, A.C. Fabian, MNRAS **249**, 352 (1991). https://doi.org/10.1093/mnras/249.2.352
77. I.M. George, T.J. Turner, H. Netzer et al., ApJS **114**, 73–120 (1998). https://doi.org/10.1086/31306710.48550/arXiv.astro-ph/9708046. [arXiv:astro-ph/9708046]
78. M. Giustini, D. Proga, A&A **630**, A94 (2019). https://doi.org/10.1051/0004-6361/201833810. [arXiv:1904.07341]
79. M. Giustini, D. Proga, IAUS **356**, 82–86 (2021). https://doi.org/10.1017/S1743921320002628. [arXiv:2002.07564]
80. J. Gofford, J.N. Reeves, F. Tombesi et al., MNRAS **430**, 60–80 (2013). https://doi.org/10.1093/mnras/sts481. [arXiv:1211.5810]
81. C.J. Grier, P. Martini, L.C. Watson et al., ApJ **773**, 90 (2013). https://doi.org/10.1088/0004-637X/773/2/90. [arXiv:1305.2447]
82. M.F. Gu, T. Holczer, E. Behar, S.M. Kahn, ApJ **641**, 1227–1232 (2006). https://doi.org/10.1086/500640. [arXiv:astro-ph/0512410]
83. M.F. Gu, S.M. Kahn, D.W. Savin et al., ApJ **563**, 462–471 (2001). https://doi.org/10.1086/323683
84. A. Gupta, S. Mathur, Y. Krongold, F. Nicastro, ApJ **772**, 66 (2013). https://doi.org/10.1088/0004-637X/772/1/66. [arXiv:1301.6139]
85. A. Gupta, S. Mathur, Y. Krongold, ApJ **798**, 4 (2015). https://doi.org/10.1088/0004-637X/798/1/4. [arXiv:1406.5968]
86. M. Gupta, M. Sikora, K. Rusinek, G.M. Madejski, MNRAS **480**, 2861–2871 (2018). https://doi.org/10.1093/mnras/sty204310.48550/arXiv.1808.07170. [arXiv:1808.07170]
87. K. Hagino, H. Odaka, C. Done et al., MNRAS **446**, 663–676 (2015). https://doi.org/10.1093/mnras/stu2095. [arXiv:1410.1640]
88. K. Hagino, H. Odaka, C. Done et al., MNRAS **461**, 3954–3963 (2016). https://doi.org/10.1093/mnras/stw1579. [arXiv:1509.05645]
89. J.P. Halpern, ApJ **281**, 90–94 (1984). https://doi.org/10.1086/162077

90. N. Higginbottom, D. Proga, ApJ **807**, 107 (2015). https://doi.org/10.1088/0004-637X/807/1/107. [arXiv:1504.03328]
91. N. Higginbottom, D. Proga, C. Knigge et al., ApJ **789**, 19 (2014). https://doi.org/10.1088/0004-637X/789/1/19. [arXiv:1402.1849]
92. N. Higginbottom, D. Proga, C. Knigge, K.S. Long, ApJ **836**, 42 (2017). https://doi.org/10.3847/1538-4357/836/1/42. [arXiv:1612.08996]
93. Hitomi Collaboration, F. Aharonian, H. Akamatsu, et al., Natur **535**, 117-121 (2016). https://doi.org/10.1038/nature18627. [arXiv:1607.04487]
94. Hitomi Collaboration, F. Aharonian, H. Akamatsu, et al., PASJ **70**, 13 (2018). https://doi.org/10.1093/pasj/psx147. [arXiv:1711.06289]
95. T. Holczer, E. Behar, S. Kaspi, ApJ **663**, 799–807 (2007). https://doi.org/10.1086/51841610.48550/arXiv.astro-ph/0703351. [arXiv:astro-ph/0703351]
96. S.S. Holt, R.F. Mushotzky, R.H. Becker et al., ApJL **241**, L13–L17 (1980). https://doi.org/10.1086/183350
97. P.F. Hopkins, M. Elvis, MNRAS **401**, 7–14 (2010). https://doi.org/10.1111/j.1365-2966.2009.15643.x. [arXiv:0904.0649]
98. Z. Igo, M.L. Parker, G.A. Matzeu et al., MNRAS **493**, 1088–1108 (2020). https://doi.org/10.1093/mnras/staa265. [arXiv:2001.08208]
99. J. Jiang, M.L. Parker, A.C. Fabian et al., MNRAS **477**, 3711–3726 (2018). https://doi.org/10.1093/mnras/sty836. [arXiv:1804.00349]
100. J. Jiang, D.J. Walton, A.C. Fabian, M.L. Parker, MNRAS **483**, 2958–2967 (2019). https://doi.org/10.1093/mnras/sty3228. [arXiv:1811.10932]
101. A. Juráňová, E. Costantini, P. Uttley, MNRAS **510**, 4225–4235 (2022). https://doi.org/10.1093/mnras/stab3731. [arXiv:2201.02640]
102. J.S. Kaastra, J.A.M. Bleeker, A&A **587**, A151 (2016). https://doi.org/10.1051/0004-6361/201527395. [arXiv:1601.05309]
103. J.S. Kaastra, R.G. Detmers, M. Mehdipour et al., A&A **539**, A117 (2012). https://doi.org/10.1051/0004-6361/201118161. [arXiv:1201.1855]
104. J.S. Kaastra, G.A. Kriss, M. Cappi et al., Sci **345**, 64–68 (2014). https://doi.org/10.1126/science.125378710.48550/arXiv.1406.5007. [arXiv:1406.5007]
105. J.S. Kaastra, M. Mehdipour, E. Behar et al., A&A **619**, A112 (2018). https://doi.org/10.1051/0004-6361/201832629. [arXiv:1805.03538]
106. J.S. Kaastra, R. Mewe, D.A. Liedahl, S. Komossa, A.C. Brinkman, A&A **354**, L83–L86 (2000). ([arXiv:astro-ph/0002345])
107. J.S. Kaastra, R. Mewe, & H. Nieuwenhuijzen, uxsa.conf 411–414 (1996)
108. J.S. Kaastra, A.J.J. Raassen, R. Mewe et al., A&A **428**, 57–66 (2004). https://doi.org/10.1051/0004-6361:200414341 0.48550/arXiv.astro-ph/0406199. [arXiv:astro-ph/0406199]
109. J.S. Kaastra, Y. Terashima, T. Kallman, et al., (2014), [arXiv:1412.1171]
110. T. Kallman, M. Bautista, ApJS **133**, 221–253 (2001). https://doi.org/10.1086/319184
111. E.S. Kammoun, J.M. Miller, M. Koss et al., ApJ **901**, 161 (2020). https://doi.org/10.3847/1538-4357/abb29f. [arXiv:2007.02616]
112. E. Kara, M. Mehdipour, G.A. Kriss et al., ApJ **922**, 151 (2021). https://doi.org/10.3847/1538-4357/ac2159. [arXiv:2105.05840]
113. S. Kaspi, W.N. Brandt, H. Netzer et al., ApJL **535**, L17–L20 (2000). https://doi.org/10.1086/312697
114. S. Kaspi, W.N. Brandt, H. Netzer et al., ApJ **554**, 216–232 (2001). https://doi.org/10.1086/321333. [arXiv:astro-ph/0101540]
115. D. Kazanas, K. Fukumura, E. Behar, I. Contopoulos, C. Shrader, AstRv **7**, 92–123 (2012). https://doi.org/10.1080/21672857.2012.11519707. [arXiv:1206.5022]
116. N. Keshet, E. Behar, ApJ **934**, 124 (2022). https://doi.org/10.3847/1538-4357/ac7c6b. [arXiv:2206.15074]
117. A.R. King, MNRAS **402**, 1516–1522 (2010). https://doi.org/10.1111/j.1365-2966.2009.16013.x. [arXiv:0911.1639]

118. A.R. King, K.A. Pounds, MNRAS **345**, 657–659 (2003). https://doi.org/10.1046/j.1365-8711.2003.06980.x. [arXiv:astro-ph/0305541]
119. A. King, K. Pounds, ARA&A **53**, 115–154 (2015). https://doi.org/10.1146/annurev-astro-082214-122316. [arXiv:1503.05206]
120. C. Knigge, J.A. Woods, J.E. Drew, MNRAS **273**, 225–248 (1995). https://doi.org/10.1093/mnras/273.2.225
121. S. Komossa, NewAR **44**, 483–485 (2000). https://doi.org/10.1016/S1387-6473(00)00084-110.48550/arXiv.astro-ph/9810105. [arXiv:astro-ph/9810105]
122. S. Komossa, S. Mathur, A&A **374**, 914–918 (2001). https://doi.org/10.1051/0004-6361:2001081910.48550/arXiv.astro-ph/0106242. [arXiv:astro-ph/0106242]
123. K.T. Korista, M.A. Bautista, N. Arav et al., ApJ **688**, 108–115 (2008). https://doi.org/10.1086/59214010.48550/arXiv.0807.0230. [arXiv:0807.0230]
124. K. Koyama, Y. Maeda, T. Sonobe et al., PASJ **48**, 249–255 (1996). https://doi.org/10.1093/pasj/48.2.249
125. S.B. Kraemer, F. Tombesi, M.C. Bottorff, ApJ **852**, 35 (2018). https://doi.org/10.3847/1538-4357/aa9ce0. [arXiv:1711.07965]
126. A. Kramida, Y. Ralchenko, J. Reader, NIST AST Team, NIST atomic spectra database version 5.10 (2022)
127. J.H. Krolik, T.R. Kallman, ApJL **320**, L5 (1987). https://doi.org/10.1086/184966
128. J.H. Krolik, G.A. Kriss, ApJ **447**, 512 (1995). https://doi.org/10.1086/17589610.48550/arXiv.astro-ph/9501089. [arXiv:astro-ph/9501089]
129. J.H. Krolik, C.F. McKee, C.B. Tarter, ApJ **249**, 422–442 (1981). https://doi.org/10.1086/159303
130. Y. Krongold, F. Nicastro, N.S. Brickhouse et al., ApJ **597**, 832–850 (2003). https://doi.org/10.1086/378639. [arXiv:astro-ph/0306460]
131. Y. Krongold, F. Nicastro, M. Elvis et al., ApJ **620**, 165–182 (2005). https://doi.org/10.1086/425293. [arXiv:astro-ph/0409490]
132. Y. Krongold, F. Nicastro, M. Elvis et al., ApJ **659**, 1022–1039 (2007). https://doi.org/10.1086/512476
133. S. Laha, M. Guainazzi, S. Chakravorty, G.C. Dewangan, A.K. Kembhavi, MNRAS **457**, 3896–3911 (2016). https://doi.org/10.1093/mnras/stw211. [arXiv:1601.06369]
134. S. Laha, M. Guainazzi, G.C. Dewangan, S. Chakravorty, A.K. Kembhavi, MNRAS **441**, 2613–2643 (2014). https://doi.org/10.1093/mnras/stu669. [arXiv:1404.0899]
135. A. Laor, ApJ **376**, 90 (1991). https://doi.org/10.1086/170257
136. K.S. Long, C. Knigge, ApJ **579**, 725–740 (2002). https://doi.org/10.1086/342879. [arXiv:astro-ph/0208011]
137. A.L. Longinotti, Y. Krongold, M. Guainazzi et al., ApJL **813**, L39 (2015). https://doi.org/10.1088/2041-8205/813/2/L39. [arXiv:1511.01165]
138. A.L. Longinotti, Y. Krongold, G.A. Kriss et al., ApJ **766**, 104 (2013). https://doi.org/10.1088/0004-637X/766/2/10410.48550/arXiv.1301.5463. [arXiv:1301.5463]
139. P. Magdziarz, O.M. Blaes, A.A. Zdziarski, W.N. Johnson, D.A. Smith, MNRAS **301**, 179–192 (1998). https://doi.org/10.1046/j.1365-8711.1998.02015.x
140. A.G. Markowitz, M. Krumpe, R. Nikutta, MNRAS **439**, 1403–1458 (2014). https://doi.org/10.1093/mnras/stt249210.48550/arXiv.1402.2779. [arXiv:1402.2779]
141. G. Matt, MNRAS **337**, 147–150 (2002). https://doi.org/10.1046/j.1365-8711.2002.05890.x. [arXiv:astro-ph/0207615]
142. G.A. Matzeu, M. Lieu, M.T. Costa et al., MNRAS **515**, 6172–6190 (2022). https://doi.org/10.1093/mnras/stac2155 [rXiv:2207.13731]
143. G.A. Matzeu, J.N. Reeves, V. Braito et al., MNRAS **472**, L15–L19 (2017). https://doi.org/10.1093/mnrasl/slx129. [arXiv:1708.03546]
144. J. Mao, J.S. Kaastra, M. Mehdipour et al., A&A **607**, A100 (2017). https://doi.org/10.1051/0004-6361/20173137810.48550/arXiv.1707.09552. [arXiv:1707.09552]
145. J. Mao, J.S. Kaastra, M. Mehdipour et al., A&A **665**, A72 (2022). https://doi.org/10.1051/0004-6361/20214263710.48550/arXiv.2204.06813. [arXiv:2204.06813]

146. M. Mehdipour, E. Costantini, A&A **625**, A25 (2019). https://doi.org/10.1051/0004-6361/20193520510.48550/arXiv.1903.11605. [arXiv:1903.11605]
147. M. Mehdipour, G.A. Kriss, E. Costantini et al., ApJL **934**, L24 (2022). https://doi.org/10.3847/2041-8213/ac822f. [arXiv:2207.09464]
148. M. Mehdipour, G.A. Kriss, E. Costantini et al., ApJL **934**, L24 (2022). https://doi.org/10.3847/2041-8213/ac822f10.48550/arXiv.2207.09464. [arXiv:2207.09464]
149. C. Mendoza, T.R. Kallman, M.A. Bautista, P. Palmeri, A&A **414**, 377–388 (2004). https://doi.org/10.1051/0004-6361:20031621. [arXiv:astro-ph/0306320]
150. J.M. Miller, ARA&A **45**, 441–479 (2007). https://doi.org/10.1146/annurev.astro.45.051806.110555. [arXiv:0705.0540]
151. J.M. Miller, E. Cackett, A. Zoghbi et al., ApJ **865**, 97 (2018). https://doi.org/10.3847/1538-4357/aadbaa. [arXiv:1808.07435]
152. M. Mizumoto, M. Nomura, C. Done, K. Ohsuga, H. Odaka, MNRAS **503**, 1442–1458 (2021). https://doi.org/10.1093/mnras/staa3282. [arXiv:2003.01137]
153. T. Minezaki, Y. Yoshii, Y. Kobayashi et al., ApJ **886**, 150 (2019). https://doi.org/10.3847/1538-4357/ab4f7b. [arXiv:1910.08722]
154. Y. Morishima, H. Sudou, A. Yamauchi, Y. Taniguchi, & N. Nakai, PASJ..tmp (2022). https://doi.org/10.1093/pasj/psac092. [arXiv:2211.02280]
155. K.D. Murphy, T. Yaqoob, MNRAS **397**, 1549–1562 (2009). https://doi.org/10.1111/j.1365-2966.2009.15025.x. [arXiv:0905.3188]
156. N. Murray, J. Chiang, S.A. Grossman, G.M. Voit, ApJ **451**, 498 (1995). https://doi.org/10.1086/176238
157. K. Nandra, MNRAS **368**, L62–L66 (2006). https://doi.org/10.1111/j.1745-3933.2006.00158.x. [arXiv:astro-ph/0602081]
158. E. Nardini, J.N. Reeves, J. Gofford et al., Sci **347**, 860–863 (2015). https://doi.org/10.1126/science.1259202. [arXiv:1502.06636]
159. M. Nenkova, M.M. Sirocky, R. Nikutta, Ž Ivezić, M. Elitzur, ApJ **685**, 160–180 (2008). https://doi.org/10.1086/590483. [arXiv:0806.0512]
160. F. Nicastro, F. Fiore, G. Matt, ApJ **517**, 108–122 (1999). https://doi.org/10.1086/30718710.48550/arXiv.astro-ph/9812392. [arXiv:astro-ph/9812392]
161. F. Nicastro, F. Fiore, G.C. Perola, M. Elvis, ApJ **512**, 184–196 (1999). https://doi.org/10.1086/30673610.48550/arXiv.astro-ph/9808316. [arXiv:astro-ph/9808316]
162. F. Nicastro, L. Piro, A. De Rosa et al., ApJ **536**, 718–728 (2000). https://doi.org/10.1086/30895010.48550/arXiv.astro-ph/0001201. [arXiv:astro-ph/0001201]
163. C. Nixon, A. King, D. Price, J. Frank, ApJL **757**, L24 (2012). https://doi.org/10.1088/2041-8205/757/2/L24. [arXiv:1209.1393]
164. H. Noda, T. Mineta, T. Minezaki, et al., (2022), [arXiv:2212.02731]
165. M. Nomura, K. Ohsuga, MNRAS **465**, 2873–2879 (2017). https://doi.org/10.1093/mnras/stw2877. [arXiv:1610.08511]
166. M. Nomura, K. Ohsuga, C. Done, MNRAS **494**, 3616–3626 (2020). https://doi.org/10.1093/mnras/staa948. [arXiv:1811.01966]
167. P. Palmeri, C. Mendoza, T.R. Kallman, M.A. Bautista, A&A **403**, 1175–1184 (2003). https://doi.org/10.1051/0004-6361:20030405
168. P. Palmeri, C. Mendoza, T.R. Kallman, M.A. Bautista, M. Meléndez, A&A **410**, 359–364 (2003). https://doi.org/10.1051/0004-6361:20031262. [arXiv:astro-ph/0306321]
169. M.L. Parker, C. Pinto, A.C. Fabian et al., Natur **543**, 83–86 (2017). https://doi.org/10.1038/nature21385. [arXiv:1703.00071]
170. M.L. Parker, G.A. Matzeu, J.H. Matthews et al., MNRAS **513**, 551–572 (2022). https://doi.org/10.1093/mnras/stac877. [arXiv:2203.14789]
171. C. Pinto, W. Alston, M.L. Parker et al., MNRAS **476**, 1021–1035 (2018). https://doi.org/10.1093/mnras/sty231. [arXiv:1708.09422]
172. G. Ponti, L.C. Gallo, A.C. Fabian et al., MNRAS **406**, 2591–2604 (2010). https://doi.org/10.1111/j.1365-2966.2010.16852.x. [arXiv:0911.1003]

173. G. Ponti, M.R. Morris, R. Terrier et al., MNRAS **453**, 172–213 (2015). https://doi.org/10.1093/mnras/stv1331. [arXiv:1508.04445]
174. K.A. Pounds, J.N. Reeves, A.R. King et al., MNRAS **345**, 705–713 (2003). https://doi.org/10.1046/j.1365-8711.2003.07006.x. [arXiv:astro-ph/0303603]
175. D. Proga, J.M. Stone, T.R. Kallman, ApJ **543**, 686–696 (2000). https://doi.org/10.1086/317154. [arXiv:astro-ph/0005315]
176. D. Proga, T.R. Kallman, ApJ **616**, 688–695 (2004). https://doi.org/10.1086/425117. [arXiv:astro-ph/0408293]
177. J.N. Reeves, V. Braito, ApJ **884**, 80 (2019). https://doi.org/10.3847/1538-4357/ab41f9. [arXiv:1909.05039]
178. J.N. Reeves, V. Braito, G. Chartas et al., ApJ **895**, 37 (2020). https://doi.org/10.3847/1538-4357/ab8cc4. [arXiv:2004.12439]
179. J.N. Reeves, P.T. O'Brien, M.J. Ward, ApJL **593**, L65–L68 (2003). https://doi.org/10.1086/378218. [arXiv:astro-ph/0307127]
180. J.N. Reeves, R.M. Sambruna, V. Braito, M. Eracleous, ApJL **702**, L187–L190 (2009). https://doi.org/10.1088/0004-637X/702/2/L18710.48550/arXiv.0908.1715. [arXiv:0908.1715]
181. C.S. Reynolds, MNRAS **286**, 513–537 (1997). https://doi.org/10.1093/mnras/286.3.51310.48550/arXiv.astro-ph/9610127. [arXiv:astro-ph/9610127]
182. C.S. Reynolds, ARA&A **59**, 117–154 (2021). https://doi.org/10.1146/annurev-astro-112420-035022. [arXiv:2011.08948]
183. G. Risaliti, E. Nardini, M. Salvati et al., MNRAS **410**, 1027–1035 (2011). https://doi.org/10.1111/j.1365-2966.2010.17503.x10.48550/arXiv.1008.5067. [arXiv:1008.5067]
184. D. Rogantini, E. Costantini, L.C. Gallo et al., MNRAS **516**, 5171–5186 (2022). https://doi.org/10.1093/mnras/stac2552. [arXiv:2209.02747]
185. D. Rogantini, M. Mehdipour, J. Kaastra et al., ApJ **940**, 122 (2022). https://doi.org/10.3847/1538-4357/ac9c01
186. R.R. Ross, A.C. Fabian, MNRAS **261**, 74–82 (1993). https://doi.org/10.1093/mnras/261.1.74
187. R.R. Ross, A.C. Fabian, MNRAS **358**, 211–216 (2005). https://doi.org/10.1111/j.1365-2966.2005.08797.x. [arXiv:astro-ph/0501116]
188. A. Różańska, I. Kowalska, A.C. Gonçalves, A&A **487**, 895–900 (2008). https://doi.org/10.1051/0004-6361:200809549
189. M. Sako, S.M. Kahn, E. Behar et al., A&A **365**, L168–L173 (2001). https://doi.org/10.1051/0004-6361:20000081. [arXiv:astro-ph/0010660]
190. R.D. Saxton, A.M. Read, P. Esquej et al., A&A **480**, 611–622 (2008). https://doi.org/10.1051/0004-6361:20079193. [arXiv:0801.3732]
191. R. Serafinelli, F. Tombesi, F. Vagnetti et al., A&A **627**, A121 (2019). https://doi.org/10.1051/0004-6361/201935275. [arXiv:1906.02765]
192. N.I. Shakura, R.A. Sunyaev, A&A **24**, 337–355 (1973)
193. X.W. Shu, T. Yaqoob, J.X. Wang, ApJS **187**, 581–606 (2010). https://doi.org/10.1088/0067-0049/187/2/581. [arXiv:1003.1790]
194. J. Silk, M.J. Rees, A&A **331**, L1–L4 (1998), ([arXiv:astro-ph/9801013])
195. C.V. Silva, P. Uttley, E. Costantini, A&A **596**, A79 (2016). https://doi.org/10.1051/0004-6361/20162855510.48550/arXiv.1607.01065. [arXiv:1607.01065]
196. S.A. Sim, K.S. Long, L. Miller, T.J. Turner, MNRAS **388**, 611–624 (2008). https://doi.org/10.1111/j.1365-2966.2008.13466.x. [arXiv:0805.2251]
197. S.A. Sim, L. Miller, K.S. Long, T.J. Turner, J.N. Reeves, MNRAS **404**, 1369–1384 (2010). https://doi.org/10.1111/j.1365-2966.2010.16396.x. [arXiv:1002.0544]
198. R.K. Smith, M.H. Abraham, R. Allured et al., SPIE **9905**, 99054M (2016). https://doi.org/10.1117/12.2231778
199. V. Springel, L. Hernquist, ApJL **622**, L9–L12 (2005). https://doi.org/10.1086/429486. [arXiv:astro-ph/0411379]
200. K.C. Steenbrugge, M. Fenovčík, J.S. Kaastra, E. Costantini, F. Verbunt, A&A **496**, 107–119 (2009). https://doi.org/10.1051/0004-6361/200810416. [arXiv:0901.3108]

201. K.C. Steenbrugge, J.S. Kaastra, D.M. Crenshaw et al., A&A **434**, 569–584 (2005). https://doi.org/10.1051/0004-6361:20047138. [arXiv:astro-ph/0501122]
202. T. Takahashi, M. Kokubun, K. Mitsuda et al., JATIS **4**, 021402 (2018). https://doi.org/10.1117/1.JATIS.4.2.021402
203. Y. Tanaka, T. Boller, L. Gallo, R. Keil, Y. Ueda, PASJ **56**, L9–L13 (2004). https://doi.org/10.1093/pasj/56.3.L9. [arXiv:astro-ph/0405158]
204. C.B. Tarter, W.H. Tucker, E.E. Salpeter, ApJ **156**, 943 (1969). https://doi.org/10.1086/150026
205. M. Tashiro, H. Maejima, K. Toda et al., SPIE **11444**, 1144422 (2020). https://doi.org/10.1117/12.2565812
206. L. Titarchuk, ApJ **434**, 570 (1994). https://doi.org/10.1086/174760
207. R. Tomaru, C. Done, J. Mao, MNRAS **518**, 1789–1801 (2023). https://doi.org/10.1093/mnras/stac3210. [arXiv:2204.08802]
208. F. Tombesi, M. Cappi, J.N. Reeves et al., A&A **521**, A57 (2010). https://doi.org/10.1051/0004-6361/200913440. [arXiv:1006.2858]
209. F. Tombesi, M. Cappi, J.N. Reeves et al., ApJ **742**, 44 (2011). https://doi.org/10.1088/0004-637X/742/1/44. [arXiv:1109.2882]
210. F. Tombesi, R.M. Sambruna, J.N. Reeves et al., ApJ **719**, 700–715 (2010). https://doi.org/10.1088/0004-637X/719/1/700. [arXiv:1006.3536]
211. E. Torresi, P. Grandi, E. Costantini, G.G.C. Palumbo, MNRAS **419**, 321–329 (2012). https://doi.org/10.1111/j.1365-2966.2011.19694.x10.48550/arXiv.1108.4890. [arXiv:1108.4890]
212. T.J. Turner, L. Miller, S.B. Kraemer, J.N. Reeves, ApJ **733**, 48 (2011). https://doi.org/10.1088/0004-637X/733/1/48. [arXiv:1103.3709]
213. D.A. Turnshek, ApJ **280**, 51–65 (1984). https://doi.org/10.1086/161967
214. C.M. Urry, P. Padovani, PASP **107**, 803 (1995). https://doi.org/10.1086/133630. [arXiv:astro-ph/9506063]
215. W. Voges, B. Aschenbach, T. Boller et al., A&A **349**, 389–405 (1999). ([arXiv:astro-ph/9909315])
216. D.J. Walton, W.N. Alston, P. Kosec et al., MNRAS **499**, 1480–1498 (2020). https://doi.org/10.1093/mnras/staa2961. [arXiv:2009.10734]
217. S. Watanabe, M. Sako, M. Ishida et al., ApJL **597**, L37–L40 (2003). https://doi.org/10.1086/379735. [arXiv:astro-ph/0309344]
218. D.R. Wilkins, L.C. Gallo, E. Costantini, W.N. Brandt, R.D. Blandford, MNRAS **512**, 761–775 (2022). https://doi.org/10.1093/mnras/stac416. [arXiv:2202.06958]
219. D.R. Wilkins, L.C. Gallo, D. Grupe et al., MNRAS **454**, 4440–4451 (2015). https://doi.org/10.1093/mnras/stv2130. [arXiv:1510.07656]
220. R.J. Weymann, S.L. Morris, C.B. Foltz, P.C. Hewett, ApJ **373**, 23 (1991). https://doi.org/10.1086/170020
221. Y. Xu, C. Pinto, S. Bianchi et al., MNRAS **508**, 6049–6067 (2021). https://doi.org/10.1093/mnras/stab2984. [arXiv:2110.06633]
222. T. Yaqoob, I.M. George, K. Nandra et al., ApJ **546**, 759–768 (2001). https://doi.org/10.1086/318315. [arXiv:astro-ph/0008471]
223. T. Yaqoob, MNRAS **423**, 3360–3396 (2012). https://doi.org/10.1111/j.1365-2966.2012.21129.x. [arXiv:1204.4196]
224. A. Zoghbi, A.C. Fabian, C.S. Reynolds, E.M. Cackett, MNRAS **422**, 129–134 (2012). https://doi.org/10.1111/j.1365-2966.2012.20587.x. [arXiv:1112.1717]
225. A. Zoghbi, A.C. Fabian, P. Uttley et al., MNRAS **401**, 2419–2432 (2010). https://doi.org/10.1111/j.1365-2966.2009.15816.x. [arXiv:0910.0367]
226. A. Zoghbi, J.M. Miller, E. Cackett, ApJ **884**, 26 (2019). https://doi.org/10.3847/1538-4357/ab3e31. [arXiv:1908.09862]
227. Y. Zu, C.S. Kochanek, S. Kozłowski, A. Udalski, ApJ **765**, 106 (2013). https://doi.org/10.1088/0004-637X/765/2/106. [arXiv:1202.3783]

Chapter 10
Charge Exchange in X-Ray Astrophysics

Liyi Gu and Chintan Shah

10.1 Introduction

The phenomenon of ion-neutral interaction was studied in early laboratory experiments. The scattering experiment of alpha particles or bare He ions, against gold atoms has led [205] to establish the classical model of atomic structure. The quantum mechanical treatment of the ion-neutral scattering problem was explored by [168, 179], among others, pioneered the theoretical spectral modeling of charge exchange using atomic orbital approximation. From 1950s, as the gas-beam technique matured, an increasing number of experiments have been put forward to measure directly the charge exchange cross sections [71, 228]. Early investigations of this process have been covered in several reviews [84, 123, 135].

It has been well known that the atom-ion charge exchange by far dominates all kinds of electron-ion collisions. For a collision of $\sim$ keV/u, the effective cross section of charge exchange between highly-ionized ions and neutrals are at least two orders of magnitude greater than the radiative recombination and electron-impact excitation cross sections. From the ion view, charge exchange provides efficient recombination on the ionization state. Unlike radiative recombination, charge exchange always ends up in an excited state, producing strong line emissions but zero continuum.

Charge exchange has been well utilised in the research of nuclear fusion. The radiative transitions from excited states of impurity ions resulting from charge exchange may constitute a substantial portion of the total energy losses; meanwhile, the charge

L. Gu (✉)
SRON Netherlands Institute for Space Research, Niels Bohrweg 4, 2333 CA Leiden, The Netherlands
e-mail: l.gu@sron.nl

C. Shah
NASA Goddard Space Flight Center, 8800 Greenbelt Rd., Greenbelt, MD 20771, USA
e-mail: chintan.shah@mpi-hd.mpg.de

Max-Planck-Institut für Kernphysik, Saupfercheckweg 1, 69117 Heidelberg, Germany

C. Bambi and J. Jiang (eds.), *High-Resolution X-ray Spectroscopy*, Springer Series in Astrophysics and Cosmology, https://doi.org/10.1007/978-981-99-4409-5_10

exchange emission can be used as a powerful tool for spectroscopic diagnostics of high-temperature plasmas (e.g., [42]).

The study of charge exchange has been extended to astrophysics, in particular the interaction between solar wind protons with neutral particles near the Sun, in the planets, and from the interstellar medium (e.g., [245]). Charge exchange has also been found to be the origin of the broad Hα emission lines observed at the shock wave of supernova remnants [45]. These lines are emitted from hot hydrogen atoms created by electron capture of hot protons in the post-shock plasma.

An important breakthrough was made in the late 1990s, when the first detection of cometary X-rays [158] and the charge exchange interpretation [53] was made. Since then, the interest on charge exchange X-ray has been activated in observations, but also in theory and in the laboratory. It has been recognized as a primary or second emission mechanism in a broad range of astrophysicobjects, as well as a time-varying background emission tl X-ray sources. In this review, we provide an overview of our current knowledge on charge exchange, with a focus on X-ray astrophysics with spectroscopic utilised In Sect. 10.2 we present the basic physical properties of charge exchange, as well as the theoretical and experimental achievements. Section 10.3 shows the observational results obtained so far. The review will close with conclusions and a short outlook in Sect. 10.4.

10.2 Plasma Modeling

10.2.1 Physics and Classic Models

Charge exchange recombination consists of the transfer of one or more electrons from (mostly) the outer atomic shell of target atom/ion M to the outer shell of an impinging/projectile atom/ion X, leading to the recombination of the impinging particle while the ionization of M. The process can be written as

$$X^{q+} + M \rightarrow X^{(q-m)+} + M^{m+}, \tag{10.1}$$

where q is the original charge of the X, and m is the number of electrons transferred. $m = 1$ can be referred as single electron capture (SEC) from the point of view of the impinging ion. $m = 2$ is designated as double electron capture (DEC).

The total cross section for this process is typically of order of 10^{-15}–10^{-14} cm^2, greater by a factor of $10^2 - 10^3$ than the cross sections of typical electron-impact radiative recombination and dielectronic recombination. The classical analytic calculations, as summarized in [219], provide order estimate of the total cross sections. They include

- Langevin model for <0.1 eV/u collision, where u stands for atomic mass unit. This model predicts an energy dependence $E^{-1/2}$ of the cross section, which is found to be in line with the Landau-Zener approximation shown in Fig. 10.1.

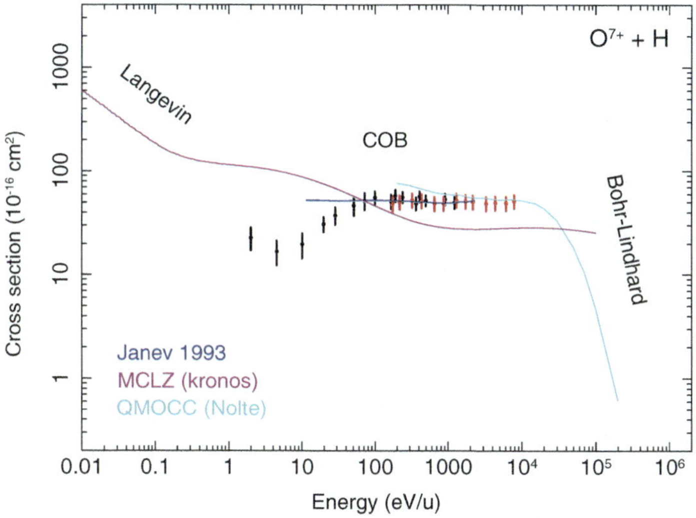

Fig. 10.1 Total cross sections for O^{7+} + H based on the experimental measurements of [267] (black) and [173] (red). The theoretical calculations from [122], multichannel Landau-Zener [177], and quantum molecular-orbital close-coupling [181] are shown in blue, purple, and cyan. The classical Langevin, over-the-barrier, and Bohr-Lindhard models are plotted in dashed lines

- Classical over-the-barrier (COB) model for intermediate collision speed. In this approximation, the cross section becomes a constant over energy.
- Bohr-Lindhard model, which predicts a $E^{-7/2}$ behavior, becomes applicable at very high energy.

These classic treatments for $O^{7+} + H$ are shown in Fig. 10.1 with a comparison to various explicit theoretical and experimental results. These simple models seem to reproduce the correct general trend as a function of energy, though apparently overestimating the absolute values.

Charge exchange operates in a semi-resonant favor when a subset of the energy levels of the projectile ion and the target overlap. If the projectile ion is highly charged, its highly excited levels overlap with the neutral's ground state, making it possible to transfer electrons onto high-n levels. The distribution of n with the most likelihood has been studied in various ways,

- Simple COB model predicts the dominant n level $n_{\max} \leq q/\sqrt{2I_M}$, where q is the initial charge and I_M is the ionization potential of the target.
- Systematic compilation of existing results by [123] reported a scaling $n_{\max} \sim q(1 + (q - 1)/\sqrt{(}2q))^{-0.5} \sim q^{3/4}$.
- More recently, [94] suggested that the n distribution is energy dependent; the capture is peaked on a single n channel at low collision energy, while it tends to be shared among several neighbour n channels at the intermediate energy range. The analytic form of the transfer is reported as Eq. A.1 of their paper. At the very high energy end, the n distributions become narrow again [251].

For the He-like triplet, the cascade contribution from large n to the formation of forbidden lines is substantially larger than that to the resonance lines, yielding a large forbidden-to-resonance line ratio.

To construct an emission spectrum from charge exchange, it is necessary to model the population of capture onto states with orbital angular momenta l in the range of (0, n-1). For collisions with $E \geq 10$ keV/u, it is expected that the l distribution function W is weighed statistically, $W_{st} = (2l + 1)/n^2$. Therefore the largest possible $l = n - 1$ will be the most populated. As the collision velocity decreases, the capture l tends to decrease towards $l = 1$ or 2 taking one of the following possible forms,

- Low energy approximation L1, $W_{L1} = (2l + 1)\,[(n-1)!]^2 / [(n + l)!\,(n-l-1)!]$,
- Modified low energy approximation or L2, $W_{L2} = l\,(l + 1)\,(2l + 1)\,(n-1)!(n-2)! / [(n + l)!(n-l-1)!]$,
- Shifted low energy approximation or L3, $W_{L3} = (2l + 3)\,[(n-1)!]^2 / [(n + l + 1)!\,(n-l-2)!]$,
- "Separable", $W_{se} = (2l + 1)/q\ e^{-l(l+1)/q}$,
- even or constant over l, $W_{ev} = 1 / n$.

For a high-speed collision, shells with large l are significantly populated, resulting in more optical and UV lines following the Yrast cascade relaxation. The low-speed charge exchange would instead yield more X-ray lines because the shells with small l become more populated.

Commonly the capture on the total spin quantum number is assumed to be statistical, and therefore the triplet-to-singlet ratio of He-like ion, produced from the H-like ion reaction, is 3. However, bias from the statistical weight is found in recent theoretical calculations (e.g., [181, 260]). These works suggested that the triplet-to-singlet ratios, and therefore the distribution on the spin quantum number, are likely to be energy-dependent.

Transitions from highly-excited levels with large n and the large forbidden-to-resonance line ratio are two basic spectral features of the charge exchange X-ray emission. The high-n Rydberg lines are usually stronger in H-like than in He-like series because the cascades in the He-like ions are subject to the singlet-to-triplet branching ratios. The UV-to-X-ray ratios of charge exchange emission are larger than those of the electron-impact emission.

10.2.2 Theoretical Calculations of Absolute Cross Sections

To produce accurate X-ray emission models, it is desirable to have absolute cross sections using state-resolved theoretical calculation for all the relevant reactions at a broad range of collision velocities. Simple approximations described above, e.g., COB, cannot reproduce accurate experiment results (e.g., [102]). Furthermore, COB-like models do not have the facility to calculate the l distribution.

For collisions with energies $E \geq 5$ keV/u, the Classical Trajectory Monte Carlo (CTMC) method is often considered to be a valid approximation (e.g., [185]). The

CTMC provides a non-perturbative solution to the classical equations of motion in the many body interactions during the collisions. However, this method does not give an adequate description of the lowering of l quantum number as the collision energy decreases below ~1 keV [21].

The Multichannel Landau-Zener approach (MCLZ) provides another flexible tool for massive calculation. MCLZ considers a slow collision between an ion and a neutral particle using a quasi-molecular configuration, therefore, it is mostly used for models with $E \leq$ a few keV/u. It studies the crossing between pre- and post-exchange potentials of the configuration and determines the final state of the capture. Since the MCLZ requires that crossing distances are fully resolved, while for a bare ion collision, the capture l degenerates with the n state, this method cannot resolve l distribution for bare ion collision [177]. Empirical l distribution functions listed in 10.2.1 must then be applied.

(Quasi-)quantum mechanical models involving close coupling are known to be accurate simulations of atomic collisions. The Atomic-Orbital Close-Coupling (AOCC) is found to be applicable between ~0.1 and 100 keV/u. It is based on bound state calculation on either center of the two reactors, assuming an infinite separation between them. The number of n and l are limited by the finite size of the close-coupling expansion.

The Molecular-Orbital Close-Coupling (MOCC) is another type of close-coupling which is optimized for low energy ($E \leq 10\,\text{keV/u}$). It models the union state of the two reactors. The molecular wave functions, as well as the coupling terms of the relevant status, are calculated and fed into the close-coupling scatter calculation. The number of n and l are limited in the same way as the atomic-orbital. A full quantum mechanical version, QMOCC, has been developed to cope with very low energy collision at $E < 10\,\text{eV/u}$ [219].

We summarize in Table 10.1 a comprehensive, though not exhaustive, list of theoretical calculations. The reactors, type of the data products (total, $n-$, and $nl-$ resolved cross sections) and the energy range indicated in the references, are all recorded. The collision energy determines the range of application; for instance, the solar wind ions collide with comet atmosphere at $E \sim 0.1$–5 keV/u, while for the charge exchange component in collisional or photo-ionization equilibrium, the relevant collisions are much less energetic with $E \sim 10$ eV/u.

10.2.3 Laboratory Measurements

Laboratory measurements provide the absolute calibration of the atomic constants. The charge exchange process has been measured in the laboratory since 1950s (e.g., [104]), using a crossed-beam setup in which a beam of ionized particles passes through a chamber filled with neutral gas. The storage ring facilities are often used for measurements at a high collision energy ($E \sim$ MeV/u), while the electron cyclotron resonances and the electron beam ion sources are applied to measurements with medium collision energy ($E \sim$ keV/u). A significant part of the early studies with

Table 10.1 Theoretical data

References	Type[a]	Ion	Energy (eV/u)
[207]	Total	$D + H^+, D^+ + H$	≤ 6
[51]	Total, *nl*	$He^{2+} - Ne^{10+} + H$, Li	10^3–10^5
[224]	Total	$Li + H^+$	10^{-4}–10
[127]	Total	$Li + H^+$	10^{-5}–10^3
[121]	Total, *nl*	$A^{A+} + H$ ($5 \leq A \leq 74$)	30–8×10^4
[241]	Total, *nl*	$B^{2+} + H$	$\sim10^{-1}$–10^4
[161]	Total, *nl*	$B^{4+} + H$	10^{-2}–2×10^5
[221]	Total	$C^+ + H$	$\leq 10^3$
[128]	Total	C, N, O, Si + H^+	$\leq 10^3$
[40]	Total	$C^{q+}, N^{q+}, O^{q+}, Ne^{q+} + H$ (q = 2,3), $O^{2+} + He$	10^{-1}–1
[39]	Total	$C^{q+}, N^{q+}, O^{q+}, Ne^{q+}, Mg^{q+}, Si^{q+}, S^{q+}, Ar^{q+} + H$ (q = 2, 3, 4)	10^{-2}–1
[109]	*nl*	$C^{2+}, N^{3+}, O^{3+}, Ne^{2+}, Ne^{3+} + H$	0.27–8.1
[92]	Total	$C^{3+}, O^{3+}, Si^{3+} + H$	$\sim$0.1–10^4
[258]	Total, *nl*	$C^{3+} + He$	$\sim10^{-4}$–10^3
[130]	Total	$C^{q+}, N^{q+}, O^{q+}, F^{q+}, Ne^{q+}, Kr^{q+}$ + H (q = 10–25 for Kr and 4–9 for rest)	600
[178]	Total, *nl*	$C^{6+}, N^{7+}, O^{8+}, Ne^{10+}, Na^{11+}, Mg^{12+}, Al^{13+}$, H_2O, CO, CO_2, OH, O	$\sim10^{-3}$–10^5
[193]	Total, *nl*	$C^{6+} + H, He$	2.7×10^3–8.3×10^3
[56]	Total, *nl*	$C^{6+} - Al^{13+} + H, He, H_2$	2×10^2–5×10^3
[154]	Total, *nl*	$N + H^+, N^+ + H$	$\sim10^{-4}$–10^3
[15]	Total	$N^{2+} + H$	2×10^{-3}–3×10^5
[110]	Total	$N^{2+} + H$	$\sim10^{-1}$–10^2
[214]	Total, *nl*	$N^{4+} + H$	$\sim$50–2×10^4
[69]	Total	$N^{4+} + H$	30–10^5
[273]	Total, *nl*	$N^{4+} + H$	$\sim$0.1–8×10^3
[225]	Total, *nl*	$N^{4+} + H$	$\sim10^{-2}$–6×10^3
[213]	Total, *nl*	$N^{5+} + H$	$\sim10^{-2}$–10^4
[102]	*n*	$N^{7+}, O^{7+} + He, CO, CO_2, H_2O$	2×10^3–4.67×10^3
[44]	Total	$O + H^+$	$\leq 10^{-1}$
[222]	Total	$O + H^+, O^+ + H$	10^{-4}–10^7
[190]	Total, *nl*	$O^+ + H_2$	1–10^4
[114]	Total, *nl*	$O^{2+} + H$	$\sim$10–3×10^4
[115]	Total	$O^{2+} + H$	$\sim3\times10^{-4}$–4
[247]	Total, *nl*	$O^{3+} + H$	0.1–10^3
[26]	Total	$O^{3+} + H$	0.1 -5×10^3
[57]	*nl*	$O^{3+} + H$	0.1–1

(continued)

Table 10.1 (continued)

References	Type[a]	Ion	Energy (eV/u)
[83]	*nl*	$O^{3+} + H$	≤ 1
[204]	Total	$O^{3+} + H$	~ 1
[259]	Total, *nl*	$O^{3+} + He$	$0.01–10^3$
[248]	Total, *nl*	$O^{3+} + H_2$	$0.1–10^4$
[261]	Total, *nl*	$O^{6+} + H$	$0.1–10^4$
[272]	Total, *nl*	$Ne^{2+} + He$	$0.1–10^4$
[165]	Total, *nl*	$Ne^{(8-10)+} + H, Mg^{(8-12)+} + H$	$10^{-3}–5 \times 10^4$
[202]	*nl*, line	$Ne^{10+} + Ne, He, H_2, CO_2, H_2O$	9
[54]	Total	$Na + H^+$	≤ 34
[6]	Total	$Mg + H^+$	≤ 40
[73]	Total, *nl*	$Mg^{12+} + H, He$	$10^3–5 \times 10^3$
[132]	total	$Si + H^+$	≤ 30
[220]	Total, *nl*	$Si^{3+} + He$	$\sim 10^{-4}–400$
[113]	Total, *nl*	$Si^{3+} + He$	$1–7 \times 10^3$
[198]	Total	$Si^{3+} + He$	$\sim 10^{-4}–10^3$
[159]	Total, *nl*	$Si^{3+} + H$	$0.01–2 \times 10^5$
[230]	Total	$Si^{4+} + He$	$10^{-3}–2.5 \times 10^3$
[249]	Total, *nl*	$Si^{4+} + He$	$\sim 10^{-4}–10^7$
[243]	Total, *nl*	$Si^{4+} + He$	$\sim 0.01–100$
[13]	Total, *nl*	$P^{2+} + H$	$\sim 100–10^5$
[129]	Total	$S + H^+$	$\leq 10^3$
[271]	Total, *nl*	$S + H^+$	$\sim 10^{-4}–10^4$
[12]	Total, *nl*	$S^{2+} + H$	$100–10^4$
[146]	Total, *nl*	$S^{3+} + H$	$2 \times 10^3–8 \times 10^3$
[11]	Total, *nl*	$S^{3+} + He$	$2 \times 10^3–5 \times 10^4$
[223]	Total, *nl*	$S^{4+} + H$	$\sim 10^{-4}–10^7$
[177]	Total, *nl*	$Fe^{25+}, Fe^{26+} + H$	$10^{-3}–10^5$
[227]	Total, *nl*	$Ge^{q+}, Se^{q+}, Br^{q+}, Kr^{q+}, Rb^{q+}, Xe^{q+}$ +H (q = 2,3,4,5)	$10^{-3}–10^6$

a: total = total cross section, nl = *nl*-resolved cross section

crossed-beam measurements contributed to the fusion research because of the charge exchange influence on plasma parameters such as charge state distribution and radiative cooling e.g., [209]. Recent interest arises for astrophysical applications with the measurements of highly charged ion collisions, including among many others, the total cross sections of bare C, N, O, and Ne charge exchanging with H_2O, He, H_2, CO_2 measured at Jet Propulsion Laboratory [88, 89], and the K-shell charge exchange between He^{2+} and H_2O using the crossed-beam facility at Kernfysisch Versneller Institute in Groningen [31, 112].

The combination of charge state determination and spectroscopic measurements becomes more available lately, providing reliable measurement on the velocity-dependent total and state-resolved cross sections. Some such measurements has been made with the cold target recoil-ion momentum spectroscopy (COLTRIMS), which has been developing since 1990s (e.g., [1]). The n states, sometimes even l levels of the captured electrons, are inferred from the measurement of the ion kinetic energy loss (the "Q" value, e.g., [70, 134, 263, 270]). A few most recent efforts are carried out in the COLTRIMS facilities in the Institute of Modern Physics of China on the Ne^{8+} and Ne^{9+} collision with He and H_2 [264], and at the COLTRIMS in combination with simultaneous X-ray spectroscopic experiments at the electron cyclotron resonance source of the University of Nevada, Reno [5].

One potential issue with beam-gas approach is that the radiative cascade from the metastable levels of the product ions might take longer than the dynamical timescale in the interaction region, making it complicated to study the X-ray spectra observed. The electron beam ion trap (EBIT) [153] devices provide a solution, as the ions are radially confined in a quasi-stationary state by the space charge potential of the electron beam and axially trapped by the electrostatic potential to the cylindrical drift tube electrodes. A useful technique for measuring charge exchange with the EBIT is the magnetic trap mode, in which the electron beam is turned off when the proper charge states are reached. The ions expand to form a larger cloud, but remain confined in the radial direction by the strong magnetic field and the potential applied to the trap electrodes. Typically, the trap potentials limit the energy of the ions stored in the EBIT. Therefore, the collision energy for the CX reaction is usually very low, less than 10–20 eV/u [21]. However, the advantage of the magnetic trapping mode is that the electron-ion collision can be safely eliminated, while the ion-neutral collision remains efficient and produces the pure charge exchange emission.

The experiments using EBITs do not measure the absolute cross sections; instead, they often record the detailed X-ray spectra of the reactions. Earlier EBIT experiments with limited spectral resolution used the parameter known as the hardness ratio, i.e., the intensity ratio of X-ray line emission from levels with $n \geq 3$ relative to $n = 2$ K-shell emission, to study the dependence on collision energies [8, 21, 22, 211]. Recent high-resolution (resolved to a few eV) X-ray spectral measurements have become possible with the introduction of X-ray microcalorimeters [17]. The high resolution is particularly important for charge exchange, which emits lines from high n levels, often in a narrow energy range. The NASA Goddard Space Flight Center X-ray Spectrometer (XRS) and EBIT Calorimeter Spectrometer (ECS) [196, 197] have been used for a number of charge exchange measurements performed at the Lawrence Livermore National Laboratory EBIT (e.g., [76, 152]).

The n-resolved line diagnostics are made possible with the EBIT plus high-resolution spectrometer, while the l-distribution of the capture, which is sensitive to the collision velocity (see Sect. 10.2.1), remains challenging since the ion-neutral collisions in the magnetic trapping mode are always mild ($\leq$10 eV). A possible solution is to measure the charge exchange photons simultaneously at longer wavelengths, e.g., in the UV and optical bands, where the radiative cascades of the Rydberg levels might provide additional constraints on the l-state distribution. In addition, efforts are

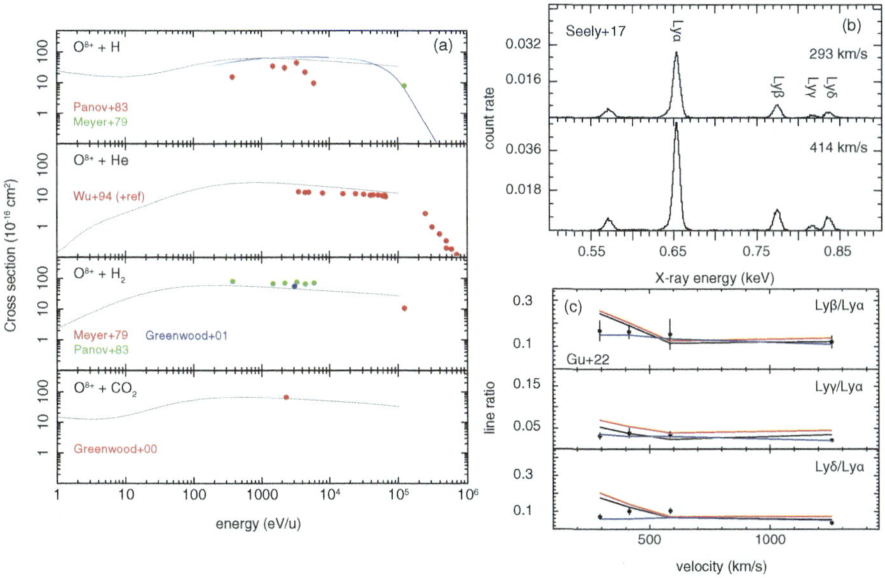

Fig. 10.2 **a** Total cross sections as a function of collision energy for O^{8+} interacting with atomic hydrogen, atomic helium, molecular hydrogen, and CO_2. The solid lines are the model values from the calculations with the MCLZ (black) and [122]. Data points are experimental results from sources indicated in the labels. **b** X-ray spectra from [208] on O^{8+} + Kr reaction for two different velocities. **c** Comparison of experimental and theoretical line ratios for O^{8+} charge exchange. The experimental data are taken from [208], and the predictions from [94], MCLZ, and [122] are shown in black, red, and blue curves

needed to combine EBIT and high-resolution calorimeters with the COLTRIMS setup to fully investigate the kinematics of charge-exchange collision and multi electron capture processes [5]. Furthermore, the development of a suitable atomic hydrogen target for EBIT measurements is ideal so that only single-electron capture capture can take place [151].

In Fig. 10.2, we show a set of historical and state-of-the-art measurements of the O^{8+} charge exchange. The total cross sections of O^{8+} with H, He, H_2, and CO_2 targets are measured in crossed-beam in the 1970s to 2000s. In panel b, the X-ray quantum microcalorimeter spectra taken with the beam-gas measurements at Oak Ridge on O^{8+} with Kr charge exchange are shown for two different velocities. The comparison of the observed and calculated line ratios is plotted in panel c. A more comprehensive list of laboratory measurement efforts made in the past can be found in Table 10.2.

Table 10.2 Experimental cross section data

References	Type[a]	Ion	Energy (eV/u)
[88]	Total	H^+, He^+, He^{2+} + H_2O, CO_2	300–7.5×10^3
[164]	nl	He^{2+}, N^{5+} + H_2	5–10^3
[143]	Total	He^{2+} + H_2, N_2, O_2, CO, CO_2	4×10^3
[142]	Total	He^{2+} + He	2×10^2–4×10^3
[212]	Total	Li^{q+} (q = 1–3)	4×10^3-3.43×10^5
[209]	Total	Li^{q+} (q = 2–3), N^{q+} (q = 2–5), Ne^{q+} (q = 3–5)	3q×10^3–1.2q×10^4
[86]	Total	B^{q+} (q = 1–5), C^{q+} (q = 1–4)	q×10^5–2.5q×10^6
	Total	C^{q+} (q = 5, 6), N^{7+}	q×10^5–2.5q×10^6
[169]	Total	B^{2+}, C^+, N^+, Mg^{2+}	8×10^2–4×10^4
[52]	Total	B^{q+} (q = 2–5), C^{q+} (q = 3, 4), N^{q+} (q = 3, 4), O^{q+} (q = 5, 6)	4q×10^3–2.5q×10^4
[192]	Total	B^{4+}	60–1.2×10^3
[82]	Total	B^{q+} (q = 2–4), C^{q+} (q = 2–4), N^{q+} (q = 2–5), O^{q+} (q = 2–5)	6q×10^3–2.3q×10^4
[173]	Total	B^{q+} (q = 2–5), C^{q+} (q = 3, 4), N^{q+} (q = 3, 4)	2.5×10^4–2×10^5
[189]	Total	C^{q+} (q = 1–4), N^{q+} (q = 1–5), O^{q+} (q = 1–5), Si^{q+} (q = 2–7)	8.6×10^3–1.65×10^6
[183]	Total	C^{2+}	5×10^2–1.4×10^3
[106]	Total	C^{3+}	0.3–3×10^3
[90]	Total, Line	C, N, O, Ne^{q+} + He, H_2, H_2O, CO_2 (q = 3-9)	7q×10^3
[89]	Total, Line	C, N, O, Ne^{q+} + H_2O, CO_2 (q = 3-10)	550–800 km s^{-1}
[74]	Line	C^{2+} + H_2	10^3–3×10^4
[188]	Total	C^{q+} (q = 3, 4), O^{q+} (q = 2–6)	10–1×10^4
[163]	nl	C^{4+}, N^{5+}, O^{6+} + H_2	5–4×10^3
[188]	Total	C^{q+} (q = 5, 6)	10–1×10^4
[206]	Total	C^{3+}	1×10^6–3.5×10^6
[48]	Total, nl	C^{q+} (q = 3, 4), N^{5+}, O^{6+}	7×10^2–4.6×10^3
[170]	Total, nl	C^{3+}	6×10^2–1.8×10^4
[187]	Total	C^{4+}, N^{5+}, O^{6+}, Ne^{8+}	100–1300 km s^{-1}
	Total	C^{q+} (q = 5, 6), N^{q+} (q = 6, 7), O^{q+} (q = 7, 8), Ne^{q+} (q = 9, 10)	100–1300 km s^{-1}
[63]	Total, nl	C^{q+} (q = 3, 4), N^{5+}, O^{6+}	2.7–13.6
[81]	Total	C^{q+} + CO (q = 3-4)	5×10^2
[77]	Total, nl	C^{4+}	100–2×10^4

(continued)

Table 10.2 (continued)

References	Type[a]	Ion	Energy (eV/u)
[112]	Total, nl	C^{4+}	2.4×10^2–4.4×10^2
[116]	Total	C^{4+} + He	2.4×10^2–4.4×10^2
[64]	Total	C^{5+} + H	0.64–1.2×10^4
[30]	Total, n	C^{5+} + H_2O	113–3.75×10^3
[19]	Line	C^{q+}, O^{q+} + CO_2 (q = 5-8)	several 100 km s^{-1}
[58]	Line	C^{6+} + He	460–3.2×10^4
[148]	Total, nl	C^{6+} + He, H_2	10^3–2.5×10^4
[9]	Line	C^{6+} + Kr	320–4.6×10^4
[149]	Line	C^{6+}, O^{8+} + Kr	500–4×10^4
[226]	Total	N^+, O^+	100–4×10^4
[75]	Total	N^{4+}	1-3×10^2
[27]	Total	N^{4+} + H , H_2	10^3-4×10^3
[118]	Total	N^{q+} + H (q = 3-5)	0.9–1.4×10^3
[266]	Total	N^{7+}, O^{7+} + H	2–1.22×10^3
[102]	Total, n	N^{7+}, O^{7+} + He, H_2O, CO_2, CO	2×10^3, 4.67×10^3
[72]	Total	O^+	100–3.6×10^4
[144]	Total	O^+ + H_2, CO_2, CO, CH_4, C_2H_2, C_2H_6, C_3H_8	200–4.5×10^3
[23]	nl	O^{3+}	45–752
[47]	Total	O^{q+} (q = 2, 3) + H	1.5×10^4
[107]	Total	O^{q+} (q = 3, 4) + H(D)	1–10^3
[173]	Total	O^{q+} (q = 3–6), Si^{q+} (q = 4–9), Fe^{q+} (q = 4–15)	2.5×10^4–2×10^5
[80]	Total	O^{q+} + CO (q = 4,5)	$1.5q\times10^3$
[105]	Total	O^{5+}	0.9–800
[29]	Total, nl	O^{6+} + H_2O	100 -7.5×10^3
[175]	Line	O^{6+} + CO	3.6×10^4
[166]	Total	O^{6+} + CO, H_2O, CO_2, CH_4, N_2, NO, N_2O, Ar	1.17×10^3–2.33×10^3
[173]	Total	O^{q+} (q = 7, 8)	2.5×10^4–2×10^5
[20]	Line	O^{8+}, O^{7+} + CO_2	15
		Ne^{10+}, Ne^{9+} + Ne	9
[14]	Total	O^{q+} + He (q = 2–5)	100–1.5×10^3
[100]	Total, n	O^{6+} + CO_2, CH_4, H_2, N_2	2.63×10^3–1.95×10^4
[185]	Line	O^{8+} + CH_4, CO_2, N_2	10
[208]	Line	O^{8+} + Kr	445–8.18×10^3
[250]	Total, n	O^{8+} + He	16–54
[176]	Total	Ne^{2+}	139–1.49×10^3
[210]	Total	Ne^{2+} + H(D)	59–949

Table 10.2 (continued)

References	Type[a]	Ion	Energy (eV/u)
[117]	Total	Ne^{q+} (q = 2–4), Ar^{q+} (q = 2–4, 6)	10^3–1.5×10^4
[201]	Total	Ne^{3+} + H	0.07–826
[3]	Total, n	Ne^{q+} + CO_2, H_2O (q = 3-5)	15q–500q
[108]	Total	Ne^{4+} + H	0.1–1006
[145]	Total	Ne^{q+} + Ne, Ar, Kr, Xe	$2q\times10^3$
[2]	Total	Ne^{6+} + CO_2, H_2O	450–2.4×10^3
[267]	Line	Ne^{8+} + He, Kr	400-900 km s^{-1}
[150]	Total, n	Ne^{8+} + He, H_2	4.6–10.9
[264]	Total, n	Ne^{8+}, Ne^{9+} + He, H_2	10^3–2.475×10^4
[184]	Line	Ne^{10+}, Ar^{18+}, Kr^{36+} + Ar	18–4×10^3
[238]	Line	Kr^{36+}, Ar^{18+}, Ne^{10+} + Ar	4×10^3
[126]	Total	Si^{q+} (q = 2–7)	5.1×10^4–2.04×10^6
[37]	Total	Si^{3+}	40–2500
[234]	Total	Si^{q+} + He (q = 3-5)	~100–1000
[235]	Total, Line	Si^{13+}, S^{15+}, Ar^{17+} + Ar, He, CH_4	10^3-7×10^4
[255]	Total	S^{2+} + H , H_2	2×10^3-8×10^3
[211]	Line	S^{16+} + CS_2	~10
[171]	Total, nl	Ar^{q+} + H, H_2, He (q = 4–6)	$q\times10^3$, $2q\times10^3$
[262]	Total, nl	Ar^{8+} + He	1.4×10^3–2×10^4
[18]	Line	Ar^{16+} + H	4×10^4
[7]	Line	Ar^{17+}, Ar^{18+} + Ar	0.25q–125q
[8]	Line	Ar^{17+}, Ar^{18+} + Ar	5–2×10^3
[22]	Line	Fe^{q+} + O_2, N_2, H_2O, CO_2 (q = 16–23)	≤10–20
[141]	Total	Cu^{q+} + H_2, Ne, Ar, Kr, Xe	2.5×10^6
[24]	Line	Ni^{19+}, Ni^{18+} + He, H_2	10
[236]	Total, Line	Kr^{q+} + He, Ne, Ar, N_2, CH_4 (q = 27–33)	2×10^3
[103]	Line	Xe^{18+}, Xe^{24+} + Na	500–4.5×10^3
[237]	Total, Line	Ta^{q+} + Ar (q = 45-49)	10^3–4×10^4

a: total = total cross section, n = *n*-resolved cross sections, nl = *nl*-resolved cross sections, line = line emissivities or ratios

10.3 Observations

10.3.1 Ionization Balance

Charge exchange is relevant for determining the ion concentration as it serves as a recombination term for ion X^{q+} with

$$X^{q+} + H(He) \rightarrow X^{(q-1)+} + H^{+}(He^{+}), \tag{10.2}$$

but also in some conditions an ionization term with

$$X^{(q-1)+} + H^{+}(He^{+}) \rightarrow X^{q+} + H(He). \tag{10.3}$$

These terms are most effective for photoionized and non-equilibrium collisional ionization plasmas where highly charged ions naturally co-exist with the neutral hydrogen and helium. As shown in [99], the charge exchange contribution could even be effective for the collisional ionization equilibrium at a low temperature ($\sim$ a few eV, Fig. 10.3). The collision induced by the thermal motion of ions has a typical velocity of km s^{-1}.

Sets of recombination and ionization rates by charge exchange are tabulated in [10, 133]. These tables have been used broadly in the ion concentration calculation, however, they lacks data of highly charged ions with $q > 4$. Moreover, a significant part of the rates are derived from theoretical calculations with classical approximations, e.g., Landau-Zener method. A systematic review and update are then needed to expand the q range as well as to improve the quality of charge exchange data for the ionization balance calculation.

In Table 10.3 we list a set of astrophysical observations made so far. For each observation, we feature the object, instrument, relevant emission features, together

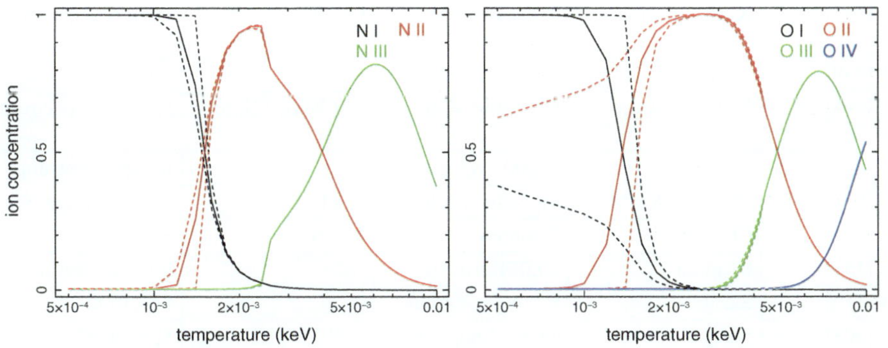

Fig. 10.3 Charge state distributions of N (left) and O (right) as a function of equilibrium temperature for the CIE plasma. The dashed lines show the calculations when the charge exchange recombination rates are changed by 50%, while the other ionization and recombination data are kept the same

Table 10.3 Observations of charge exchange in X-ray

References	Detector	Sources	Relevant ion(s)	Note
[158]	ROSAT	C/Hyakutake 1996 B2	–	Image
[186]	BeppoSAX	C/1995 O1 (Hale-Bopp)	–	Image, spectrum
[155]	ROSAT, EUVE	P/Encke 1997	–	Image
[59]	Chandra	Mars	–	Spectrum, low S/N
[140]	Chandra	C/1999 T1	–	Image, spectrum
[252]	Chandra	Earth geocorona	O VII, O VIII	Spectrum
[218]	XMM-Newton	Heliosphere	C, O, Ne, Mg	Spectrum
[156]	Chandra	2P/Encke 2003	–	Image, spectrum
[62]	XMM-Newton	Mars	C, N, O, Ne	High res. spectrum
[254]	Swift	9P/Tempel 1 (deep impact)	–	Timing, spectrum
[139]	Chandra	sample of 4 comets	–	Spectrum
[78]	Suzaku	Earth geocorona	–	Spectrum
[60]	Chandra	Venus	–	Spectrum, low S/N
[28]	Chandra	sample of 8 comets	–	Spectrum
[217]	XMM-Newton	Earth geocorona	O VII	Timing, spectrum
[256]	Chandra	73P/2006, C/1999 S4	–	Spectrum
[46]	Chandra	8p/Tuttle, 17p/Holmes	–	Spectrum
[67]	Suzaku	Earth geocorona	O VII variation	Timing, spectrum
[138]	XMM-Newton	Heliosphere	O VII, O VIII	Spectrum
[125]	Suzaku	Cygnus Loop SNR	O VII at 0.7 keV	Spectrum
[160]	XMM-Newton	M82	O VII, Ne IX, Mg XI	High res. spectrum
[136]	Suzaku	M82	O VIII, Ne X	Spectrum
[162]	XMM-Newton	9 star-forming galaxies	O VII	High res. spectrum
[239]	Chandra	Carina Nebula	–	Spectrum
[195]	XMM-Newton	Colliding-wind stars	Mg XI & XII, Si XIII	High res. spectrum
[43]	Swift	C/2007 N3 (Lulin)	–	Image, spectrum
[91]	Chandra, XMM	Pl. Nebula A30	C VI	Spectrum, low S/N
[66]	Chandra	5 comets	Mg, Si	Spectrum
[157]	Chandra	103P/Hartley 2	–	Image, spectrum
[119]	Suzaku	Earth geocorona	–	Spectrum
[203]	Suzaku	Cygus Loop SNR	O VII triplet	Spectrum
[216]	Chandra	2 comets	–	Image, spectrum
[55]	Suzaku	M82	Ne X	Spectrum

Table 10.3 (continued)

References	Detector	Sources	Relevant ion(s)	Note
[93]	XMM-Newton	sample of clusters	S XVI at 3.5 keV	Spectrum
[97]	Hubble	NGC 1275 AGN	Ne X, S XV	Spectrum, 2σ
[94]	XMM-Newton	C/2000 WM1	–	High res. spectrum
[178]	XMM-Newton	C/2000 WM1	–	High res. spectrum
[98]	XMM-Newton	sample of clusters	O VIII	High res. spectrum
[268]	XMM-Newton	Earth geocorona	–	Spectrum
[167]	XMM-Newton	NGC 5548 AGN	N VII	High res. spectrum
[96]	XMM-Newton	NGC 5548 AGN	O VI	High res. spectrum
[265]	XMM-Newton	M51	O VII, O VIII	High res. spectrum
[244]	Chandra	Perseus cluster	–	Spectrum
[4]	Hitomi	Perseus cluster	S XVI	High res. spectrum
[242]	XMM-Newton	Cygnus Loop SNR	–	High res. spectrum
[120]	RXTE	EXO 1745-248 burst	Ti, Cr, Fe, Co	Spectrum
[111]	Hitomi	Perseus cluster	S XVI, Fe XVII	High res. spectrum
[232]	XMM-Newton	G296.1-0.5 SNR	–	High res. spectrum
[231]	XMM-Newton	N132D SNR	–	Low significance
[137]	XMM-Newton	J0453.6-6829 SNR	O VII	High res. spectrum
[200]	XMM-Newton	1E 0102.2-7219 SNR	–	High res. spectrum
[194]	XMM-Newton	NGC 4636	O VII	High res. spectrum
[34]	XMM-Newton	Jupiter	–	High res. spectrum
[215]	Chandra	SWCX	O VII, O VIII	Spectrum
[257]	XQC	SWCX	C, N, O	High res. spectrum
[269]	XMM-Newton	M82	N, O, Ne	High res. spectrum

with a few remarks on the quality of the detection (e.g, using image only, with timing, and using spectrum with normal or high spectral resolution).

10.3.2 Solar System Objects

On 25 March 1996, comet C/1996 B2 (Hyakutake) had a close encounter with Earth at a distance of ~0.1 AU. Rosat took the opportunity to search for X-ray signal from this unusual event. While astrnomers expected very low luminosity because comets are too cold to emit X-ray, the Rosat image showed a surprisingly bright source from a crescent-shaped region on the sunward side of comet [158]. Follow-up studies of Rosat archival data proved that X-ray emission is common in comets. One year later after the Rosat discovery, [53] solved the mystery: the X-ray emission originates from the charge exchange between highly charged heavy solar wind ions and neutral gas around the comet atmosphere. The spatial and spectral properties predicted by the charge exchange model has been found to be consistent with Rosat data of other comets, as well as the later Chandra and XMM-Newton observations with better spatial and spectral resolutions [61].

That the 1996 Rosat data established the charge exchange study of solar system objects may seem odd, as a number of earlier studies (e.g., [87, 101]) already pointed out that charge exchange is non-negligible in the solar wind interaction. However, the early works tended to underestimate or overlook the charge exchange contribution in the X-ray band. The Rosat discovery greatly expanded the field, leading to a number of dedicated astrophysical observations, but also stimulating relevant laboratory measurements and theoretical calculations. A new focus of the laboratory and theoretical studies is to reproduce the X-ray observations, with highly-charged ions (C, N, O, Ne, etc) and cometary neutrals colliding at a speed of several hundred kilometer per second [17].

10.3.2.1 Comets

In the context of solar wind charge exchange, comets are often used as natural probes of the latitude- and time-dependent properties of the solar wind ions, which produce emission features that can be recognized in the X-ray spectra. The solar wind study with comets benefits from the facts that the comets approach the Sun in a much broader range of latitudes than the planets, and their occurrence in a proper distance range is quasi-constant on year basis. The representative X-ray observations of comets are summarized in Table 10.3.

The solar winds are found to be largely bimodal at solar minimum: the fast wind with velocity $\sim$700 km s^{-1}, characterized by low density and low ionization, launched probably from the open magnetic fields regions at high solar latitudes. The slow wind, blowing at $v \sim 400$ km s^{-1} from the solar equatorial region with a closed magnetic field, is often found to be denser and more ionized. When the Sun moves towards its maximum, the fast and slow winds become more mixed up.

Line diagnostics with the cometary X-ray spectra provide useful constraints on the speed and the chemical composition of the solar wind (e.g., [28, 139]). In addition, the X-ray morphology of comets has been used to infer the shape and the spatial structure of bow shocks [253]. Finally, [178] derived a constraint on the neutral composition of the cometary atmosphere using a global fit to the *XMM-Newton* grating spectrum.

Two typical comet spectra are shown in Fig. 10.4. The *Chandra* observation of C/1999 S4 reveals a solar wind composition of C, N, and O in fully striped and hydrogen-like states. The *XMM-Newton* grating spectrum of C/2000 WM1 (linear) further resolves the strong individual C, N, and O lines, posing even tighter constraints on the solar wind abundance and even the chemistry of the cometary atmosphere. These data showcase the cometary X-ray as a powerful solar wind probe. However, most of the X-ray spectra are taken with a low CCD-like resolution of $\sim$100 eV,

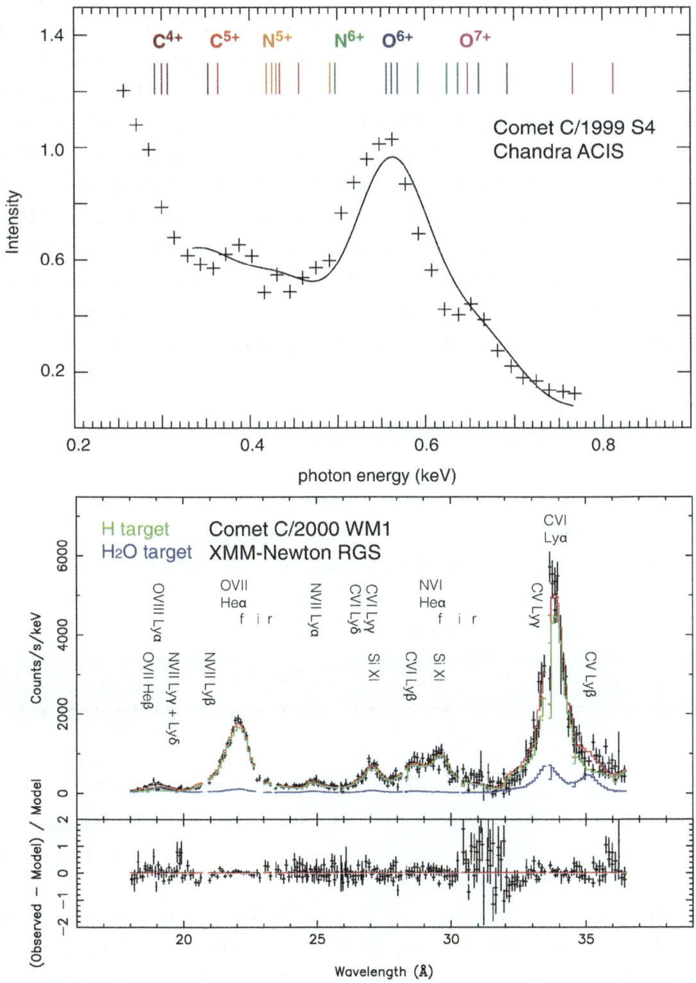

Fig. 10.4 (Upper) Comet C/1999 S4 *Chandra* CCD spectrum taken on 14 July 2000, compared with the model of C, N, O ions charge exchanging with CO_2 measured at EBIT [19]. (Lower) Comet C/2000 WM1 *XMM Newton* grating spectrum fit with the SPEX *cx* model taking into account the hydrogen and water targets

making it impossible to study the solar wind dynamics using comets. A further advance in the X-ray study of comets requires observations using high-resolution X-ray spectrometers, in particular the microcalorimeters onboard XRISM [233] and Athena [180] missions.

10.3.2.2 Planets

Jupiter is known as a bright X-ray source from the early Einstein observations [172]. The origin of the X-ray has been settled just recently: the emission from the equatorial region is likely the scattering of solar X-ray in the dense planetary atmosphere, while the X-ray in the aurorae can be a mixture of two components, the bremsstrahlung radiation dominating > 2 keV by energetic electrons precipitating from the magnetosphere, and the charge exchange in soft X-ray between the in-falling ions and the atmosphere c neutrals [25, 33–36, 65, 85, 131, 182]. The ion diagnostics using charge exchange provides an opportunity to understand where the ions come from: the X-ray spectra would be C- and O- rich if the solar wind is the main source of the ions, and S- and O is rich if the bulk population originates from the Io plasma torus falling into the Jovian magnetosphere.

The *XMM-Newton* reflection grating spectrometer (RGS) spectrum of Jupiter in 2003 shown in Fig. 10.5 show so-far the best example of the charge exchange diagnostics with X-ray spectroscopy of planets. It shows a clear separation of the charge exchange emission, mostly in oxygen lines, from the low-latitude disk component with Fe and O lines. The charge exchange lines are broadened by a velocity of $\sim$5000 km s^{-1}, implying that we are witnessing a bulk of accelerated particles precipitating into the Jovian polar area.

Mars and Venus are possible charge exchange emitters, even though the detection was made at a lower significance than Jupiter [60, 62]. The scatter of solar X-rays

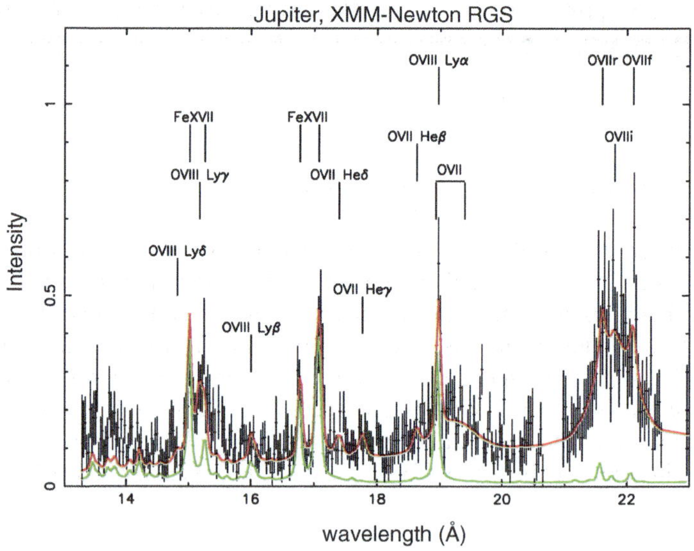

Fig. 10.5 The *XMM-Newton* grating spectrum of Jupiter. A phenomenological fit to the spectrum is plotted in red, and the disk component is shown in green. The emission from X-ray aurorae is the difference between the two

in the planetary atmosphere is constantly bright in, even dominating, the overall soft X-ray band, making it challenging for charge exchange observation. No significant charge exchange signal has yet been detected in other non-Earth planets.

10.3.2.3 Geocorona and Heliosphere

Charge exchange emission is generated when the highly ionized solar wind interacts with neutral or near-neutral gas in the Earth's exosphere and the heliosphere. This component is commonly known as the solar wind charge exchange (SWCX), which can affect any X-ray observation so far. Therefore, it is treated as a troublesome background, consisting of multiple emission lines (e.g., C, N, and O) in the soft X-ray band. Concerning its temporal, spectral, and spatial variations, it is often challenging to fully eliminate the SWCX from the extrasolar sources.

Chandra observation of the dark moon has been considered direct evidence for the nearby SWCX since it shows charge exchange-type spectrum [252]. Similarly, [215] reported an enhancement of O VII and O VIII lines from SWCX in the *Chandra* observation of molecular cloud MBM 12, and [78] observed a transient increase of SWCX flux by comparing several spectra of a *Suzaku* blank field in the direction of the north ecliptic pole.

In a CCD resolution, the X-ray spectra from the geocoronal and heliospheric SWCX could sometimes resemble those from the local hot bubble around the solar system, making it difficult to isolate one from another in the spectrum. One solution is to separate them based on directional variation: the local hot bubble is isotropic while the SWCX might increase in the focusing cone of heliospheric helium caused by the inflow of interstellar medium. By scanning the sky through the helium cone using the diffuse X-rays from the local galaxy (DXL) mission, [79] separated the SWCX and the local hot bubble emission based on their spatial distributions. Their results showed that the SWCX emission contributes about 40% of the 1/4 keV X-ray flux in the Galactic plane.

A second approach is to distinguish SWCX using high-resolution X-ray spectroscopy. The X-ray quantum calorimeter (XQC, $\sim$6 eV FWHM), onboard the University of Wisconsin-Madison/Goddard Space Flight Center sounding rocket, has measured several times the diffuse X-ray background below 1 keV [257]. As shown in Fig. 10.6, the high-resolution spectroscopy allows the separation of the SWCX from the 0.1 keV local hot bubble and the 0.2 keV Milky Way halo components. It could be used to further isolate the charge exchange from fast solar wind and slow wind components.

10.3.3 Astrophysical Objects

While the charge exchange from the solar system objects is well established, the detection in extrasolar astrophysical objects remains somewhat controversial though

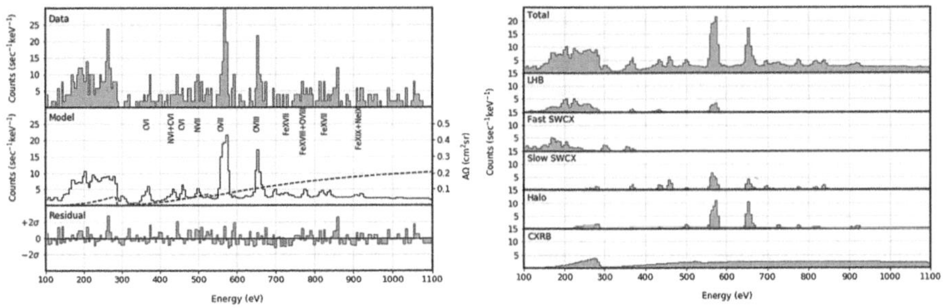

Fig. 10.6 (Left) X-ray quantum calorimeter spectrum of the diffuse soft X-ray background. The model is shown in the middle, and the residual is plotted at the bottom. The effective area of the detector is shown in the dashed line. (Right) The model is divided into components: LHB—local hot bubble; Fast/slow SWCX—fast/slow solar wind charge exchange; Halo—Milky way gas halo; and CXRB—cosmic X-ray background

evidence is accumulating in particular in the past decade. A general issue is that many claims are made with the CCDs on *Chandra*, *XMM-Newton*, and *Suzaku* with 50–100 eV resolutions, or with dispersive gratings (e.g., RGS) of better but source-dependent resolutions. A revisit with non-dispersive high resolution X-ray spectrometers (e.g., XRISM and Athena) is needed to verify these features.

10.3.3.1 Stars and Supernova Remnants

Pollock [195] discussed a scenario that colliding wind ejected from binary stars might lead to ion-ion collision and charge exchange. The ion-ion charge exchange can be realized only when the collisions occur at a large velocity ($\geq$2000 km s^{-1}), which indeed resembles some of the observed terminal shock speeds of the colliding stellar winds. Pollock [195] suggested that the charge exchange recombination might explain the low Mg XII/Mg XI ratios observed in systems including WR22, WR140, and ζ Pup.

The wind interaction condition might apply to other types of stellar objects. Iwakiri [120] reported the detection of a 6.6 keV emission feature in the spectrum of a neutron star X-ray binary EXO 1745-248 40 h after its superburst. It is a broad feature with an equivalent width of 4.3 keV, which is much larger than that of the standard ionized Fe line formed by the reflection in the disk. A spectral analysis with the RXTE data implied that the feature can be a set of charge exchange lines, likely emitted from the interaction of a failed wind launched by the superburst with the neutrals in the neutron star disk.

The interaction between supernova remnants and ambient neutral clouds might generate charge exchange emission [124]. It is expected that this emission would be visible around the edge of the remnants, where interstellar neutrals pass through the shock front and collide with the post-shock hot ions. [125] reported a detection of charge exchange emission at the outer shell of the Cygnus Loop, which shows an

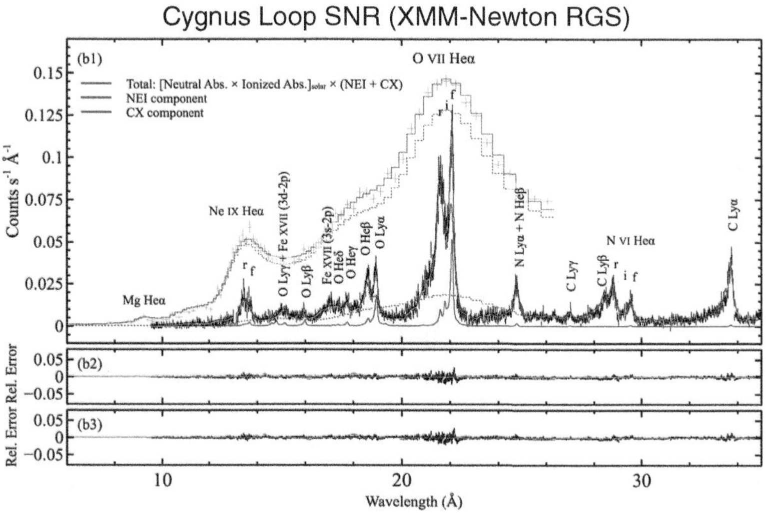

Fig. 10.7 *XMM-Newton* grating spectrum of a bright knot in the Cygnus Loop supernova remnant, fit with the charge exchange plus non-equilibrium collisional ionization model. The neutral and ionized absorbers are taken into account

enhanced feature at the energy of highly excited O VII lines in the *Suzaku* spectrum [203]. As plot in Fig. 10.7, [242] took an *XMM-Newton* RGS spectrum of a bright X-ray spot on the outer shell, and showed that the charge exchange component, together with an ionized absorber, are indeed required to explain the forbidden-to-resonance line ratio of O VII. Similarly, the RGS spectra of G296.1-0.5 [232] and J0453.6-6829 [231] show high forbidden-to-resonance ratios that can be explained by charge exchange.

The limited imaging capability of *XMM-Newton* RGS allows brief mapping of the ion-neutral interaction region in the supernova remnants. [242] pioneered this method by differentiating the O VII forbidden line and resonance line monochromatic images, and discovered that the charge exchange lines can be emitted from a compact region where a dense neutral cloud is also found.

10.3.3.2 Interstellar Medium and Star Forming Regions

The north polar spur is a large ridge of X-ray and radio-emitting gas in the northern sky. Similar to some of the supernova remnant spectra discussed above, the north polar spur also shows large forbidden-to-resonance ratios for He-like triplets of O and Ne [147, 174], which might potentially indicate a charge exchange origin. However, this scenario is far from conclusive since the forbidden-to-resonance ratio could also be explained by others atomic physics including an ionized absorber. As suggested in

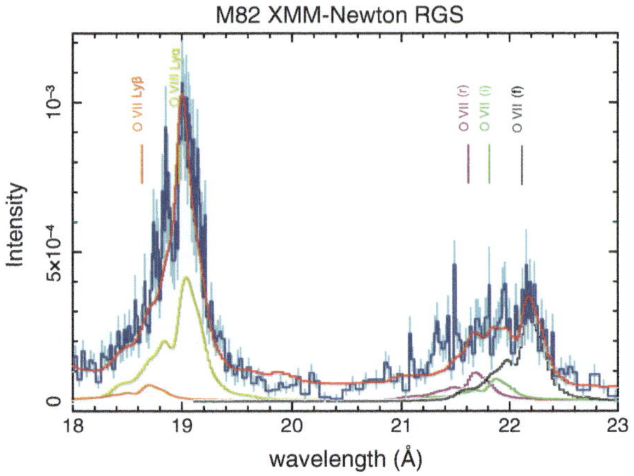

Fig. 10.8 Part of the *XMM-Newton* grating spectrum of the M82 star-forming galaxy. The charge exchange lines from O VII and O VIII are plotted in colors indicated by the label

[95], the ionized absorption might be responsible for most of the observed line ratios while the charge exchange contributes no more than 10% to the forbidden lines.

The star-forming regions in normal or starburst galaxies are another locations where interactions between hot gas and cold clouds are expected. By modeling the *Chandra* spectrum of the Carina star-forming complex with multiple thermal components, [239] reported several line-like excesses in the residual in $0.5 - 2.0$ keV. The distribution of these line residuals is spatially correlated with the cold star-forming filaments, where the gas-filament charge exchange might be responsible for these line emissions.

There have been claims for the detection of charge exchange X-ray in several starburst galaxies. M82 is the best-studied object in this category. The X-ray spectroscopic works using *XMM-Newton* and *Suzaku* data showed that the thermal model is insufficient in particular for the O VII and O VIII lines, while adding an additional charge exchange component could fix most of the problem [55, 136, 160, 199, 240, 269]. A recent study using the RGS spectrum is plotted in Fig. 10.8. [162] presented a sample study of nine star-forming galaxies with *XMM-Newton* RGS, and showed that most of these galaxies are undergoing charge exchange as indicated by the forbidden-to-resonance line ratios. Therefore charge exchange can be a class property of star-forming regions and galaxies. Naturally, the ion-neutral interaction could occur between the starburst-driven outflows and the cold interstellar medium.

In the quiescent galaxies, charge exchange might still occur at various locations including the interfaces between the nucleus-driven outflows and the circumnuclear neutral clouds. [265] reported a possible detection of charge exchange component which contributes significantly to the N, O, and Ne lines, in particular the He-like forbidden lines, in the RGS spectrum of M51 nucleus. The interaction between the radio jet and the cold interstellar medium might be responsible for the emission. Yet another possibility is that these lines originate from photoionization by the AGN outbursts in the past.

10.3.3.3 Active Galactic Nuclei

AGNs host multiple gas inflows and outflows of a broad range of ionization states, making it another potential target of the charge exchange search. [96] reported a detection of unidentified features at 18.4 Å in the grating spectra of Seyfert I AGN NGC 5548. By stacking all observations, this feature is seen at $> 5\sigma$ significance taking into account the look-elsewhere effect. As indicated in Fig. 10.9, the wavelength of the anomaly coincides with the high-n transitions from He- and Li-like oxygen, therefore, it is likely a charge exchange line. The authors suggested that the interaction could happen within the same outflow by mixing the partially ionized and neutral layers, or between the outflow with the neutral close environment. Another report by [97] showed that the Hubble STIS spectrum of the radio galaxy NGC 1275 can be modeled by including three weak lines at 1223.6 Å, 1242.4 Å, and 1244.0 Å, each with a significance of 2–3 σ. These features can be explained by a mixture of charge exchange between highly ionized hydrogen, neon, and sulfur with neutral matter, indicating for an outflow with $v \sim 3400$ km s^{-1}.

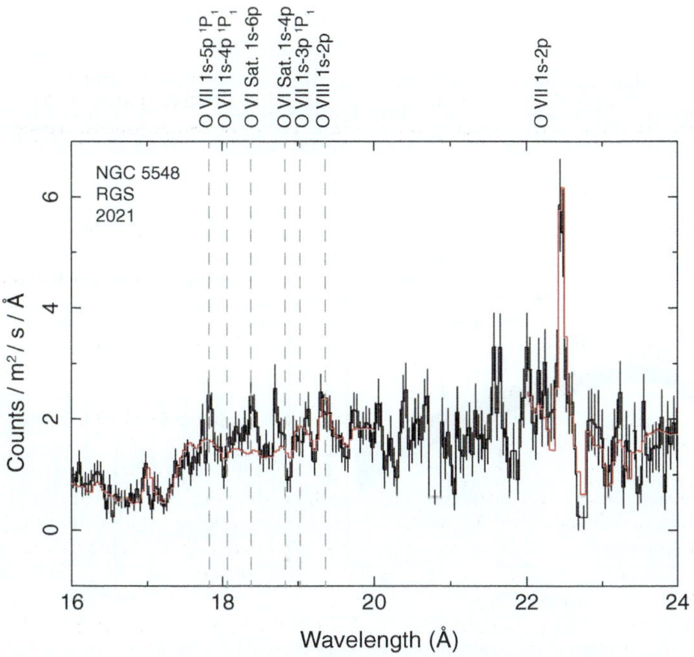

Fig. 10.9 *XMM-Newton* RGS spectrum of NGC 5548 AGN taken in January 2021 fit with the photoionization model. A set of excess features are seen around 18.4 Å indicating the possible charge exchange emission

10.3.3.4 Clusters of Galaxies

Though the hot plasmas are the dominating baryonic component in clusters of galaxies, cold gas clouds do exist, and they are often observed in and near the central galaxies, as well as in the wake of member galaxies during their infall to the center [50]. These neutral structures are immersed in the giant pool of hot, highly ionized plasma, and their interfaces are candidates for charge exchange. [68] and [244] suggested that the charge exchange can account for a part of the Hα and soft X-ray emission from bright filaments in the Perseus cluster.

During its brief lifetime, the micro-calorimeter onboard *Hitomi* observed the Perseus cluster with a $\sim$5 eV resolution in the 2–10 keV band. As shown in Fig. 10.10, there are indeed hints for charge exchange in the *Hitomi* spectrum. The high-*n* Rydberg transitions are found to be 1.6σ and 2.4σ significance for S XVI and Fe XXV. The detection remains challenging even for *Hitomi*, since the charge exchange in Perseus cluster is overwhelmed by the thermal emission by two orders of magnitude in flux. The S XVI charge exchange is in particular interesting since it nearly overlaps in energy with the mysterious 3.5 keV line detected in a large sample of clusters by [32, 38]. The 3.5 keV line was originally proposed as an evidence for the radiative decay of sterile neutrino, a theoretical form of dark matter. As shown in [4], the Hitomi spectrum of the Perseus cluster prefers the S XVI charge exchange scenario, although the data quality is insufficient to rule out the dark matter line. In addition, [98] reported a 2.8σ detection of a line feature at 14.82 Å, which is likely a O VIII charge exchange line. The line flux measured with the RGS is in good agreement with the possible *Hitomi* detection.

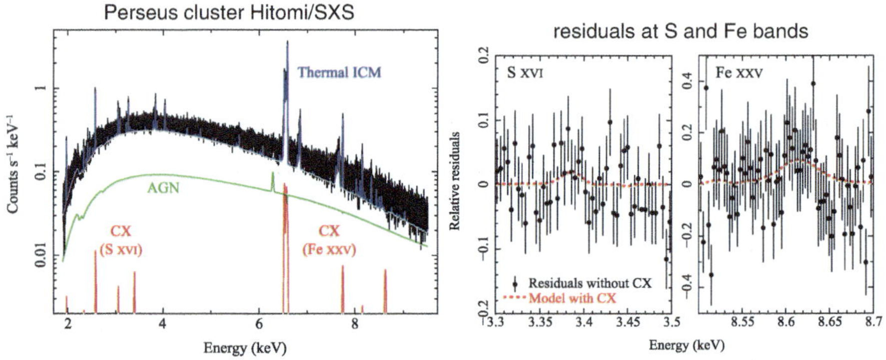

Fig. 10.10 (Left) *Hitomi* spectrum of the Perseus cluster fit with a sum of thermal, AGN, and charge exchange components. (Right) Residuals of the fit in the S XVI 3.2–3.6 keV and Fe XXV 8.3 – 8.8 keV bands with the thermal plus AGN model. The red curve in each panel shows the model change by including the charge exchange component

10.4 Ending Remarks

The introduction of the charge exchange process to general X-ray astronomy was made in 1996 by the discovery of X-ray from a comet. It has been broadly understood now that, in parallel to the electron impact and photon induced X-ray, the ion-neutral interaction can also efficiently generate X-ray in astrophysical objects. In the past two decades after the cometary discovery, X-ray astronomers have spotted charge exchange-like X-rays in a variety of objects, ranging from the neighbour planets and heliosphere to the interstellar medium, galaxies, supermassive black holes, and clusters of galaxies on the cosmological scales. Charge exchange emission is no doubt the best tool for detecting an interface, because the normal electron-impact X-ray flux emitted from the physically-thin interface is likely much dimmer. It has the potential to change the research landscape on the interactions of (1) AGNs and the host galaxies; (2) galactic outflow and interstellar medium; (3) infalling galaxies and the cluster.

Charge exchange produces only line emission, making it a science case tailored for high-resolution spectroscopy. With low-resolution spectra it often remains challenging to disentangle charge exchange from a thermal or photoionized component. Once fully resolved, charge exchange lines provide diagnostics on (1) the collision speed between ions and neutrals; (2) the chemical composition and ionization state of the hot ions; (3) relation with the Hα-emitting region; and (4) the neutral species. Spatially-resolved spectroscopy can further constrain the interaction region.

New astrophysical discoveries could feedback to the theoretical and laboratory studies of charge exchange. Substantial work remains to remove the tension between the theoretical calculations of cross sections and the corresponding laboratory measurements. The Li-like and ion species with more than three electrons are still poorly covered in the current calculations and experiments. The l- and S-distributions of the electron capture are crucial for an accurate line model, but they remain less certain than the n-distribution. Finally, the multi-electron charge exchange remain to be addressed. A consistent and continuous effort will be required to ensure that the atomic data are ready for the next generation of high-resolution X-ray data to be obtained with XRISM and Athena.

Acknowledgements SRON is supported financially by NWO, the Netherlands Organization for Scientific Research. C.S. acknowledges support from NASA under award number 80GSFC21M0002 and Max-Planck-Gesellschaft (MPG).

References

1. M.A. Abdallah, W. Wolff, H.E. Wolf et al., Phys. Rev. A **58**, 2911 (1998). https://doi.org/10.1103/PhysRevA.58.2911
2. O. Abu-Haija, A. Hasan, A. Kayani et al., EPL (Europhys. Lett.) **93**, 13003 (2011). https://doi.org/10.1209/0295-5075/93/13003

3. O. Abu-Haija, J. Harris, A. Hasan et al., Int. J. Mass Spectrom. **385**, 19 (2015). https://doi.org/10.1016/j.ijms.2015.05.002
4. F.A. Aharonian, H. Akamatsu, F. Akimoto et al., ApJL **837**, L15 (2017). https://doi.org/10.3847/2041-8213/aa61fa
5. R. Ali, P. Beiersdorfer, C.L. Harris et al., Phys. Rev. A **93**, 012711 (2016). https://doi.org/10.1103/PhysRevA.93.012711
6. R.J. Allan, R.E.S. Clegg, A.S. Dickinson et al., MNRAS **235**, 1245 (1988). https://doi.org/10.1093/mnras/235.4.1245
7. F.I. Allen, C. Biedermann, R. Radtke et al., J. Phys: Conf. Ser. **58**, 188 (2007). https://doi.org/10.1088/1742-6596/58/1/038
8. F.I. Allen, C. Biedermann, R. Radtke et al., Phys. Rev. A **78**, 032705 (2008). https://doi.org/10.1103/PhysRevA.78.032705
9. V.M. Andrianarijaona, D. Wulf, D. McCammon et al., Nucl. Inst. Methods Phys. Res. B **350**, 122 (2015). https://doi.org/10.1016/j.nimb.2015.01.040
10. M. Arnaud, R. Rothenflug, A&AS **60**, 425 (1985)
11. M.C. Bacchus-Montabonel, Chem. Phys. **228**, 181 (1998). https://doi.org/10.1016/S0301-0104(97)00359-5
12. M.C. Bacchus-Montabonel, K. Amezian, Zeitschrift fur Physik D Atoms Molec. Clust. **25**, 323 (1993). https://doi.org/10.1007/BF01437298
13. M.-C. Bacchus-Montabonel, Y.S. Tergiman, Chem. Phys. Lett. **422**, 122 (2006). https://doi.org/10.1016/j.cplett.2006.02.019
14. J.P. Bangsgaard, P. Hvelplund, J.O.P. Pedersen et al., Phys. Scrip. Vol. T **28**, 91 (1989). https://doi.org/10.1088/0031-8949/1989/T28/016
15. P. Barragán, L.F. Errea, L. Méndez et al., Phys. Rev. A **70**, 022707 (2004). https://doi.org/10.1103/PhysRevA.70.022707
16. P. Barragán, L.F. Errea, L. Méndez et al., Astrophys. J. **636**, 544 (2006). https://doi.org/10.1086/497884
17. P. Beiersdorfer, ARA&A **41**, 343 (2003). https://doi.org/10.1146/annurev.astro.41.011802.094825
18. P. Beiersdorfer, M. Bitter, M. Marion et al., Phys. Rev. A **72**, 032725 (2005). https://doi.org/10.1103/PhysRevA.72.032725
19. P. Beiersdorfer, K.R. Boyce, G.V. Brown et al., Science **300**, 1558 (2003). https://doi.org/10.1126/science.1084373
20. P. Beiersdorfer, C.M. Lisse, R.E. Olson et al., ApJL **549**, L147 (2001). https://doi.org/10.1086/319143
21. P. Beiersdorfer, R.E. Olson, G.V. Brown et al., PRL **85**, 5090 (2000). https://doi.org/10.1103/PhysRevLett.85.5090
22. P. Beiersdorfer, L. Schweikhard, P. Liebisch et al., Astrophys. J. **672**, 726 (2008). https://doi.org/10.1086/522617
23. J.P.M. Beijers, R. Hoekstra, R. Morgenstern, J. Phys. B Atomic Molec. Phys. **29**, 1397 (1996). https://doi.org/10.1088/0953-4075/29/7/020
24. G.L. Betancourt-Martinez, P. Beiersdorfer, G.V. Brown et al., ApJL **868**, L17 (2018). https://doi.org/10.3847/2041-8213/aaef82
25. A. Bhardwaj, R.F. Elsner, G.R. Gladstone et al., J. Geophys. Res. (Space Phys.) **111**, A11225 (2006). https://doi.org/10.1029/2006JA011792
26. S. Bienstock, T.G. Heil, A. Dalgarno, Phys. Rev. A **27**, 2741 (1983). https://doi.org/10.1103/PhysRevA.27.2741
27. F.W. Bliek, G.R. Woestenenk, R. Hoekstra et al., Phys. Rev. A **57**, 221 (1998). https://doi.org/10.1103/PhysRevA.57.221
28. D. Bodewits, D.J. Christian, M. Torney et al., A&A **469**, 1183 (2007). https://doi.org/10.1051/0004-6361:20077410
29. D. Bodewits, R. Hoekstra, Phys. Rev. A **76**, 032703 (2007). https://doi.org/10.1103/PhysRevA.76.032703
30. D. Bodewits, R. Hoekstra, Atoms **7**, 17 (2019). https://doi.org/10.3390/atoms7010017

31. D. Bodewits, A.G.G.M. Tielens, R. Morgenstern et al., Nucl. Inst. Methods Phys. Res. B **235**, 358 (2005). https://doi.org/10.1016/j.nimb.2005.03.205
32. A. Boyarsky, O. Ruchayskiy, D. Iakubovskyi et al., PRL **113**, 251301 (2014). https://doi.org/10.1103/PhysRevLett.113.251301
33. G. Branduardi-Raymont, A. Bhardwaj, R.F. Elsner et al., Planet. Space Sci. **55**, 1126 (2007). https://doi.org/10.1016/j.pss.2006.11.017
34. G. Branduardi-Raymont, A. Bhardwaj, R.F. Elsner et al., A&A **463**, 761 (2007). https://doi.org/10.1051/0004-6361:20066406
35. G. Branduardi-Raymont, R.F. Elsner, M. Galand et al., J. Geophys. Res. (Space Phys.) **113**, A02202 (2008). https://doi.org/10.1029/2007JA012600
36. G. Branduardi-Raymont, R.F. Elsner, G.R. Gladstone et al., A&A **424**, 331 (2004). https://doi.org/10.1051/0004-6361:20041149
37. H. Bruhns, H. Kreckel, D.W. Savin et al., Phys. Rev. A **77**, 064702 (2008). https://doi.org/10.1103/PhysRevA.77.064702
38. E. Bulbul, M. Markevitch, A. Foster et al., Astrophys. J. **789**, 13 (2014). https://doi.org/10.1088/0004-637X/789/1/13
39. S.E. Butler, A. Dalgarno, Astrophys. J. **241**, 838 (1980). https://doi.org/10.1086/158395
40. S.E. Butler, T.G. Heil, A. Dalgarno, Astrophys. J. **241**, 442 (1980). https://doi.org/10.1086/158357
41. C.N. Cabello, L.F. Errea, L. Fernández et al., J. Phys. B Atomic Molec. Phys. **36**, 307 (2003). https://doi.org/10.1088/0953-4075/36/2/311
42. P.G. Carolan, B.P. Duval, A.R. Field et al., Phys. Rev. A **35**, 3454 (1987). https://doi.org/10.1103/PhysRevA.35.3454
43. J.A. Carter, D. Bodewits, A.M. Read et al., A&A **541**, A70 (2012). https://doi.org/10.1051/0004-6361/201117950
44. G. Chambaud, B. Levy, P. Millie et al., J. Phys. B Atomic Molec. Phys. **13**, 4205 (1980). https://doi.org/10.1088/0022-3700/13/21/014
45. R.A. Chevalier, J.C. Raymond, ApJL **225**, L27 (1978). https://doi.org/10.1086/182785
46. D.J. Christian, D. Bodewits, C.M. Lisse et al., ApJS **187**, 447 (2010). https://doi.org/10.1088/0067-0049/187/2/447
47. D.A. Church, H.M. Holzscheiter, PRL **49**, 643 (1982). https://doi.org/10.1103/PhysRevLett.49.643
48. D. Ciric, A. Brazuk, D. Dijkkamp et al., J. Phys. B Atomic Molec. Phys. **18**, 3629 (1985). https://doi.org/10.1088/0022-3700/18/17/022
49. N.J. Clarke, P.C. Stancil, B. Zygelman et al., J. Phys. B Atomic Molec. Phys. **31**, 533 (1998). https://doi.org/10.1088/0953-4075/31/3/019
50. C.J. Conselice, J.S. Gallagher, R.F.G. Wyse, AJ **122**, 2281 (2001). https://doi.org/10.1086/323534
51. K.R. Cornelius, K. Wojtkowski, R.E. Olson, J. Phys. B Atomic Molec. Phys. **33**, 2017 (2000). https://doi.org/10.1088/0953-4075/33/11/304
52. D.H. Crandall, R.A. Phaneuf, F.W. Meyer, Phys. Rev. A **19**, 504 (1979). https://doi.org/10.1103/PhysRevA.19.504
53. T.E. Cravens, Geophys. Res. Lett. **24**, 105 (1997). https://doi.org/10.1029/96GL03780
54. H. Croft, A.S. Dickinson, J. Phys. B Atomic Molec. Phys. **29**, 57 (1996). https://doi.org/10.1088/0953-4075/29/1/011
55. R.S. Cumbee, L. Liu, D. Lyons et al., MNRAS **458**, 3554 (2016). https://doi.org/10.1093/mnras/stw527
56. R.S. Cumbee, P.D. Mullen, D. Lyons et al., Astrophys. J. **852**, 7 (2018). https://doi.org/10.3847/1538-4357/aa99d8
57. A. Dalgarna, T.G. Heil, S.E. Butler, Astrophys. J. **245**, 793 (1981). https://doi.org/10.1086/158853
58. X. Defay, K. Morgan, D. McCammon et al., Phys. Rev. A **88**, 052702 (2013). https://doi.org/10.1103/PhysRevA.88.052702
59. K. Dennerl, A&A **394**, 1119 (2002). https://doi.org/10.1051/0004-6361:20021116

60. K. Dennerl, Planet. Space Sci. **56**, 1414 (2008). https://doi.org/10.1016/j.pss.2008.03.008
61. K. Dennerl, Space Sci. Rev. **157**, 57 (2010). https://doi.org/10.1007/s11214-010-9720-5
62. K. Dennerl, C.M. Lisse, A. Bhardwaj et al., A&A **451**, 709 (2006). https://doi.org/10.1051/0004-6361:20054253
63. D. Dijkkamp, D. Ciric, A. de Boer et al., J. Phys. B Atom. Molec. Phys. **18**, 4763 (1985). https://doi.org/10.1088/0022-3700/18/24/017
64. I.N. Draganić, D.G. Seely, C.C. Havener, Phys. Rev. A **83**, 054701 (2011). https://doi.org/10.1103/PhysRevA.83.054701
65. W.R. Dunn, G. Branduardi-Raymont, L.C. Ray et al., Nat. Astron. **1**, 758 (2017). https://doi.org/10.1038/s41550-017-0262-6
66. I. Ewing, D.J. Christian, D. Bodewits et al., Astrophys. J. **763**, 66 (2013). https://doi.org/10.1088/0004-637X/763/1/66
67. Y. Ezoe, K. Ebisawa, N.Y. Yamasaki et al., PASJ **62**, 981 (2010). https://doi.org/10.1093/pasj/62.4.981
68. A.C. Fabian, J.S. Sanders, R.J.R. Williams et al., MNRAS **417**, 172 (2011). https://doi.org/10.1111/j.1365-2966.2011.19034.x
69. C.A. Feickert, R.J. Blint, G.T. Surratt et al., Astrophys. J. **286**, 371 (1984). https://doi.org/10.1086/162609
70. D. Fischer, B. Feuerstein, R.D. Du Bois et al., J. Phys. B Atom. Molec. Phys. **35**, 1369 (2002). https://doi.org/10.1088/0953-4075/35/5/318
71. W.L. Fite, R.T. Brackmann, Phys. Rev. **112**, 1141 (1958). https://doi.org/10.1103/PhysRev.112.1141
72. W.L. Fite, A.C.H. Smith, R.F. Stebbings, Proc. R. Soc. Lond. Ser. A **268**, 527 (1962). https://doi.org/10.1098/rspa.1962.0156
73. M.R. Fogle, M.S. Pindzola, J. Phys. B Atom. Molec. Phys. **53**, 095203 (2020). https://doi.org/10.1088/1361-6455/ab7524
74. M. Fogle, D. Wulf, K. Morgan et al., Phys. Rev. A **89**, 042705 (2014). https://doi.org/10.1103/PhysRevA.89.042705
75. L. Folkerts, M.A. Haque, C.C. Havener et al., Phys. Rev. A **51**, 3685 (1995). https://doi.org/10.1103/PhysRevA.51.3685
76. M. Frankel, P. Beiersdorfer, G.V. Brown et al., Astrophys. J. **702**, 171 (2009). https://doi.org/10.1088/0004-637X/702/1/171
77. W. Fritsch, C.D. Lin, J. Phys. B Atom. Molec. Phys. **17**, 3271 (1984). https://doi.org/10.1088/0022-3700/17/16/010
78. R. Fujimoto, K. Mitsuda, D. McCammon et al., Prog. Theor. Phys. Suppl. **169**, 71 (2007). https://doi.org/10.1143/PTPS.169.71
79. M. Galeazzi, M. Chiao, M.R. Collier et al., Nature **512**, 171 (2014). https://doi.org/10.1038/nature13525
80. H. Gao, V.H.S. Kwong, Astrophys. J. **567**, 1272 (2002). https://doi.org/10.1086/338589
81. H. Gao, V.H. Kwong, Phys. Rev. A **69**, 052715 (2004). https://doi.org/10.1103/PhysRevA.69.052715
82. L.D. Gardner, P.M. Koch, J.E. Bayfield et al., Phys. Rev. A **21**, 1397 (1980). https://doi.org/10.1103/PhysRevA.21.1397
83. M. Gargaud, R. McCarroll, L. Opradolce, A&A **208**, 251 (1989)
84. H.B. Gilbody, Phys. Scr. **24**, 712 (1981). https://doi.org/10.1088/0031-8949/24/4/006
85. G.R. Gladstone, J.H. Waite, D. Grodent et al., Nature **415**, 1000 (2002). https://doi.org/10.1038/4151000a
86. T.V. Goffe, M.B. Shah, H.B. Gilbody, J. Phys. B Atom. Molec. Phys. **12**, 3763 (1979). https://doi.org/10.1088/0022-3700/12/22/021
87. T.I. Gombosi, M. Horanyi, K. Kecskemety et al., Astrophys. J. **268**, 889 (1983). https://doi.org/10.1086/161011
88. J.B. Greenwood, A. Chutjian, S.J. Smith, Astrophys. J. **529**, 605 (2000). https://doi.org/10.1086/308254

89. J.B. Greenwood, I.D. Williams, S.J. Smith et al., ApJL **533**, L175 (2000). https://doi.org/10.1086/312615
90. J.B. Greenwood, I.D. Williams, S.J. Smith et al., Phys. Rev. A **63**, 062707 (2001). https://doi.org/10.1103/PhysRevA.63.062707
91. M.A. Guerrero, N. Ruiz, W.-R. Hamann et al., Astrophys. J. **755**, 129 (2012). https://doi.org/10.1088/0004-637X/755/2/129
92. N.L. Guevara, E. Teixeira, B. Hall et al., Phys. Rev. A **83**, 052709 (2011). https://doi.org/10.1103/PhysRevA.83.052709
93. L. Gu, J. Kaastra, A.J.J. Raassen et al., A&A **584**, L11 (2015). https://doi.org/10.1051/0004-6361/201527634
94. L. Gu, J. Kaastra, A.J.J. Raassen, A&A **588**, A52 (2016). https://doi.org/10.1051/0004-6361/201527615
95. L. Gu, J. Mao, E. Costantini et al., A&A **594**, A78 (2016). https://doi.org/10.1051/0004-6361/201628609
96. L. Gu, J. Mao, J.S. Kaastra et al., A&A **665**, A93 (2022). https://doi.org/10.1051/0004-6361/202244075
97. L. Gu, J. Mao, C.P. O'Dea et al., A&A **601**, A45 (2017). https://doi.org/10.1051/0004-6361/201730596
98. L. Gu, J. Mao, J. de Plaa et al., A&A **611**, A26 (2018). https://doi.org/10.1051/0004-6361/201731861
99. L. Gu, C. Shah, R. Zhang, Sensors **22**, 752 (2022). https://doi.org/10.3390/s22030752
100. J. Han, L. Wei, B. Wang et al., ApJS **253**, 6 (2021). https://doi.org/10.3847/1538-4365/abde44
101. M. Harwit, F. Hoyle, Astrophys. J. **135**, 875 (1962). https://doi.org/10.1086/147331
102. A.A. Hasan, F. Eissa, R. Ali et al., ApJL **560**, L201 (2001). https://doi.org/10.1086/324058
103. V.G. Hasan, S. Knoop, R. Morgenstern et al., J. Phys: Conf. Ser. **58**, 199 (2007). https://doi.org/10.1088/1742-6596/58/1/041
104. J.B. Hasted, Proc. R. Soc. Lond. Ser. A **205**, 421 (1951). https://doi.org/10.1098/rspa.1951.0038
105. C.C. Havener, M.S. Huq, H.F. Krause et al., Phys. Rev. A **39**, 1725 (1989). https://doi.org/10.1103/PhysRevA.39.1725
106. C.C. Havener, A. Müller, P.A. Zeijlmans van Emmichoven et al., Phys. Rev. A **51**, 2982 (1995). https://doi.org/10.1103/PhysRevA.51.2982
107. C.C. Havener, M.P. Nesnidal, M.R. Porter et al., Nucl. Inst. Methods Phys. Res. B **56**, 95 (1991). https://doi.org/10.1016/0168-583X(91)95980-R
108. C.C. Havener, R. Rejoub, C.R. Vane et al., Phys. Rev. A **71**, 034702 (2005). https://doi.org/10.1103/PhysRevA.71.034702
109. T.G. Heil, S.E. Butler, A. Dalgarno, Phys. Rev. A **27**, 2365 (1983). https://doi.org/10.1103/PhysRevA.27.2365
110. B. Herrero, I.L. Cooper, A.S. Dickinson et al., J. Phys. B Atom. Molec. Phys. **28**, 711 (1995). https://doi.org/10.1088/0953-4075/28/4/019
111. Hitomi Collaboration, F. Aharonian, H. Akamatsu, et al., PASJ **70**, 12 (2018). https://doi.org/10.1093/pasj/psx156
112. R. Hoekstra, J.P.M. Beijers, A.R. Schlatmann et al., Phys. Rev. A **41**, 4800 (1990). https://doi.org/10.1103/PhysRevA.41.4800
113. P. Honvault, M.C. Bacchus-Montabonel, M. Gargaud et al., Chem. Phys. **238**, 401 (1998). https://doi.org/10.1016/S0301-0104(98)00331-0
114. P. Honvault, M.C. Bacchus-Montabonel, R. McCarroll, J. Phys. B Atom. Molec. Phys. **27**, 3115 (1994). https://doi.org/10.1088/0953-4075/27/14/042
115. P. Honvault, M. Gargaud, M.C. Bacchus-Montabonel et al., A&A **302**, 931 (1995)
116. M. Hoshino, L. Pichl, Y. Kanai et al., Phys. Rev. A **75**, 012716 (2007). https://doi.org/10.1103/PhysRevA.75.012716
117. B.A. Huber, Zeitschrift für Physik A Hadrons and Nuclei **299**, 307 (1981). https://doi.org/10.1007/BF01441273

118. M.S. Huq, C.C. Havener, R.A. Phaneuf, Phys. Rev. A **40**, 1811 (1989). https://doi.org/10.1103/PhysRevA.40.1811
119. K. Ishikawa, Y. Ezoe, Y. Miyoshi et al., PASJ **65**, 63 (2013). https://doi.org/10.1093/pasj/65.3.63
120. W.B. Iwakiri, M. Serino, T. Mihara et al., PASJ **73**, 1405 (2021). https://doi.org/10.1093/pasj/psab085
121. R.K. Janev, D.S. Belic, B.H. Bransden, Phys. Rev. A **28**, 1293 (1983). https://doi.org/10.1103/PhysRevA.28.1293
122. R.K. Janev, R.A. Phaneuf, H. Tawara et al., At. Data Nucl. Data Tables **55**, 201 (1993). https://doi.org/10.1006/adnd.1993.1021
123. R.K. Janev, H. Winter, Phys. Rep. **117**, 265 (1985). https://doi.org/10.1016/0370-1573(85)90118-8
124. S. Katsuda, H. Tsunemi, K. Mori et al., ApJ **756**, 49 (2012). https://doi.org/10.1088/0004-637X/756/1/49
125. S. Katsuda, H. Tsunemi, K. Mori et al., Astrophys. J. **730**, 24 (2011). https://doi.org/10.1088/0004-637X/730/1/24
126. H.J. Kim, R.A. Phaneuf, F.W. Meyer et al., Phys. Rev. A **17**, 854 (1978). https://doi.org/10.1103/PhysRevA.17.854
127. M. Kimura, C.M. Dutta, N. Shimakura, Astrophys. J. **430**, 435 (1994). https://doi.org/10.1086/174418
128. M. Kimura, J.P. Gu, G. Hirsch et al., Phys. Rev. A **55**, 2778 (1997). https://doi.org/10.1103/PhysRevA.55.2778
129. M. Kimura, J.-P. Gu, G. Hirsch et al., Phys. Rev. A **56**, 1892 (1997). https://doi.org/10.1103/PhysRevA.56.1892
130. M. Kimura, T. Iwai, Y. Kaneko et al., J. Phys. Soc. Jpn. **53**, 2224 (1984). https://doi.org/10.1143/JPSJ.53.2224
131. T. Kimura, R.P. Kraft, R.F. Elsner et al., J. Geophys. Res. (Space Phys.) **121**, 2308 (2016). https://doi.org/10.1002/2015JA021893
132. M. Kimura, A.B. Sannigrahi, J.P. Gu et al., Astrophys. J. **473**, 1114 (1996). https://doi.org/10.1086/178221
133. J.B. Kingdon, G.J. Ferland, ApJS **106**, 205 (1996). https://doi.org/10.1086/192335
134. S. Knoop, D. Fischer, Y. Xue et al., J. Phys. B Atom. Molec. Phys. **41**, 195203 (2008). https://doi.org/10.1088/0953-4075/41/19/195203
135. H. Knudsen, Phys. Electron. Atom. Collis.: ICPEAC XII, 657 (1982)
136. S. Konami, K. Matsushita, T.G. Tsuru et al., PASJ **63**, S913 (2011). https://doi.org/10.1093/pasj/63.sp3.S913
137. Y. Koshiba, H. Uchida, T. Tanaka et al., PASJ **74**, 757 (2022). https://doi.org/10.1093/pasj/psac033
138. D. Koutroumpa, R.K. Smith, R.J. Edgar et al., Astrophys. J. **726**, 91 (2011). https://doi.org/10.1088/0004-637X/726/2/91
139. V.A. Krasnopolsky, J. Geophys. Res. (Space Phys.) **111**, A12102 (2006). https://doi.org/10.1029/2006JA012003
140. V.A. Krasnopolsky, D.J. Christian, V. Kharchenko et al., Icarus **160**, 437 (2002). https://doi.org/10.1006/icar.2002.6965
141. A. Kumar, D. Misra, A.H. Kelkar et al., J. Phys. B Atom. Molec. Phys. **39**, 331 (2006). https://doi.org/10.1088/0953-4075/39/2/008
142. T. Kusakabe, S. Kitamuro, Y. Nakai et al., Plasma Fus. Res. **7**, 2401062–2401062 (2012). https://doi.org/10.1585/pfr.7.2401062
143. T. Kusakabe, Y. Miyamoto, M. Kimura et al., Phys. Rev. A **73**, 022706 (2006). https://doi.org/10.1103/PhysRevA.73.022706
144. T. Kusakabe, H. Nakanishi, A. Iida et al., J. Phys. B Atom. Molec. Phys. **34**, 4809 (2001). https://doi.org/10.1088/0953-4075/34/23/323
145. T. Kusakabe, T. Shirai, Plasma Fus. Res. **9**, 3401119–3401119 (2014). https://doi.org/10.1585/pfr.9.3401119

146. M. Łabuda, Y.S. Tergiman, M.-C. Bacchus-Montabonel et al., Chem. Phys. Lett. **394**, 446 (2004). https://doi.org/10.1016/j.cplett.2004.07.045
147. R. Lallement, Space Sci. Rev. **143**, 427 (2009). https://doi.org/10.1007/s11214-008-9428-y
148. A.C.K. Leung, T. Kirchner, Phys. Rev. A **93**, 052710 (2016). https://doi.org/10.1103/PhysRevA.93.052710
149. A.C.K. Leung, T. Kirchner, Phys. Rev. A **97**, 062705 (2018). https://doi.org/10.1103/PhysRevA.97.062705
150. A. Leung, T. Kirchner, Atoms **7**, 15 (2019). https://doi.org/10.3390/atoms7010015
151. M.A. Leutenegger, P. Beiersdorfer, G.L. Betancourt-Martinez et al., Rev. Sci. Instrum. **87**, 11E516 (2016). https://doi.org/10.1063/1.4959919
152. M.A. Leutenegger, P. Beiersdorfer, G.V. Brown et al., PRL **105**, 063201 (2010). https://doi.org/10.1103/PhysRevLett.105.063201
153. M.A. Levine, R.E. Marrs, J.R. Henderson et al., Phys. Scrip. Vol. T **22**, 157 (1988). https://doi.org/10.1088/0031-8949/1988/T22/024
154. C.Y. Lin, P.C. Stancil, J.P. Gu et al., Phys. Rev. A **71**, 062708 (2005). https://doi.org/10.1103/PhysRevA.71.062708
155. C.M. Lisse, D. Christian, K. Dennerl et al., Icarus **141**, 316 (1999). https://doi.org/10.1006/icar.1999.6176
156. C.M. Lisse, D.J. Christian, K. Dennerl et al., Astrophys. J. **635**, 1329 (2005). https://doi.org/10.1086/497570
157. C.M. Lisse, D.J. Christian, S.J. Wolk et al., Icarus **222**, 752 (2013). https://doi.org/10.1016/j.icarus.2012.09.025
158. C.M. Lisse, K. Dennerl, J. Englhauser et al., Science **274**, 205 (1996). https://doi.org/10.1126/science.274.5285.205
159. C.H. Liu, L. Liu, J.G. Wang, Phys. Rev. A **90**, 012708 (2014). https://doi.org/10.1103/PhysRevA.90.012708
160. J. Liu, S. Mao, Q.D. Wang, MNRAS **415**, L64 (2011). https://doi.org/10.1111/j.1745-3933.2011.01079.x
161. C.-H. Liu, J.-G. Wang, Eur. Phys. J. D **71**, 168 (2017). https://doi.org/10.1140/epjd/e2017-80020-3
162. J. Liu, Q.D. Wang, S. Mao, MNRAS **420**, 3389 (2012). https://doi.org/10.1111/j.1365-2966.2011.20263.x
163. G. Lubinski, Z. Juhász, R. Morgenstern et al., J. Phys. B Atom. Molec. Phys. **33**, 5275 (2000). https://doi.org/10.1088/0953-4075/33/23/302
164. G. Lubinski, Z. Juhász, R. Morgenstern et al., PRL **86**, 616 (2001). https://doi.org/10.1103/PhysRevLett.86.616
165. D. Lyons, R.S. Cumbee, P.C. Stancil, ApJS **232**, 27 (2017). https://doi.org/10.3847/1538-4365/aa8d16
166. J.R. Machacek, D.P. Mahapatra, D.R. Schultz et al., Astrophys. J. **809**, 75 (2015). https://doi.org/10.1088/0004-637X/809/1/75
167. J. Mao, J.S. Kaastra, M. Mehdipour et al., A&A **612**, A18 (2018). https://doi.org/10.1051/0004-6361/201732162
168. H.S.W. Massey, R.A. Smith, Proc. R. Soc. Lond. Ser. A **142**, 142 (1933). https://doi.org/10.1098/rspa.1933.0161
169. R.W. McCullough, W.L. Nutt, H.B. Gilbody, J. Phys. B Atom. Molec. Phys. **12**, 4159 (1979). https://doi.org/10.1088/0022-3700/12/24/021
170. R.W. McCullough, F.G. Wilkie, H.B. Gilbody, J. Phys. B Atom. Molec. Phys. **17**, 1373 (1984). https://doi.org/10.1088/0022-3700/17/7/020
171. R.W. McCullough, S.M. Wilson, H.B. Gilbody, J. Phys. B Atom. Molec. Phys. **20**, 2031 (1987). https://doi.org/10.1088/0022-3700/20/9/016
172. A.E. Metzger, J.L. Luthey, D.A. Gilman et al., J. Geophys. Res. **88**, 7731 (1983). https://doi.org/10.1029/JA088iA10p07731
173. F.W. Meyer, R.A. Phaneuf, H.J. Kim et al., Phys. Rev. A **19**, 515 (1979). https://doi.org/10.1103/PhysRevA.19.515

174. E.D. Miller, T. Hiroshi, M.W. Bautz, et al., PASJ **60**, S95 (2008). https://doi.org/10.1093/pasj/60.sp1.S95
175. K.A. Miller, W.W. Smith, T. Ehrenreich et al., Astrophys. J. **742**, 130 (2011). https://doi.org/10.1088/0004-637X/742/2/130
176. T. Mroczkowski, D.W. Savin, R. Rejoub et al., Phys. Rev. A **68**, 032721 (2003). https://doi.org/10.1103/PhysRevA.68.032721
177. P.D. Mullen, R.S. Cumbee, D. Lyons et al., ApJS **224**, 31 (2016). https://doi.org/10.3847/0067-0049/224/2/31
178. P.D. Mullen, R.S. Cumbee, D. Lyons et al., Astrophys. J. **844**, 7 (2017). https://doi.org/10.3847/1538-4357/aa7752
179. R.S. Mulliken, J. Chem. Phys. **7**, 20 (1939). https://doi.org/10.1063/1.1750319
180. K. Nandra, D. Barret, X. Barcons, et al., (2013), arXiv:1306.2307
181. J.L. Nolte, P.C. Stancil, H.P. Liebermann et al., J. Phys. B Atom. Molec. Phys. **45**, 245202 (2012). https://doi.org/10.1088/0953-4075/45/24/245202
182. M. Numazawa, Y. Ezoe, T. Ohashi et al., PASJ **73**, 894 (2021). https://doi.org/10.1093/pasj/psab053
183. W.L. Nutt, R.W. McCullough, H.B. Gilbody, J. Phys. B Atom. Molec. Phys. **11**, L181 (1978). https://doi.org/10.1088/0022-3700/11/6/005
184. S. Otranto, N.D. Cariatore, R.E. Olson, Phys. Rev. A **90**, 062708 (2014). https://doi.org/10.1103/PhysRevA.90.062708
185. S. Otranto, R.E. Olson, P. Beiersdorfer, J. Phys, Conf. Ser. **58**, 165 (2007). https://doi.org/10.1088/1742-6596/58/1/032
186. A. Owens, T. Oosterbroek, A. Orr et al., Earth Moon Planet. **77**, 293 (1997). https://doi.org/10.1023/A:1006253028962
187. M.N. Panov, A.A. Basalaev, K.O. Lozhkin, Phys. Scrip. Vol. T **3**, 124 (1983). https://doi.org/10.1088/0031-8949/1983/T3/025
188. R.A. Phaneuf, I. Alvarez, F.W. Meyer et al., Phys. Rev. A **26**, 1892 (1982). https://doi.org/10.1103/PhysRevA.26.1892
189. R.A. Phaneuf, F.W. Meyer, R.H. McKnight, Phys. Rev. A **17**, 534 (1978). https://doi.org/10.1103/PhysRevA.17.534
190. L. Pichl, Y. Li, H.-P. Liebermann et al., J. Chem. Phys. **118**, 4872 (2003). https://doi.org/10.1063/1.1545777
191. L. Pichl, Y. Li, H.-P. Liebermann et al., Phys. Rev. A **69**, 062715 (2004). https://doi.org/10.1103/PhysRevA.69.062715
192. M. Pieksma, C.C. Havener, Phys. Rev. A **57**, 1892 (1998). https://doi.org/10.1103/PhysRevA.57.1892
193. M.S. Pindzola, M. Fogle, J. Phys. B Atom. Molec. Phys. **48**, 205203 (2015). https://doi.org/10.1088/0953-4075/48/20/205203
194. C. Pinto, A.C. Fabian, A. Ogorzalek et al., MNRAS **461**, 2077 (2016). https://doi.org/10.1093/mnras/stw1444
195. A.M.T. Pollock, Astron. Nachr. **333**, 351 (2012). https://doi.org/10.1002/asna.201211668
196. F.S. Porter, B.R. Beck, P. Beiersdorfer et al., Can. J. Phys. **86**, 231 (2008). https://doi.org/10.1139/P07-147
197. F.S. Porter, P. Beiersdorfer, G.V. Brown et al., J. Low Temp. Phys. **151**, 1061 (2008). https://doi.org/10.1007/s10909-008-9788-4
198. D. Rabli, M. Gargaud, R. McCarroll, J. Phys. B Atom. Molec. Phys. **34**, 1395 (2001). https://doi.org/10.1088/0953-4075/34/8/305
199. P. Ranalli, A. Comastri, L. Origlia et al., MNRAS **386**, 1464 (2008). https://doi.org/10.1111/j.1365-2966.2008.13128.x
200. A.P. Rasmussen, E. Behar, S.M. Kahn et al., A&A **365**, L231 (2001). https://doi.org/10.1051/0004-6361:20000231
201. R. Rejoub, M.E. Bannister, C.C. Havener et al., Phys. Rev. A **69**, 052704 (2004). https://doi.org/10.1103/PhysRevA.69.052704

202. M. Rigazio, V. Kharchenko, A. Dalgarno, Phys. Rev. A **66**, 064701 (2002). https://doi.org/10.1103/PhysRevA.66.064701
203. S.R. Roberts, Q.D. Wang, MNRAS **449**, 1340 (2015). https://doi.org/10.1093/mnras/stv319
204. E. Roueff, A. Dalgarno, Phys. Rev. A **38**, 93 (1988). https://doi.org/10.1103/PhysRevA.38.93
205. F.R.S. Rutherford, London, Edinburgh. Dublin Philos. Mag. J. Sci. **21**(125), 669–688 (1911)
206. M.M. Sant'Anna, W.S. Melo, A.C.F. Santos et al., J. Phys. B Atom. Molec. Phys. **33**, 353 (2000). https://doi.org/10.1088/0953-4075/33/3/305
207. D.W. Savin, Astrophys. J. **566**, 599 (2002). https://doi.org/10.1086/324218
208. D.G. Seely, V.M. Andrianarijaona, D. Wulf et al., Phys. Rev. A **95**, 052704 (2017). https://doi.org/10.1103/PhysRevA.95.052704
209. W. Seim, A. Muller, I. Wirkner-Bott et al., J. Phys. B Atom. Molec. Phys. **14**, 3475 (1981). https://doi.org/10.1088/0022-3700/14/18/020
210. B. Seredyuk, H. Bruhns, D.W. Savin et al., Phys. Rev. A **75**, 054701 (2007). https://doi.org/10.1103/PhysRevA.75.054701
211. C. Shah, S. Dobrodey, S. Bernitt et al., Astrophys. J. **833**, 52 (2016). https://doi.org/10.3847/1538-4357/833/1/52
212. M.B. Shah, T.V. Goffe, H.B. Gilbody, J. Phys. B Atom. Molec. Phys. **11**, L233 (1978). https://doi.org/10.1088/0022-3700/11/7/008
213. N. Shimakura, M. Kimura, Phys. Rev. A **44**, 1659 (1991). https://doi.org/10.1103/PhysRevA.44.1659
214. E.Y. Sidky, W. Fritsch, C.D. Lin, Phys. Rev. A **59**, 1994 (1999). https://doi.org/10.1103/PhysRevA.59.1994
215. R.K. Smith, R.J. Edgar, P.P. Plucinsky et al., Astrophys. J. **623**, 225 (2005). https://doi.org/10.1086/428568
216. B. Snios, V. Kharchenko, C.M. Lisse et al., Astrophys. J. **818**, 199 (2016). https://doi.org/10.3847/0004-637X/818/2/199
217. S.L. Snowden, M.R. Collier, T. Cravens et al., Astrophys. J. **691**, 372 (2009). https://doi.org/10.1088/0004-637X/691/1/372
218. S.L. Snowden, M.R. Collier, K.D. Kuntz, Astrophys. J. **610**, 1182 (2004). https://doi.org/10.1086/421841
219. P.C. Stancil, Spectroscop. Chall. Photoion. Plasmas **247**, 3 (2001)
220. P.C. Stancil, N.J. Clarke, B. Zygelman et al., J. Phys. B Atom. Molec. Phys. **32**, 1523 (1999). https://doi.org/10.1088/0953-4075/32/6/015
221. P.C. Stancil, C.C. Havener, P.S. Krstić et al., Astrophys. J. **502**, 1006 (1998). https://doi.org/10.1086/305937
222. P.C. Stancil, D.R. Schultz, M. Kimura et al., Astron. Astrophys. Supplem. **140**, 225 (1999). https://doi.org/10.1051/aas:1999419
223. P.C. Stancil, A.R. Turner, D.L. Cooper et al., J. Phys. B Atom. Molec. Phys. **34**, 2481 (2001). https://doi.org/10.1088/0953-4075/34/12/313
224. P.C. Stancil, B. Zygelman, Astrophys. J. **472**, 102 (1996). https://doi.org/10.1086/178044
225. P.C. Stancil, B. Zygelman, N.J. Clarke et al., J. Phys. B Atom. Molec. Phys. **30**, 1013 (1997). https://doi.org/10.1088/0953-4075/30/4/020
226. R.F. Stebbings, W.L. Fite, D.G. Hummer, J. Chem. Phys. **33**, 1226 (1960). https://doi.org/10.1063/1.1731360
227. N.C. Sterling, P.C. Stancil, A&A **535**, A117 (2011). https://doi.org/10.1051/0004-6361/201117584
228. P.M. Stier, C.F. Barnett, Phys. Rev. **103**, 896 (1956). https://doi.org/10.1103/PhysRev.103.896
229. S. Suzuki, N. Shimakura, J.-P. Gu et al., Phys. Rev. A **60**, 4504 (1999). https://doi.org/10.1103/PhysRevA.60.4504
230. R. Suzuki, A. Watanabe, H. Sato et al., Phys. Rev. A **63**, 042717 (2001). https://doi.org/10.1103/PhysRevA.63.042717

231. H. Suzuki, H. Yamaguchi, M. Ishida et al., Astrophys. J. **900**, 39 (2020). https://doi.org/10.3847/1538-4357/aba524
232. Y. Tanaka, H. Uchida, T. Tanaka et al., Astrophys. J. **933**, 101 (2022). https://doi.org/10.3847/1538-4357/ac738f
233. M. Tashiro, H. Maejima, K. Toda et al., Proc. SPIE **10699**, 1069922 (2018). https://doi.org/10.1117/12.2309455
234. H. Tawara, K. Okuno, C.W. Fehrenbach et al., Phys. Rev. A **63**, 062701 (2001). https://doi.org/10.1103/PhysRevA.63.062701
235. H. Tawara, P. Richard, U.I. Safronova et al., Phys. Rev. A **64**, 042712 (2001). https://doi.org/10.1103/PhysRevA.64.042712
236. H. Tawara, P. Richard, U.I. Safronova et al., Phys. Rev. A **65**, 042509 (2002). https://doi.org/10.1103/PhysRevA.65.042509
237. H. Tawara, P. Richard, U.I. Safronova et al., Can. J. Phys. **80**, 821 (2002). https://doi.org/10.1139/p02-014
238. H. Tawara, E. Takács, T. Suta et al., Phys. Rev. A **73**, 012704 (2006). https://doi.org/10.1103/PhysRevA.73.012704
239. L.K. Townsley, P.S. Broos, Y.-H. Chu et al., ApJS **194**, 16 (2011). https://doi.org/10.1088/0067-0049/194/1/16
240. T.G. Tsuru, M. Ozawa, Y. Hyodo et al., PASJ **59**, 269 (2007). https://doi.org/10.1093/pasj/59.sp1.S269
241. A.R. Turner, D.L. Cooper, J.G. Wang et al., Phys. Rev. A **68**, 012704 (2003). https://doi.org/10.1103/PhysRevA.68.012704
242. H. Uchida, S. Katsuda, H. Tsunemi et al., Astrophys. J. **871**, 234 (2019). https://doi.org/10.3847/1538-4357/aaf8a6
243. N. Vaeck, M.-C. Bacchus-Montabonel, E. Baloïtcha et al., Phys. Rev. A **63**, 042704 (2001). https://doi.org/10.1103/PhysRevA.63.042704
244. S.A. Walker, P. Kosec, A.C. Fabian et al., MNRAS **453**, 2480 (2015). https://doi.org/10.1093/mnras/stv1829
245. M.K. Wallis, Planet. Space Sci. **21**, 1647 (1973). https://doi.org/10.1016/0032-0633(73)90156-6
246. J.G. Wang, B. He, Y. Ning et al., Phys. Rev. A **74**, 052709 (2006). https://doi.org/10.1103/PhysRevA.74.052709
247. J.G. Wang, P.C. Stancil, A.R. Turner et al., Phys. Rev. A **67**, 012710 (2003). https://doi.org/10.1103/PhysRevA.67.012710
248. J.G. Wang, P.C. Stancil, A.R. Turner et al., Phys. Rev. A **69**, 062702 (2004). https://doi.org/10.1103/PhysRevA.69.062702
249. J.G. Wang, A.R. Turner, D.L. Cooper et al., J. Phys. B Atom. Molec. Phys. **35**, 3137 (2002). https://doi.org/10.1088/0953-4075/35/14/310
250. F. Wang, M.-M. Wang, Can. J. Phys. **90**, 283 (2012). https://doi.org/10.1139/p2012-017
251. B.J. Wargelin, P. Beiersdorfer, G.V. Brown, Can. J. Phys. **86**, 151 (2008). https://doi.org/10.1139/P07-125
252. B.J. Wargelin, M. Markevitch, M. Juda et al., Astrophys. J. **607**, 596 (2004). https://doi.org/10.1086/383410
253. R. Wegmann, K. Dennerl, A&A **430**, L33 (2005). https://doi.org/10.1051/0004-6361:200400124
254. R. Willingale, P.T. O'Brien, S.W.H. Cowley et al., Astrophys. J. **649**, 541 (2006). https://doi.org/10.1086/506315
255. S.M. Wilson, T.K. McLaughlin, R.W. McCullough et al., J. Phys. B Atom. Molec. Phys. **23**, 2969 (1990). https://doi.org/10.1088/0953-4075/23/17/009
256. S.J. Wolk, C.M. Lisse, D. Bodewits et al., Astrophys. J. **694**, 1293 (2009). https://doi.org/10.1088/0004-637X/694/2/1293
257. D. Wulf, M.E. Eckart, M. Galeazzi et al., Astrophys. J. **884**, 120 (2019). https://doi.org/10.3847/1538-4357/ab41f8

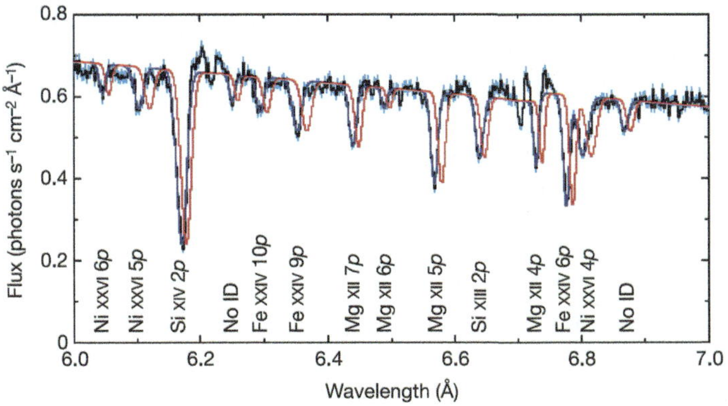

Fig. 11.1 Blueshifted absorption lines from an accretion disk wind in the black hole LMXB GRO J1655–40 from [196]. The black curve with blue error bars is the *Chandra* HETGS spectrum, while the red curve is the best fit model adjusted to zero Doppler shift. High-resolution spectra like these allow sensitive diagnostics of the dynamics, temperature, ionization, and even density of absorbing gas

CCD spectra had revealed absorption lines in some bright X-ray sources [e.g., 68, 149], but without the resolution to measure Doppler shifts, it was not possible to determine whether those lines represented extended atmospheres or outflows. With grating spectra provided by *Chandra* [21, 26] and *XMM-Newton* [54], it became clear that they were produced in gas flowing away from the compact object (at least in many cases, but see Sect. 11.3). These outflows typically have blueshifts of a few hundred to a few thousand km s^{-1} (Fig. 11.1; see also Table 11.1), which can complicate the determination of an accretion disk origin when the companion star is a strong source of stellar winds. Therefore, most of our observational knowledge about disk winds in XRBs (and hence our focus here) comes from LMXBs, since their companion stars do not exhibit strong winds. Disk winds are often envisioned as biconical, roughly equatorial outflows (see Fig. 11.2). They are also typically highly ionized, such that hydrogen- and helium-like iron tend to dominate the spectra with strong lines at $\sim$7 keV and 6.7 keV, respectively.

It was already evident from simulations of accretion disk winds [e.g., 12] that photoionized accretion disk winds would play a significant role in both the physics of accretion onto compact objects and efforts to understand them, and this prediction was quickly confirmed with early high-resolution X-ray observations. For instance, in the 2000 discovery of P-Cygni lines in Circinus X-1, [20] noted the likely connection to the high luminosity of the source. Extending this suggestion, in 2002 [157] estimated the mass loss rate in the accretion disk wind in GRS 1915+105, finding it to be comparable to the accretion rate onto the black hole [see also 113]. In the two decades since, high-resolution observations of X-ray binaries have significantly enhanced our understanding of the connections between accretion and ejection processes, and in 2016 [78] argued that winds could eject more than half the accreted matter during

Table 11.1 List of disk atmosphere and wind detections in neutron star and black hole X-ray binaries. This is an update of [57]; winds reported since then are indicated by **. The quoted absorption column densities (N^{H}_{Gal}) are solely meant to be indicative and taken from [108]. Upper limits for outflow velocities typically mean that no gratings data were available. Source classification: B = BeXRB, D = dipper, E = eclipser, MP/SP = accreting millisecond/slow X-ray pulsar, HE = high-Eddington source, SL = outflow inferred based on a single significant line detection

Source	P_{orb}	N^{H}_{Gal} (10^{21} cm^{-2})	v_{out} (10^3 km^{-2})	Class	References
Neutron star LMXBs					
**1RXS J180408.9–342058[a]	?	2	26	SL	[48]
**GX340+0[b]	?	2	12	SL, HE	[195]
**4U 1820-30	0.19h	1.3	1.2		[40]
**IGR J17062–6143	0.63h	1	14	MP	[53, 312]
XB1916–053	0.83h	2	atm	D	[19, 60, 118, 128, 302, 333]
1A 1744–361	1.62h	3	atm	D	[95]
4U 1323–62	2.93h	10	?	D	[7, 17, 34]
**XTE J1710–281	3.28h	2	0.8	D, E	[261]
EXO 0748−676	3.82h	0.9	$\lesssim 1$	D, E	[60, 250, 315]
XB 1254–690	3.93h	2	atm	D	[18, 60, 61, 117]
MXB 1659–298	7.12h	2	$\lesssim 0.1$	D, E	[60, 247, 283]
AX J1745.6−2901	8.35h	0.9	$\lesssim 1$	D, E	[116, 246]
**IGR J17591–2342	8.80h	10	2.8	MP	[228]
X1624–490	20.9h	20	atm	D	[60, 119, 237, 326]
IGR J17480–2446	21.3h	5.5	1–3	MP	[193]
**Swift J1858.6–0814	21.3h	2	2	HE	[23]
GRO J1744–28	11.8d	10	8	SL, SP	[51]
Cir X–1	16.6d	20	2	HE	[20, 44, 119, 275, 279]
GX13+1	24.1d	10	out	HE	[45, 63, 168, 284, 305, 308]
Neutron star IMXBs and HMXBs					
Swift J0243.6+6124	1.2d	7	66	B, SP, HE	[314]
Her X-1	1.7d	0.15	$0.2-1$	IMXB, SP	[145]
1A 0535+262	111d	4.5	$1.5-3.0$	B	[267]

(continued)

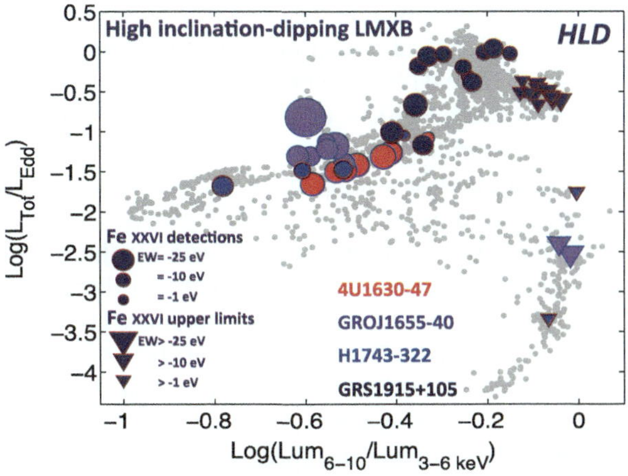

Fig. 11.3 Observations of accretion disk winds in high inclination black hole X-ray binaries from [249]. Background gray points show the hardness-intensity trajectories of outbursts. Circles show wind detections, with symbol size scaling with absorption line equivalent width. Triangles indicate non-detections, again with symbol size proportional to the upper limit on any line equivalent width

above. For instance, [202] noted that absorption lines from accretion disk winds were stronger during softer (e.g., disk-dominated) states. This pattern was reflected in observations of H1743–322 [200] and GRO J1655–40 [196] as well as archival observations of GRS 1915+105. These latter observations were published in detail in [220], who characterized the state dependence as broad emission lines in harder states (consistent with reflection) and narrow absorption lines in softer states.

If accretion disk winds vary by state, it is natural to wonder whether there is a connection between winds and radio jets [given the well-characterized link between jets and X-ray hard states, e.g., 76, 77, 80]. This connection must be complex, because both [157, 200] reported non-zero radio flux coincident with detections of absorption lines. Reference [202] suggested that an anticorrelation between winds and jets might relate to changes in the wind density and its effect on the magnetic field configuration in the inner disk. In contrast, [220] interpreted the state dependence of accretion disk winds as evidence of a material tradeoff between winds and jets, in which mass loss in winds from the outer accretion disk (see Sect. 11.2.1.3) effectively reduces the amount of matter fed into the jet launched from the inner disk.

The picture of this anticorrelation is substantially enhanced with additional data. Reference [249] performed a systematic search for accretion disk wind absorption lines (specifically Fe XXV and Fe XXVI) in X-ray observations of black hole X-ray binaries. In systems seen at high inclination, they found that iron absorption lines were preferentially detected in softer states, generally at higher luminosity (see Fig. 11.3), matching the early trend described in [202, 220]. Interestingly, they detected no significant hot wind absorption in any system at low inclination, which is consistent with the idea of disk winds as equatorial outflows [e.g., 18, 60]. The only apparent counterexample is a warm absorber in GX 339-4 [198], but this is apparent only in Ne, O, and Mg below 1.5 keV and its origin is not entirely clear [but see, 167]. See Table 11.1 for an updated census of winds in BH XRBs.

While [249] has often been cited as evidence that X-ray absorbing accretion disk winds are never found in X-ray hard states or that jets and such winds cannot coexist, the historical record described above requires a more nuanced view. Reference [200] found strong radio activity coincident with X-ray absorption lines in H1743–322, and [115] reported similar results at high luminosity in GRS 1915+105, V404 Cyg, GX 13+1, Sco X-1, and Cir X-1. Thus it seems clear that any interaction between winds and jets—whether mediated by mass loss or changes in the magnetic field configuration—must at the very least be gradual, so that both outflows can be observed simultaneously.

Regardless of the precise timescale for the transition to wind-dominated states, Fig. 11.3 [taken from 249] suggested that there is a real trend in the observability of absorption lines in black hole outbursts. The discovery of X-ray absorption lines in *Chandra* HETGS observations of 4U 1630–47 during its 2012–2013 outburst lent additional credibility to the idea that winds can be reliably detected in soft states of high-inclination systems [216, 218]; see also [191, 303]. But if that fact is generally accepted, the apparent absence of winds in hard states is still not fully understood, partly because several processes can affect their observability, including (1) overionization, (2) thermodynamics, and (3) time evolution.

Overionization: Implicit in our discussion of ionization balance and the definition of the ionization parameter ξ (Eq. 11.4) is the shape of the ionizing spectrum. Consider two identical shells of gas equidistant from X-ray sources with luminosity L; let one be illuminated by a blackbody spectrum and the other by a power law spectrum. The gas shells will have the same ionization parameter but very different ion fractions. These depend on the integral of the photon rate spectrum over the ionization cross section $\sigma_i(E)$ of the relevant charge state, i.e.:

$$\Phi_i = \int_{\chi_i}^{\infty} \frac{L_E}{E} \sigma_i(E) dE, \tag{11.7}$$

where E is energy, L_E is the monochromatic luminosity and χ_i is the ionization threshold for charge state i. For a given luminosity, a harder/shallower spectrum will generally ionize each charge state shell more effectively. For ionization parameters of interest here, $\log(\xi) \approx 4 - 5$, a harder ionizing spectrum will typically lead to weaker iron absorption lines.

The idea behind "overionization" as an explanation for the apparent absence of winds in spectrally hard states [e.g., 199, 216, 222, 241, 246, 281] is that winds might be present during these states but—given their harder spectra—too ionized to produce detectable absorption lines. The difficulty with overionization is that for a given luminosity, a harder spectrum may have **fewer ionizing photons** than a softer spectrum due to the factor of E^{-1} in Eq. 11.7. Moreover, from Fig. 11.3, much of the hard state at issue actually has lower luminosity than the soft states where winds are ubiquitous. Accordingly, one should expect a given parcel of gas to have a *lower* ionization parameter during the hard state, which can compensate for the shift in the ionization balance.[4] Even power law spectra at higher luminosities have

[4] Given the differences in luminosity, a proper statistical analysis of the presence or absence of winds in different outburst states needs to account for the 7 keV sensitivity of each individual observation.

repeatedly proven insufficient to fully ionize disk winds present in soft states [222, 246]. Thus, though the details of ionization are certainly important, overionization is not likely to be the primary driver of the state dependence of hot winds in black hole X-ray binaries [64, reached a similar conclusion based on models of thermally driven winds]. However, overionization may play a more important role in optical winds [e.g., 209].

Thermodynamics: In order to detect Fe XXV and Fe XXVI absorption lines (or any other particular line series) from our shell, it needs to maintain an appropriate temperature and ionization parameter over the course of an observation. This is not particularly likely if the shell is not in a stable thermal and ionization equilibrium. The question is then whether—given all the relevant heating, cooling, and ionization processes—such an equilibrium is more or less likely during hard states than soft states. This has been studied extensively, and while it is still necessary to account for the actual radiation field in any specific observation, the broad conclusion is fairly robust that gas capable of producing hot iron absorption lines is probably thermally unstable during spectrally hard states [e.g., 13, 29–31, 241].

Typically thermal stability is diagnosed by inspecting *stability curves*: the set of points corresponding to thermal equilibrium in a plot of T vs ξ/T [152]. The stability curve partitions the parameter space: to the left of the curve, the cooling rate exceeds the heating rate, while the reverse is true to the right of the curve. Our shell will only be in a thermodynamically stable equilibrium where the slope of the stability curve is positive (as this is where thermally perturbed gas will return to equilibrium).[5] Consider the example shown in Fig. 11.4. Reference [246] observed the neutron star AX J1745.6–2901 on numerous occasions with *XMM-Newton* and *NuSTAR*, finding absorption lines—likely from an accretion disk atmosphere—during soft states and none during hard states. Reference [13] used their constructed hard and soft SEDs to compute stability curves for this absorbing gas.

As shown, they found that the absorber lay on the stable portion of the stability curve during the soft state, but that the same gas would be thermodynamically unstable during the hard state. Thus, during the hard state, the same gas would rapidly evolve to a much higher or much lower temperature and ionization than observed in the soft state. As hard spectra often render $\log(\xi) = 3 - 5$ gas thermodynamically unstable, this is a very compelling explanation[6] of X-ray visibility trends for accretion disk winds. It is worth noting that [111] performed radiation hydrodynamic simulations of disk winds, including a self-consistent ionization and frequency structure, including radiative driving (see below) and thermal instabilities but still found

[5] As discussed by [[256], see also [101]], this analysis is appropriate for steady illumination of an optically thin medium; alternative stability curves give similar results in the case of time dependence or non-negligible attenuation.

[6] One important complicating factor noted by [64] is that state changes in the broadband spectrum lead to changes in the Compton temperature (see Sect. 11.2.1.3). One consequence is that thermally driven winds may be launched from different locations in the disk during hard and soft states. If the value nR^2 is not preserved in the wind during a state transition, then it becomes much more difficult to predict the expected temperature and ionization of the wind in the hard state. Robust conclusions about overionization thermal instability may therefore depend on the inferred driving mechanism.

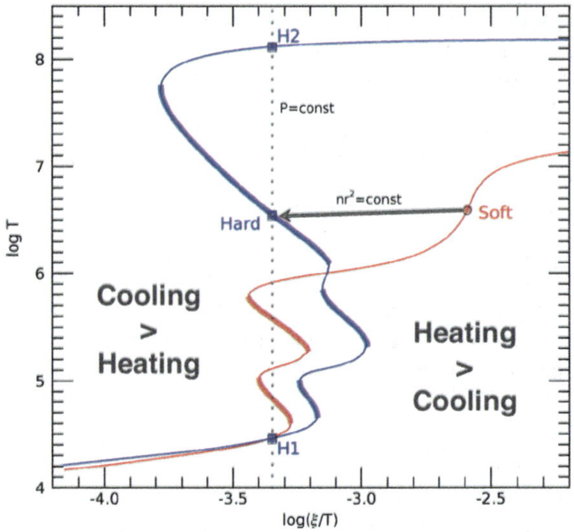

Fig. 11.4 Stability curves for a disk atmosphere in AX J1745.6–2901, adapted from [13]. The red curve represents thermal equilibrium for gas illuminated by the soft spectrum observed by *XMM-Newton* and *NuSTAR* [246]. The best-fit parameters for the absorber are shown as a red circle, which lies on the stability curve where the slope is positive. The blue curve, on the other hand, corresponds to thermal equilibrium for gas illuminated by the hard state. Adjusting for the luminosity of the hard state, the same disk atmosphere would be thermodynamically unstable (see the point labeled "Hard," where the slope is negative). Gas under these conditions would rapidly evolve to one of the points labeled "H1" or "H2"

that hot winds should be detectable in hard states at a high inclination ($i > 75°$). The source of this discrepancy is not clear, since [199] placed tight upper limits on absorption line equivalent widths in hard states of H1743–322.

Nevertheless, thermodynamic instabilities also offer an appealing way to account for the fact that winds are rarely detected as X-ray absorption lines during hard states, but that they *are* detectable in the optical and infrared during these states. For example, [208] detected P-Cygni profiles of optical emission lines coincident with radio jet emission during the 2015 outburst of V404 Cygni, and [211] found similar results during likely hard states of V4641 Sgr. Reference [209] observed cold optical winds throughout the rising and decaying hard states in MAXI J1820–070 (but not in the soft state) and suggested that these outflows are a common feature of accreting black hole systems. Interestingly, [274] found evidence for a relatively steady (i.e., state-independent) wind in NIR spectra of the same outburst. But finally, [210] have reported the simultaneous detection of X-ray and optical signatures of accretion disk winds in V404 Cyg; they argue that the data are consistent with a multiphase outflow, i.e., hot and cold gas coexisting in a single wind [see also, 320]. Therefore disappearance of X-ray absorption lines need not indicate the disappearance of the wind itself.

Time Evolution: In other words, the fact that accretion disk winds may be very difficult to observe in the X-ray during spectrally hard states does not mean that they are not present during these states. Indeed, [293] argued that massive accretion disk winds shape the lightcurves of black hole outbursts, including during the hard state. As noted above, optical and IR spectra of black hole disk winds in hard states help to reveal how the high-energy signatures of these outflows may be shaped by thermodynamic instability.

But there are also several definitive instances of state-dependent evolution of hot accretion disk winds that cannot be explained by ionization or thermodynamics. For example, two observations of an accretion disk wind in the 2005 outburst of GRO J1655–40 were inconsistent with a static wind [222]. In addition, [94] showed that overionization and thermal instabilities could not account for the disappearance of the accretion disk wind in the 2012–2013 outburst of 4U 1630–47 [see, 216, 218].

While these examples do not pertain directly to the apparent absence of winds in most spectrally hard states, they provide compelling evidence that the physical properties of accretion disk winds evolve significantly over the course of black hole outbursts. Theoretically, this does not come as a surprise: depending on the driving or launch mechanism of the wind, the mass loss rate is expected to change with accretion state (e.g., the geometry and spectrum of the accretion flow). For instance, [249] noted that if the wind is driven by irradiation or Compton heating of the outer accretion disk (i.e., "thermal driving"), any changes in the illumination of the outer disk—such as might be caused by an increase in the disk scale height during hard states—could lead to less effective wind driving. As discussed above and below, changes in the broadband X-ray spectrum are also expected to affect thermal wind launching, introducing a time-dependent $\dot{M}_w$ [see also, 64]. But magnetohydrodynamic (MHD) winds may also evolve during outbursts: [88] inferred changes in the wind density profile in different accretion states, though the theoretical basis for those changes is not understood.

In summary, there is strong evidence that photoionization, thermodynamics, and true physical time dependence all shape the appearance of accretion disk winds in X-ray binaries. Now we turn to the driving mechanisms, which (in addition to being important in their own right), will shed some additional light on the physical time dependence of winds.

11.2.1.3 Wind Driving Mechanisms

From the preceding discussion, it should be clear that efforts to interpret the results of [220, 249] and understand the mass loss rate in winds, we need to understand the physical processes that launch these outflows from the accretion disk in the first place. There has been a significant effort in recent years to infer these mechanisms based on observations [e.g., 87, 88, 130, 134, 196, 216, 223, 225, 282, 295]. Following decades of literature, we shall consider three main driving mechanisms: (1) radiation pressure, (2) thermal driving, and (3) MHD processes. It is common in observational papers to consider these mechanisms separately, though (a) hybrid winds are allowed and often distinct from the main categories [64, 111, 252, 253, 281, 319], and (b) all three mechanisms may be operating at some level in the disk at any given moment. Here we briefly discuss each mechanism in turn and discuss the overlap between them as it arises. Efforts to determine the (dominant) driving mechanisms of winds are typically a process of elimination, so where possible, we focus on observational diagnostics of and constraints on the driving mechanism.

Radiation Pressure: In a radiation pressure driven wind, photons scatter off atoms or electrons in the vicinity of the compact object, accelerating the gas until it reaches some terminal speed in excess of the local escape velocity. Here the process that accelerates the wind is the transfer of momentum from the radiation field to the gas [265]. This mechanism seems intuitively relevant for X-ray binaries, which often reach high luminosities and therefore might be expected to launch radiatively-driven winds efficiently.

The reality is rather more complicated. Briefly, radiation driving requires a source of opacity for scattering, and is therefore especially sensitive to the ionization state of the gas [316]. For a fully ionized gas, that opacity is due to electron scattering. At lower temperatures, there is significantly more opacity in atomic lines, so that resonant scattering or "line driving" [e.g., 213] is much more effective at launching winds from accretion disks. This is especially the case for the UV, where the line opacity is particularly high. To be quantitative, the line force can exceed that due to electron scattering by $\lesssim 2000\times$ [1, 27, 96].

In detailed radiation hydrodynamic simulations, [255] found that radiation pressure could only effectively drive winds in LMXBs when the central luminosity is within a factor of $\sim$2 of the Eddington limit,[7] but that such winds are best described as thermally driven (see below) with assistance from electron scattering [see also, 111]. Line driving was not effective at ionization parameters $\log(\xi) \gtrsim 3$ due to the reduction in UV opacity. Thus when we encounter a wind absorption line system with a measured high ionization parameter in X-ray binaries, we typically disregard radiation pressure as a viable driving mechanism for that particular outflow.

Practically speaking, this criterion rules out radiation pressure as the source of winds in most black hole and neutron star X-ray binaries, with a few notable exceptions. For example, [308] argued that the wind in the neutron star GX 13+1 (see below) could be explained as a radiatively-driven wind launched at high Eddington ratio, though [298] recently modeled *Chandra* HETGS 1st- and 3rd-order spectra of this source as a radiative-thermal wind. Reference [254] argued that line driving is the most plausible origin for disk winds in CVs [see also 239, and references therein].

Thermal Driving: If the accretion disk is sufficiently heated by X-rays from the inner accretion flow, the average thermal velocity of the irradiated disk surface can exceed the local escape velocity, especially in the outer disk where matter is less

[7] Reference [308] argued for the effectiveness of radiation-driven winds at similar luminosities on the idea that for the purposes of gas kinetics, radiation pressure reduces the effective mass of the compact object, and that the gas is effectively unbound at high Eddington ratios.

tightly bound to the central source [12]. Here we briefly follow [259] to cover the important points. Matter in the disk can be heated to the Compton temperature

$$T_C = \frac{1}{4k_B} \frac{\int_0^\infty E L_E dE}{\int_0^\infty L_E dE}, \tag{11.8}$$

where k_B is the Boltzmann constant; T_C is often in the range 10^6-10^8 K for typical XRB spectra. The characteristic size scale for a thermally-driven or *Compton heated* wind is the Compton radius R_{IC}, where the local escape velocity equals the thermal velocity at the Compton temperature:

$$R_{IC} = \frac{9.8 \times 10^{17}}{T_C} \frac{M}{M_\odot} \text{ cm}, \tag{11.9}$$

where M is the mass of the central object and T_C is given in K. The Compton radius is often comparable to the size of the accretion disk, but hydrodynamic simulations [323] have indicated that thermal winds can be launched from radii $R \gtrsim 0.1 R_{IC}$; the inclusion of radiation pressure may enable thermal-radiative winds to be launched from $\sim$ anywhere in the disk at high Eddington ratio [64]. Finally, in order for the wind to be heated fast enough to overcome gravity, the luminosity must exceed a critical threshold

$$L_{\rm cr} = \frac{288 L_{\rm Edd}}{\sqrt{T_C}}, \tag{11.10}$$

where $L_{\rm Edd}$ is the Eddington limit and again T_C is given in K. Because of the appearance of the luminosity and spectral shape in these equations, thermally-driven winds are naturally dependent on the accretion state, with a complex tradeoff between the increase in luminosity and decrease in spectral hardness (and T_C) as sources head from hard states to soft states [109].

With the exception of observations at a high Eddington ratio, the primary diagnostic of thermal driving is whether or not the X-ray absorber can be located outside the Compton radius (or the equivalent $0.1-0.2 R_{IC}$). If so, then thermal driving could be the dominant mechanism responsible for the wind. There are several ways to estimate the location of a wind. References [134, 196, 202] used density-sensitive line features to measure the density of a wind in GRO J1655–40; if L, n, and ξ are known, one can solve Eq. 11.4 to determine R, though this formulation implicitly assumes a point source emitter [223]. Reference [221] reported a similarly complex absorber in GRS 1915+105; [309] found a large photoionization radius ($\gtrsim 10^5 r_g$), consistent with a radiative/thermal wind. Reference [139] found strong emission lines and P-Cygni profiles at high luminosity in the 2015 outburst of V404 Cyg; ionization analysis implied a large radius and a wind driven by radiation or irradiation. Some variability arguments [e.g., 219, 224] have enabled estimates of the location of X-ray absorbing gas and led to the inference of thermally driven winds. Reference [220] assumed that the blueshift of absorption lines in GRS 1915+105 represented the local escape velocity, which placed the absorption lines in the outer accretion

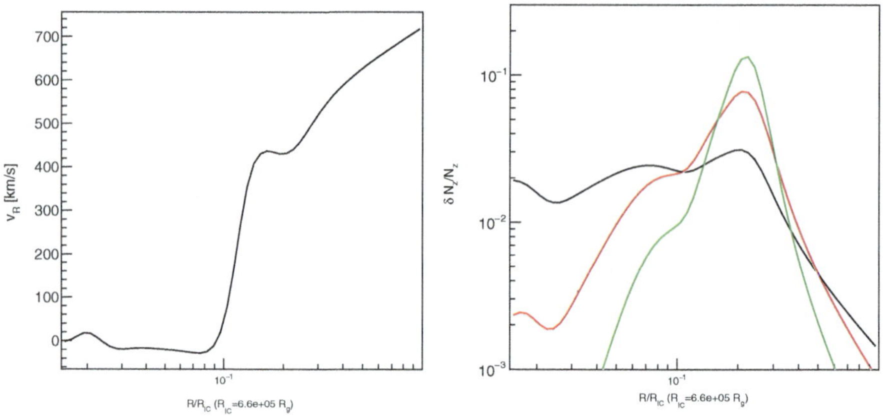

Fig. 11.5 Calculations from [297] showing the radial velocity profile (left) and fractional ionic column densities (i.e., scaled to the total; right) for a thermal-radiative wind in H1743–322. These profiles are computed with Monte Carlo radiative transfer using a radiation hydrodynamic simulation as input. In the right panel, red and green indicate Fe XXVI and Fe XXV, respectively, while the black curve shows the total column density. The blueshift and ionization are strong functions of position, so comparing these models to data permits estimates of the location and turbulent velocity in the wind (see their Fig. 11.9)

disk. Similarly, the velocity profile of observed absorption lines may be compared to predictions of models based on different launch mechanisms [297, 298, 308, Fig. 11.5;].

Studies of thermally-driven winds have benefited enormously from recent advances in simulations, which have facilitated direct comparisons between models and observations [109–112, 166, 281, 297, 298]. As a result, we also have a deeper understanding of the connection between atomic processes and the bulk behavior of winds. For example, [112] pointed out that the thermodynamic instability discussed in Sect. 11.2.1.2 is where winds are accelerated most efficiently: when plasma becomes thermally unstable, its temperature rapidly approaches the Compton temperature [see also, 12]. The sudden increase in the gas temperature provides a significant boost to thermal wind launching. Reference [67] found that winds may undergo several stages of acceleration if the irradiating flux is sufficient to drive the thermal equilibrium curve through multiple steep or unstable branches. Reference [66] demonstrated that winds themselves may represent a significant source of disk irradiation if they scatter photons down onto the disk [see also, 136]. In short, cutting edge theoretical and observational studies of thermal winds continue to reveal the rich relationships between accretion and ejection processes in X-ray binaries.

MHD Winds: The same can be said for studies of winds driven by magnetohydrodynamic processes. The virtue of this driving mechanism was put concisely by [253]: "one of the reasons for favoring magnetic fields as an explanation for mass outflows from accretion disks is the fact that magnetic fields are very likely crucial for the existence of all accretion disks." One recurring theme in studies of disk winds

11.2.2 Neutron Stars

Having reviewed the basics of disk atmospheres and disk winds in the context of black hole XRBs we now turn to neutron star systems. Their accretion properties are overall very similar, which is not surprising because their gravitational potentials are not too different. Nevertheless, neutron stars are fundamentally different objects than black holes, given that they have a solid surface (versus an event horizon) and a magnetic field anchored to it. Both these properties can impact the surrounding accretion flow, and outflows, in various ways.

Observationally, thermal radiation from the hot neutron star surface, or a boundary/spreading layer where matter from the disk splashes into the star, provides an additional emission component in X-ray spectra. Physically, this additional reservoir of soft X-ray photons can Compton-cool the corona [24], illuminate the accretion disk [6] and perhaps also affect the thermal stability criteria of the disk (Sect. 11.2.1.2).

The presence of a surface magnetic field can also have a significant impact. The inner disk may be truncated and plasma channeled onto the magnetic poles of the neutron star so that it manifests itself as an X-ray pulsar [14, 322]. The magnitude of this effect depends both on the pressure exerted by the accretion flow (which scales with the mass-accretion rate), and the strength of the surface magnetic field. Neutron stars in LMXBs typically have weak magnetic fields ($\lesssim 10^9$ G), apart from a few exceptions where the magnetic field is stronger [$\simeq 10^{10-12}$ G; 5]. X-ray pulsars in HMXBs, on the other hand, have strong magnetic fields ($\simeq 10^{12-13}$ G). The stellar magnetic field may also interact with that of the accretion disk and thereby impact the production of both jets and disk winds, where again the strength of the surface magnetic field determines the magnitude of the effect [236, 269, 313].

Other differences between neutron stars and black holes that can affect their accretion and ejection properties are that neutron stars, by definition, are less massive than black holes [231]. In the context of disk winds, for instance, mass is one of the parameters that determine where in the disk a thermal wind can be launched (cf. Eq. 11.9). Moreover, neutron stars appear to spin (much) slower than black holes in XRBs.[10] Finally, likely due to observational biases, neutron star LMXBs are clustered towards smaller orbital periods than their black hole analogues [5]. In reviewing our current observational knowledge of disks and winds in neutron star XRBs, we will draw parallels with black hole systems to gauge where their different properties come into play. Table 11.1 gives an overview of the properties of the disk atmospheres and winds observed in neutron star (and black hole) XRBs.

[10] The fastest rotating neutron star known in a XRB has a spin frequency of 599 Hz [55], which translates into a dimensionless spin parameter of $a \simeq 0.3$ [50].

11.2.2.1 Observations of Disk Atmospheres and Winds in Neutron Star Systems

For nearly two dozen neutron star XRBs ($\simeq$10% of the currently known population), ionized absorbers have been detected in X-ray spectra (Table 11.1). Many of these systems are known to be viewed at high inclination, as evidenced by the detection of eclipses and/or dips (resulting from the central X-ray source being obscured by either the companion star or the accretion stream, respectively). For several neutron stars, the X-ray aborption lines are blue-shifted, showing that plasma is outflowing with velocities in the range of $\simeq(1-30) \times 10^3$ km s^{-1} ($\simeq$0.004–0.09c; Table 11.1). Similar to black hole XRBs, there are some neutron star systems that show wind signatures at other wavelengths but not in X-rays. Examples are the very bright source Sco X-1 [wind feature in the nIR; 10], and X2127+119 in the globular cluster source M15 [wind feature in the UV; 120].

Although disk winds have so far been detected in only a modest fraction ($\simeq$10%) of neutron star LMXBs, it is likely that many disk winds go undetected, e.g., due to inclination or ionization effects (Sect. 11.2.1.2). Indeed, there is indirect evidence of (highly) non-conservative mass transfer in several neutron star LMXBs, as inferred from the accelerated orbital decay of X-ray pulsars [55, 171] and modeling the disk spectral energy distribution in other systems [107, 248]. In black hole systems there is also indirect evidence for non-conservative mass transfer, as inferred from transient decay light curves [293], or modeling the outburst properties [334]. Disk winds are an attractive explanation for the severe mass loss that these systems must be experiencing [66].

Similar to the phenomenology seen in black hole systems, the observability of the ionized absorbers in neutron star LMXBs shows a dependence on accretion state. This is, for instance, clearly demonstrated by studies of three high-inclination eclipsing (transient) neutron star low-mass ray binaries across different states: MXB 1659–28 [245], AX J1745.6–2901 [see Fig. 11.8; 246], EXO 0748–676 [250]. In all cases, the absorber is clearly seen during soft states, but not during hard states. For the former, the lines are suggested to be at rest (velocities $\lesssim$ 200 km s^{-1}), hence indicating the absorber is a static ionised atmosphere rather than an outflowing wind. Indeed, for this system the location of the absorbing plasma was inferred to be (well) within the Compton radius, hence eliminating the possibility of a thermal wind [245]. For the other two sources no velocity shifts were detected at CCD energy resolution, which sets an upper limit on the velocity of the absorbing plasma of $\lesssim 10^3$ km s^{-1}. This does not rule out a (slow) wind.

Whereas the above studies point towards a state dependence for detecting the ionized absorber in X-rays, there are a few (tentative) X-ray wind signatures found for neutron stars during hard states. This includes two LMXB pulsars, IGR J17062–6143 [53, 312] and IGR J17591–2342 [228]. The former accretes at a very low Eddington ratio ($\simeq$0.1%) and has a very compact orbit (0.6 h), hence its disk wind may be driven by magnetic processes (see Sect. 11.2.2.2). The latter has a wider orbit (8.8 h) and accreted at only $\simeq$1% of the Eddington rate during its outburst [154].

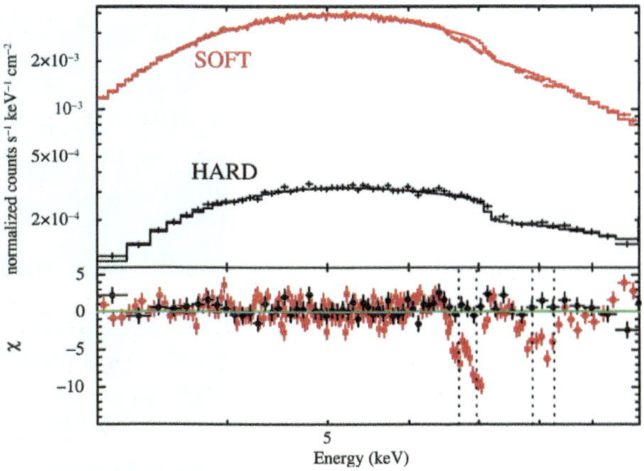

Fig. 11.8 Comparison between the XMM-Newton (CCD) spectra of the neutron star LMXB AX J1745.6–2901 taken in hard (black) and soft (red) states. This clear state-dependence of the observability of the ionized absorber is similar to the behavior seen in black holes. Figure from [246]

During this time it was in a hard X-ray spectral state [154, 169] and it displayed (unusually luminous) radio jet emission [271].

11.2.2.2 Wind Driving Mechanisms for Neutron Star Systems

Accreting neutron stars can launch disk winds via the same physical mechanisms as discussed for black holes (i.e., radiative, thermal or magnetic processes). In neutron star systems there are, however, additional means to launch disk winds. Firstly, winds may be driven by the interaction of the disk and stellar magnetic field [269]. Secondly, many neutron stars display thermonuclear bursts, short but intense flashes of Eddington-limited radiation [92] that interact with the accretion flow and can potentially also launch a temporary disk wind [49].

Disk winds from disk-magnetosphere interaction. Early work on the interaction of a magnetized star with α-type disks revealed that outflows may be launched at the disk-magnetosphere boundary [102, 162, 269, 310]. In these non-relativistic simulations, conically-shaped winds flow out of the inner disk because in-falling plasma compresses the neutron star magnetosphere, causing the field lines to inflate due to the differential rotation between the disk and the star [269, see Fig. 11.9]. Such a conical thin wind with a half-opening angle of $\theta = 30 - 40°$ and velocities up to $v = 0.1c$ [269] should be detectable with high-resolution spectroscopy in low-inclination systems that harbor a magnetic neutron star.

More recently, the first suites of GRMHD simulations of accretion onto magnetic neutron stars have been performed. The first calculations for neutron stars with dipolar magnetic fields [236] were further extended to include more complex magnetic fields [46], motivated by the recent evidence of multi-polar magnetic fields from pulse-profile modeling [e.g., 268]. These GRMHD simulations have not confirmed the

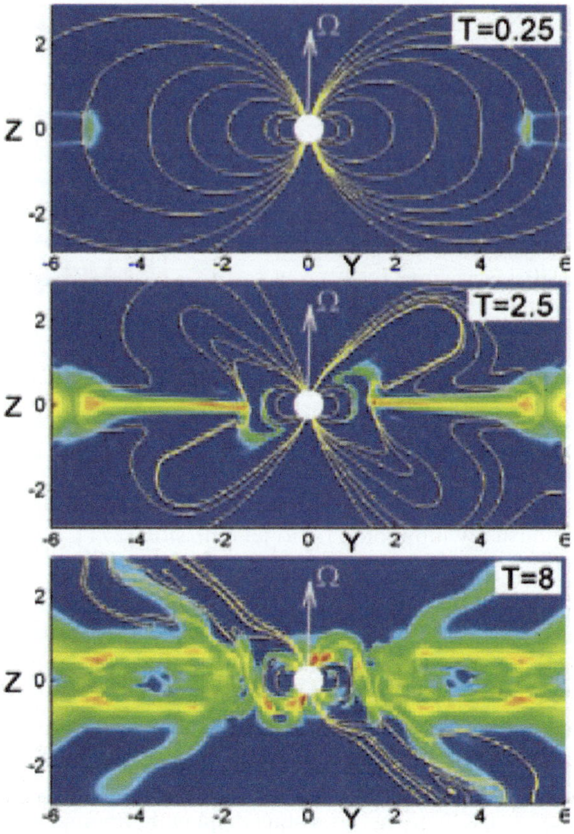

Fig. 11.9 Non-relativistic 3D MHD simulations that show the formation of conical winds from accretion onto a magnetized star [269]. Shown are the distribution of matter flux (colors) and sample magnetic field lines at different times (in the YZ plane). The direction of the stellar magnetic moment and angular velocity of the rotation is indicated by arrows

presence of the conical wind patterns seen by earlier non-relativistic calculations [269, 270], but instead find that collimated jet-type outflows are launched.

Observationally, many of the disk wind detections in neutron star systems are consistent with thermal driving [57]. There appear to be, however, interesting exceptions. For instance, there are a few (potential) disk wind detections in neutron star LMXBs that have very short orbital periods that should not allow them to drive thermal winds. Prime examples are 4U1820–30 [40] and IGR J17062–6143 [53, 312], with orbital periods of 11.4 and 37.4 min, respectively. As their accretion luminosity is well below Eddington, these systems may be, by elimination, good candidates of magnetically-driven winds. In this respect, it is interesting to note that IGR J17062–6143 is an LMXB pulsar [287] and its accretion disk appears to be truncated by its magnetic field [53]. Several other pulsars with longer orbital periods have recently been reported to show disk winds. This includes the weakly magnetic (LMXB) pulsar IGR J17591–2342 [228] and the strongly magnetic pulsars GRO J1744–28 [51],

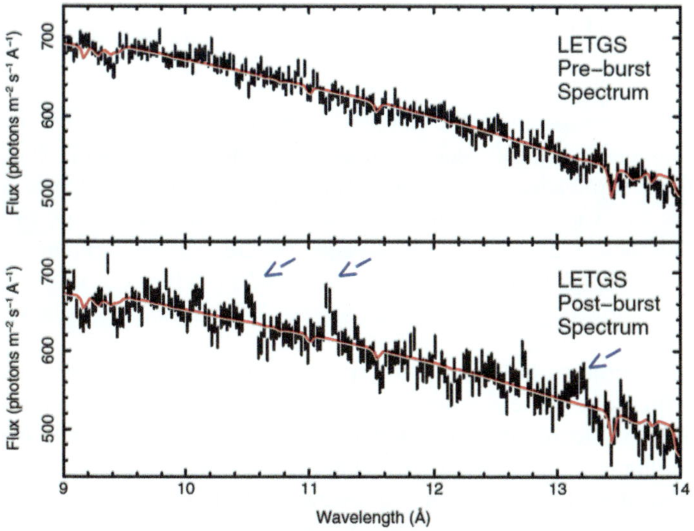

Fig. 11.10 Temporary disk wind detected in the short-period neutron star system SAX J1808.4–3658 [242]. The top and bottom panels show the gratings data recorded before and after an energetic thermonuclear burst, respectively. The emergence of narrow, blue-shifted emission lines of Ne X and Fe XXIV suggest that the intense radiation of the burst may have induced a disk wind [49]

Swift J0243.6–6124 [314], Her X-1 [146]. It is interesting to note that the blueshifts of several of these pulsars (IGR J17062–6143, Swift J0243.6–6124, GRO J1744–28) stand out as high compared to those found in other neutron star and black hole XRBs where thermal driving might be the dominant process (see Table 11.1). This is another reason to suspect that magnetic processes might be involved in driving the disk winds in some neutron star XRBs

Burst-driven disk winds. Another interesting observational development is that there are several neutron star LMXBs for which there is evidence of temporary winds related to bursts [see Fig. 11.10 52, 242, 286]. Their connection to bursts suggests that these disk winds are purely radiatively driven, or perhaps reveal the impact that radiation can have on an existing (magnetic or thermally driven) wind (see Sect. 11.2.1.3). These examples all concerned particularly energetic bursts, which may have promoted wind detection. Perhaps with future instrumentation (see Sect. 11.5), the impact of bursts on disk winds can be studied more routinely. It has not been explored in numerical simulations yet what impact bursts may have on driving or enhancing disk winds.

11.3 Geometry

> ...the X-ray luminosity of the source is determined by the wind from the star, i.e., the highest luminosity states coincide with the strongest wind. It also confirms the identification of the disk or accretion structure close to the compact object as the origin for the fluorescent Kα line and the blackbody, and it suggests that these features and the associated accretion flow structure (i.e., the accretion disk) are anticorrelated with the wind from the companion [133].

In addition to outflows, high-resolution spectroscopy of X-ray binaries has proven to be a powerful probe of the geometry of accreting systems, from the innermost stable circular orbits of accretion disks to their ionized outer atmospheres, from the hot accretion columns of white dwarfs to relativistic jets and the clumpy ionized winds from massive stars. In this section, we review a selection of these results.

11.3.1 Inner Accretion Flows

11.3.1.1 Inner Disks, Reflection, and Spin

X-ray spectral diagnostics of accretion disks in strong gravitational fields—whether thermal emission or fluorescent iron lines—rely on the relativistic broadening of emission from the inner flow, a combination of Doppler boosting and gravitational redshifts [e.g., 182, 266]. This broadening may be used to measure the inner radius of the accretion disk (or the spin of a black hole in cases where this radius corresponds to the innermost stable circular orbit). Because accurate measurements of the underlying continuum are essential, it is tempting to think of reflection and black hole spin measurements as largely the domain of telescopes with low or moderate resolution but significant hard X-ray sensitivity, like *NuSTAR*, *RXTE*, and *INTEGRAL*. But this is inaccurate both historically and in practice: high-resolution spectroscopy can play critical roles in revealing the structure of broad iron lines in X-ray binaries and the spins of black holes.

First, high-resolution spectra are ideal for disentangling broad and narrow lines, so they can lead to more powerful and robust probes of broad line profiles even if reflection is primarily the domain of larger missions with broader energy ranges [190]. Indeed, one of the first detections of a relativistically-broadened iron line in an XRB was secured with the help of the *Chandra* HETGS: [192] reported a composite Fe line profile in Cyg X-1, definitively demonstrating the existence of both a narrow and a broadened component. Interestingly, [300] argue that microcalorimeter observations of such broad lines can constrain the precession of the inner accretion disk. High-resolution spectra have played a similarly important role in uncovering the origin of broad lines in neutron star X-ray binaries. Reference [for example 32] used grating spectra to show that there was no narrow line emission in Ser X-1, effectively requiring a relativistic origin for the line. Though spin is not a factor because neutron star spacetime is approximately Schwarzschild, these broad lines can therefore probe the physical scale of the inner disk [and provide upper limits on the neutron star radius, e.g., 25, 165, and references therein].

Second, to the extent that spin and reflection diagnostics are sensitive to the underlying continuum, they are also sensitive to the inferred level of interstellar and ionized absorption (see Sect. 11.4 and discussion in [226]). For missions like *NuSTAR* and *RXTE* with little or no soft X-ray sensitivity, indirect constraints on interstellar absorption are degenerate with thermal emission from the accretion disk as well as the low energy tail of any power-law component. High-resolution spectra of absorption edges facilitate direct measurements of the column densities of gas and dust in the ISM [e.g., 157]. Reference [227] performed a detailed analysis of observations of Cyg X-1 with *Chandra*, *Suzaku*, and *RXTE* and demonstrated both a composite line at high resolution and the importance of soft X-ray constraints on neutral and ionized absorption for modeling the continuum and broad lines. Again, the presence of coordinated or contemporaneous high-resolution spectra can enhance the robustness of reflection and spin diagnostics.

11.3.1.2 Central Engines and Boundary Layers

Beyond spin and relativistic diagnostics of the inner edge of the accretion disk, high-resolution spectra can provide other insights into the nature of the inner accretion flows around compact objects.

One powerful probe not available in black hole systems is emission from the boundary layer (or the accretion column in the case of magnetically-driven accretion). Reference [212] gives a broad overview of the many subclasses of accreting white dwarf systems, which we will not review in detail here. We do note, however, that expected line properties generally vary significantly depending on the geometry of the emission region, and therefore we should not be overly quick to generalize between different classes of CVs, e.g., from dwarf novae to polars (where the configuration of the accretion flow is very different).

As in black hole and neutron star X-ray binaries, the spectral behavior of CVs can change with accretion rate. In dwarf novae, for example, high-resolution spectra typically show strong emission lines in quiescence. *XMM-Newton* RGS spectra of OY Car, for example, indicated a multi-temperature plasma with components at $\sim$3 keV and $\sim$7 keV [262]. *Chandra* HETGS spectra have revealed similar series of strong lines [e.g., 234, 240, 290]. According to these sources the lines are consistent with an origin in the boundary layer itself, i.e., plasma settling onto the white dwarf from the accretion disk [see 212, and references therein for a more extensive discussion of this point].

Reference [106, 263] discussed the presence of strong fluorescent Fe lines even in quiescent CVs, which likely indicate reflection off the white dwarf surface. Reference [105] performed Monte Carlo simulations of reflection in magnetic CVs to study

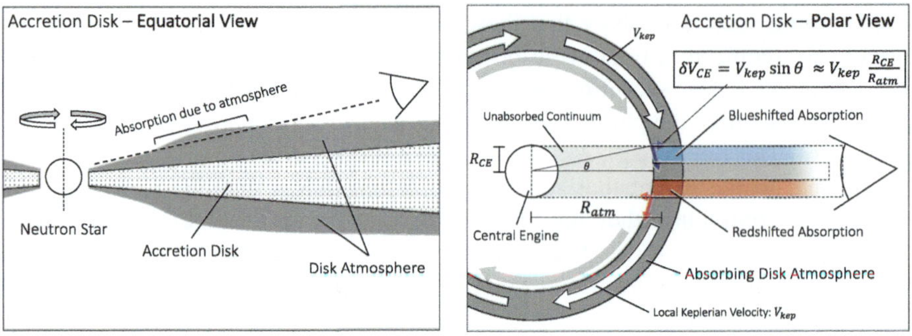

Fig. 11.11 Cartoon illustrating geometric velocity broadening from [301]. A disk atmosphere sufficiently close to the center will exhibit velocity broadening that depends on the physical size of the central engine

the dependence of the fluorescence features on parameters like white dwarf mass, spin, viewing angle, abundances, and the specific accretion rate. But there are also notable changes in the emission components in dwarf novae between outburst and quiescence. Reference [263] found that the Fe XXV line was significantly broadened during outburst in some systems, suggesting outflowing material near the white dwarf surface. Indeed, [179] argued that the *Chandra* LETGS spectrum of SS Cyg in outburst comprised continuum emission from the boundary layer, absorbed and scattered by an outflowing wind.

From the preceding discussion, it is clear that the physical scale of the inner accretion flow—be it a black hole, neutron star, or white dwarf—is of central importance for accretion diagnostics. While relativistic modeling and ionization analysis are common methods for quantifying the size and location of emitting regions, [301] developed a novel high-resolution spectroscopic method for measuring the size scale of the central engine in accreting compact objects using gravitationally-redshifted absorption lines from a rotating disk atmosphere. As shown in Fig. 11.11, the velocity width of these absorption lines will depend in part on the size of the X-ray emitting region. They were able to place 3σ upper limits of $< 90\ GM/c^2$ on the size of the central engine in the short-period neutron star LMXB XTE J1710–053. Though it has only been applied in a small number of systems to date, it seems ideal for microcalorimeters on next-generation X-ray missions like *XRISM* and *Athena*.

11.3.2 Outer Limits

11.3.2.1 Outer Accretion Disks

Even in cases where the relevant gas is not close enough to the central engine to determine its size scale, emission and absorption lines can provide detailed information about the structure and behavior of accretion disks. One well-established example of outer accretion flow diagnostics comes from dipping LMXBs, where early *XMM-Newton* and *Chandra* observations revealed the existence of ionized absorption and emission lines in a number of high-inclination systems [e.g., 17, 41, 42, 123, 237, 283, 284]. These observations implied the existence of hot and warm gas extending well above and below the accretion disk midplane at relatively large radii: the extended accretion disk corona (ADC) [e.g., 33, 34, 118, 278]. It is now understood [see 315, and references therein] that dipping phenomena are caused by transient obscuration of the central X-ray source by structures in the outer accretion disk, such as the disk rim, accretion stream, or the bulge at the disk-stream impact point. Reference [8] used *Chandra*/*XMM-Newton* grating spectra of Cyg X-2 to argue that the dips at orbital phase 0.35 could be attributed to structures opposite the accretion stream impact point. More recently, [257] used X-ray eclipses to map hot gas above the disk in EXO 0748–676. They found two separate ionization components; they interpreted the higher ionization emitter as the extended disk atmosphere, and the cooler gas as clumpy dense gas related to the accretion stream impact itself. Furthermore, to the extent that the ADC or other disk atmosphere are azimuthally asymmetric, any associated emission/absorption features can be expected to vary with orbital phase, a fact used by [326] to map the ionization structure and viewing geometry of the "Big Dipper" 4U 1624–490.

Similar techniques can be applied to similar benefit to study the structure of disks and their environments directly. For example, high-resolution X-ray spectra have illuminated the behavior of superorbitally-precessing accretion disks in both low-mass and high-mass X-ray binaries. Reference [122] calculated the expected recombination line emission from an accretion disk irradiated by a central X-ray source. Based on those calculations, [122] argued that photoionized emission lines seen in *XMM-Newton* RGS observations of Her X-1 likely originated in the disk atmosphere or the illuminated face of the secondary star. Drawing on 170 ks of *Chandra* grating spectra, [121] inferred cool, dense material distributed throughout the disk in Her X-1. Reference [146] divided the emission lines in Her X-1 into three groups, attributed to (1) the outer accretion disk, (2) the disk/magnetosphere boundary, and (3) X-ray reflection from the accretion curtain. This multi-zone structure echoes similar results from [220] in LMC X-4, where the spectral variability properties of different line groups seen by the *Chandra* and *XMM-Newton* gratings could be traced to the inner disk, illuminated outer disk, or the photoionized stellar wind from the massive donor.

There has also been significant interest in modeling and interpreting narrow Fe Kα lines. Reflection from distant cold gas, for example, is sensitive to the geometry, ionization, dynamics, and abundances in the accretion flow [97].[11] Reference [318] reported a sensitive measurement of the iron line in GX 301–2, including a detection of the "Compton shoulder," a faint feature on the red wing of the Fe Kα line due to backscattering of 6.4 keV photons in a Comptonizing medium. In cases where significant scattering is expected, the Compton shoulder can be used to measure the

[11] With upcoming microcalorimeter missions, it may even be possible to use Fe Kα lines to measure the mass functions of X-ray binaries [47].

column density, electron temperature, orientation, abundances, and Compton scattering optical depth [e.g., 177]. Reference [299] argued that the Fe Kα line equivalent widths and column densities are consistent with a spherical distribution of reprocessing material around the compact object, though [304] concluded the opposite for a smaller sample of X-ray binaries. Reference [299] also reported a decrease of iron line equivalent width with luminosity in X-ray binaries (the so-called "X-ray Baldwin effect") that they interpreted as the result of photoionization. Finally, they noted that these lines are ubiquitous in HMXBs but rare in LMXBs, which suggests that illuminated stellar companions and their winds may play a significant role in producing Kα complexes in X-ray binaries. Follow-up work on HMXBs by [100] used *XMM-Newton* to confirm both the X-ray Baldwin effect and the contribution of stellar winds.

11.3.2.2 Stellar Winds

It is clear that stellar winds are important for studies of accretion geometry via fluorescent and recombination lines, but there is also an enormous body of work devoted to the study of stellar winds in their own right. For the purposes of this chapter, we focus our attention on massive stars in XRB systems (particularly Vela X-1 and Cyg X-1, which have received the most attention in recent decades). By way of demonstrating the impact of stellar wind physics for XRB studies, [99] noted that the velocity of winds from supergiant donors can have a significant impact on the HMXB class. In other words, stellar winds play an important role in determining the overall behavior of X-ray binaries! Stellar wind physics also has implications for binary evolution scenarios: the difficulty of producing a system like Cyg X-1 (with a rapidly-spinning $21M_{\odot}$ black hole in close orbit with a $41M_{\odot}$ O star with enhanced He and Fe abundances) requires significant adjustments to prescriptions for wind mass loss rates from the black hole progenitor [215, and references therein].

In Vela X-1, the presence of ionized circumstellar gas was already apparent well before the era of high-resolution X-ray spectroscopy. For example, in 1986, [214] used a series of *Tenma* spectra to infer a moderately ionized absorber responsible for an iron edge and emission line. Around the same time, it was becoming clear how complex the structure of such winds would be in a binary system, with accretion wakes and photoionization wakes, as well as a possible accretion stream as sources of large-scale inhomogeneity and clumps on smaller scales [16] [see also, 69, 71]. Winds of massive stars were expected to be clumpy due to instabilities associated with line driving [163, 164], but advances in computation made it possible to simulate the development of shocks in these winds [e.g., 230] and to show how they fragment into clumps [e.g., 229, 288, and references therein]. Reference [272] argued that the wind in Vela X-1 had to be inhomogeneous in order to resolve discrepancies between the mass loss rates inferred from optical and X-ray spectra, while [276] drew a similar conclusion based on fluorescent line fluxes. Clumps have also been invoked to explain variability on timescales of hours [90], absorption dips [103], and associated line variability [206]. In turn, variability measurements can be used to infer the statistical properties of clumps [70].

High-resolution X-ray spectral studies of Cyg X-1 have been similarly informative. A number of early *Chandra* observations [e.g., 174, 192, 277] showed strong absorption lines at different orbital phases, while modeling by [81] indicated that the absorption lines had an extended red wing that was more pronounced in lines from more highly-ionized ions. Reference [206] found that the line profile actually varies with orbital phase, appearing as symmetric absorption lines at superior conjunction but exhibiting full P-Cygni profiles at inferior conjunction. Reference [205] used orbital phase comparisons to deduce that the X-ray absorbing portion of the wind must be gravitationally focused: tidal and centrifugal effects produce a distorted, asymmetric wind, as shown by [72, 84] and inferred from optical spectra for Cyg X-1 by [98]. Reference [103] described a *Chandra* HETGS spectrum with a model including two wind absorption components, which they associated with spherical and focused regions of the wind. Reference [133] similarly divided the stellar wind in Cyg X-3 into a "nebular" component and a "wind" component [see also, 233, 289, and references therein]. In Cyg X-1, short-lived dips are commonly observed when the line of sight intersects the densest region of the wind, such that flux-resolved spectra can reveal the ionization structure of clumps in the wind [114]. Careful measurements of line velocities (based on reference energies from an electron beam ion trap) indicate that the clumps are likely stratified, with denser gas shielding the center and back of the clumps from ionizing radiation.

In short, as is often the case, the presence of a compact object provides ample opportunities for studying the structure and behavior of stellar winds, while also substantially influencing them. A detailed review of the stellar winds of isolated stars is outside the scope of this chapter, but for a discussion of high-resolution X-ray spectroscopy and the solar wind, we refer the interested reader to Chapter 12 of this volume, by L. Gu.

11.3.2.3 Jets

Finally, we turn briefly to spectroscopic constraints on relativistic jets from X-ray binaries. There are several examples of X-ray jets in X-ray binaries in the literature [e.g., 36, 74, 285], but in these cases the X-ray emission arises when jet ejecta interact with the local environment. The best-known example of a spectroscopically-identified jet is SS 433. which exhibits Doppler-shifted lines from its precessing jet. Optical line shifts were attributed to the jet in 1979 [2, 75, 189], and Doppler-shifted X-ray emission lines had been associated with the jet within a decade [321]; [for a review of the X-ray literature, see, 172, and references therein]. Currently, the prevailing interpretation is that the X-ray emission lines arise in hot thermal plasma in the jet at distances of $\sim 10^{12-13}$ cm from the compact object [e.g., 22, 148, 161, 172, 185]. The advent of high-resolution X-ray spectroscopy has produced a number of insights into the X-ray jet from this important microquasar, including a $\sim 5\times$ smaller opening angle in the X-ray than in the optical [173] and a large overabundance of

Ni [suggesting the donor star may have been enriched by the supernova that created the compact object; 172].

It is clear that the advance to microcalorimeters will further advance our understanding of the dynamics and structure of jets. But there are also several results from lower resolution spectra that suggest a similar cause for optimism. For example, [188] found hot iron line emission in *Chandra* CCD spectra of the jet at distances of $>10^{17}$ cm from the compact object [long after adiabatic cooling should have rendered the gas undetectable in X-rays; 173]. This result implies that there is continued heating of the atomic component of the jet long past the initial acceleration region. More recently, [187] identified a time lag in *NuSTAR* observations SS 433 that—due to its narrow energy range–could only be attributed to atomic processes. Sensitive observations of this lag with next-generation facilities will reveal its origin in more detail.

More broadly, diagnostics of baryonic jets represent an open question for high-resolution X-ray spectra. In particular, it is notable that there are so **few** detections of atomic transitions clearly associated with outflowing jets. Reference [59] reported detecting relativistically Doppler-shifted emission lines in 4U 1630–47 using *XMM-Newton* CCD spectra, though *Chandra* grating observations of 4U 1630–47 in different accretion states did not show similar emission lines [218]. If confirmed, this would have represented only the second clear case of baryons in a jet from an XRB out of a population of dozens. Next-generation spectra may resolve the question of baryonic jets, but it remains to be seen whether they do so by detecting faint relativistic lines or by revealing baryonic jets to be intrinsically rare.

11.4 ISM

In this section, we briefly lay out how high-resolution X-ray spectroscopy can be leveraged to study the ISM. For a more extensive overview, the reader is referred to the recent review of Costantini & Corrales [39].

Bright X-ray binaries provide an excellent backdrop to study the composition and structure of the ISM at high spectral resolution. Gas along the line of sight to these sources leads to photoelectric edges in X-ray spectra. By comparing the optical depth in these edges to the continuum absorption, it is possible to: (1) measure ISM and stellar abundances [e.g., 126, 157], (2) infer the contributions of interstellar gas vs. absorption local to the binary [e.g., 126, 127], (3) quantify the depletion of interstellar gas into dust grains [e.g., 128, 244, see below], and (4) map the spatial distribution of cold, warm, and hot interstellar gas [e.g., 93, 232, 244].

In addition to photoelectric edges, the X-ray spectra of X-ray binaries are imprinted with extinction features from solid dust particles present in the ISM. If X-ray photons interact with electrons inside a dust grain (rather than gas phase atoms), the energy levels can be modified by the effect of neighboring atoms, resulting in an interference pattern referred to as "X-ray absorption fine structure" (XAFS). The details of such features depend on the energy of electrons and the complexity of the compound, hence allowing us to study the physical properties of interstellar dust such as its chemical composition, the size of the dust grains [38] and their crystallinity [331, 332]. Bright X-ray binaries provide sufficiently high-quality data to detect these extinction features and can reveal potential differences in the dust properties of the ISM in different Galactic environments [156, 158]. The chemical composition of interstellar dust appears to be dominated by carbon and silicates, hence the most prominent extinction features are of Si, Mg, O, and Fe (which are the main constituents of silicates). These features have been studied in detail with current X-ray missions [40, 157, 243, 311, 331, 332]. Future missions like *Athena* with higher sensitivity may be able to study similar structures and compositions using edges of less abundant species.

Another aspect of interstellar dust that is relevant for high-resolution X-ray spectroscopy is dust scattering halos. These halos consist of X-rays from the binary that are scattered into our line of sight by intervening dust screens, and have been observed around a number of XRBs on a variety of angular scales [3, 125, 131, 155, 159, 183, 280, 294, 325]. In addition to practical considerations [e.g., the size of the dust scattering halo compared to the spatial resolution of the instrument, 226, 227], through XAFS the structure and distribution of dust grains affect both the intensity profile of dust scattering halos and the shape and depth of photoelectric edges [39, and references therein]. Future missions that combine sensitivity with spatial and spectral resolution will therefore have the best opportunity to probe the properties of interstellar dust.

11.5 Future and Outlook

Finally, we turn to the future: what can we hope to learn from high-resolution spectroscopy of X-ray binaries with next-generation facilities? As of this writing, there is particular interest in microcalorimeters on *XRISM* and *Athena*. The improved sensitivity and spectral resolution of these facilities will be a game changer for all the XRB studies discussed in this chapter. In particular, the ability to compute absorption and emission diagnostics as described in the preceding sections will be greatly enhanced by the reduced time to detect line features, enabling faster constraints on the behavior and geometry of gas in variable systems and more robust detections of weaker lines. Since at the time of writing, the *Athena* mission is in a re-definition stage, the discussion below is focused on *XRISM*.

The launch of *XRISM* [292] is anticipated in August 2023. It carries a soft X-ray spectrometer, Resolve, which offers a constant <7 eV FWHM spectral resolution over the entire 0.3–12 keV bandpass and an effective area of 160 (210) cm^{-2} at 1 (6) keV. The spectral resolution of Resolve is a factor 20–40 higher than the CCD instruments on board *Chandra* and *XMM-Newton*, while it has substantially increased the collecting area and bandpass over the grating instruments on these missions. Whereas previous studies have shown that hot plasmas in X-ray binaries can have very high velocities, some of the gas may have much smaller characteristic

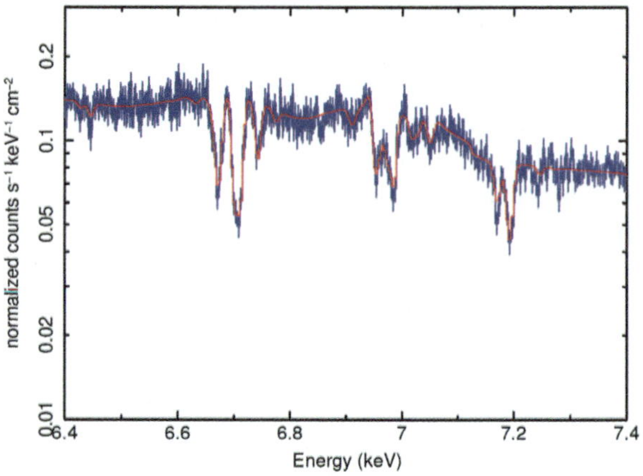

Fig. 11.12 Simulated *XRISM*/Resolve spectrum of the black hole XRB GRS 1915+105 (50 ks) based on the best *Chandra*/HETG model [197] shown in Fig. 11.7. Reproduced with permission from [327]

motions of the order of $\simeq$100 km s^{-1} and hence remain unresolved with current instrumentation.

Disk winds. For X-ray binaries, it remains to be established what the relative importance of thermal and magnetic processes is in driving their disk winds. The increased spectral resolution of new instrumentation will make it possible to separate complexes of lines (Fig. 11.12). This, in turn, will allow us to accurately measure velocities, to separate distinct wind zones, and to measure launching radii. These are all key indicators of the underlying launch mechanism, as discussed in Sect. 11.2.1.3. Also, by virtue of a broad pass band, density-sensitive lines from intermediate charge states become accessible; these enable direct measurements of the density so that the absorption radius can be derived from the photoionization parameter. Wind launch radii can also be measured independently from the velocity broadening of the emission components in lines that have a P-Cygni profile. Furthermore, high-resolution spectrometers may shed new light on whether or not disk winds are intrinsically transient: if the apparent disappearance in the hard state is due to lower column densities and higher ionization, the increased sensitivity and resolution of *XRISM*/Resolve compared to current instrumentation may detect hot X-ray winds in hard states.

In this context, it is worth noting that X-ray polarimetry missions will also contribute to advancing studies of XRB disk winds, albeit less directly. For instance, *IXPE* has already started to put constraints on the magnetic field topology in disks [150, 151], which facilitates further advancement in numerical simulations of disk accretion and any resulting outflows.

Geometry. We may also expect to see large strides in understanding the accretion geometry in X-ray binaries. For instance, the scale height of the outer disk and the extent to which it might be flared, can potentially be revealed by studying narrow emission lines in high-inclination dipping X-ray binaries. For stellar winds, the highly increased velocity resolution enabled by *XRISM* will allow the separation of lines produced close to the compact object from those excited in the larger stellar wind. Moreover, it will allow a precise detection of velocity shifts as a function of the orbital phase. Improved measures of the ionization, density and temperature of the accreting material will enhance detailed comparisons to hydrodynamic simulations. Furthermore, the increased sensitivity of next-generation missions will facilitate studies of wind structure and clumping in many more systems and on short timescales [160].

ISM. High-resolution X-ray spectroscopy studies of the ISM will also take a leap with new spectrometers. For instance, *XRISM* provides access to dense environments of the Galaxy ($N_H \simeq 1-10 \times 10^{22}$ cm^{-2}) near the Galactic Center or molecular clouds, which is currently uncharted territory. *XRISM* will allow to compare the spectral features of interstellar dust (especially for heavier elements such as Ca, S, and Fe) to material absorption measured in the lab [327].

As detailed in the pages above, the strength of high-resolution spectroscopy lies in the power to illuminate the physical conditions in gas and dust and, accordingly, reveal the physical processes that produce them. In X-ray binaries, these processes—accretion and ejection, ionization, and so on—vary from sub-second timescales to years and decades. Building on existing high-resolution X-ray spectra, next-generation facilities will rewrite our understanding of accretion physics by probing shorter timescales with superior constraints on the ionization and dynamics of gas and dust around compact objects.

References

1. D.C. Abbott, The theory of radiatively driven stellar winds. II. The line acceleration. ApJ **259**, 282 (1982). https://doi.org/10.1086/160166
2. G.O. Abell, B. Margon, A kinematic model for SS433. Nature **279**, 701 (1979). https://doi.org/10.1038/279701a0
3. M.D. Audley, F. Nagase, K. Mitsuda, L. Angelini, R.L. Kelley, Dust-scattered haloes of high-mass X-ray binaries observed with ASCA, in *New Century of X-ray Astronomy*, ed. by H. Inoue, H. Kunieda. Astronomical Society of the Pacific Conference Series, vol. 251 (2001), p. 334
4. M.C. Baglio, P. D'Avanzo, S. Campana, P. Goldoni, N. Masetti, T. Muñoz-Darias et al., 1RXS J180408.9-342058: an ultra compact X-ray binary candidate with a transient jet. A&A **587**, A102 (2016). https://doi.org/10.1051/0004-6361/201527147. arXiv:1601.05091
5. A. Bahramian, N. Degenaar, Low-Mass X-ray binaries (2022). https://doi.org/10.48550/arXiv.2206.10053. arXiv:2206.10053
6. D.R. Ballantyne, Reflection spectra from an accretion disc illuminated by a neutron star X-ray burst. MNRAS **351**, 57 (2004). https://doi.org/10.1111/j.1365-2966.2004.07767.x. arXiv: astro-ph/0402603
7. M. Bałucińska-Church, T. Dotani, T. Hirotsu, M.J. Church, Neutral absorber dips in the periodic burster LMXB XB 1323–619 from Suzaku. A&A **500**, 873 (2009). https://doi.org/10.1051/0004-6361/200811215. arXiv: 0905.0618
8. M. Bałucińska-Church, N.S. Schulz, J. Wilms, A. Gibiec, M. Hanke, R.E. Spencer et al., Dipping in CygnusX-2 in a multi-wavelength campaign due to absorption of extended ADC emission. A&A **530**, A102 (2011). https://doi.org/10.1051/0004-6361/201015931. arXiv:1104.1715
9. R. Bandyopadhyay, T. Shahbaz, P.A. Charles, M.H. van Kerkwijk, T. Naylor, Infrared spectroscopy of low-mass X-ray binaries. MNRAS **285**, 718 (1997). https://doi.org/10.1093/mnras/285.4.718

10. R.M. Bandyopadhyay, T. Shahbaz, P.A. Charles, T. Naylor, Infrared spectroscopy of low-mass X-ray binaries - II. MNRAS **306**, 417 (1999). https://doi.org/10.1046/j.1365-8711.1999.02547.x. arXiv: astro-ph/9901327
11. M.A. Bautista, T.R. Kallman, The XSTAR atomic database. ApJS **134**, 139 (2001). https://doi.org/10.1086/320363
12. M.C. Begelman, C.F. McKee, G.A. Shields, Compton heated winds and coronae above accretion disks. I. Dynamics. ApJ **271**, 70 (1983). https://doi.org/10.1086/161178
13. S. Bianchi, G. Ponti, T. Muñoz-Darias, P.-O. Petrucci, Photoionization instability of the Fe K absorbing plasma in the neutron star transient AX J1745.6-2901. MNRAS **472**, 2454 (2017). https://doi.org/10.1093/mnras/stx2173. arXiv: 1709.00860
14. L. Bildsten, D. Chakrabarty, J. Chiu, M.H. Finger, D.T. Koh, R.W. Nelson et al., Observations of accreting pulsars. ApJS **113**, 367 (1997). https://doi.org/10.1086/313060. arXiv: astro-ph/9707125
15. R.D. Blandford, D.G. Payne, Hydromagnetic flows from accretion disks and the production of radio jets. MNRAS **199**, 883 (1982). https://doi.org/10.1093/mnras/199.4.883
16. J.M. Blondin, T.R. Kallman, B.A. Fryxell, R.E. Taam, Hydrodynamic simulations of stellar wind disruption by a compact X-ray source. ApJ **356**, 591 (1990). https://doi.org/10.1086/168865
17. L. Boirin, M. Méndez, M. Díaz Trigo, A.N. Parmar, J.S. Kaastra, A highly-ionized absorber in the X-ray binary 4U 1323-62: a new explanation for the dipping phenomenon. A&A **436**, 195 (2005). https://doi.org/10.1051/0004-6361:20041940. arXiv: astro-ph/0410385
18. L. Boirin, A.N. Parmar, Discovery of narrow X-ray absorption features from the low-mass X-ray binary X 1254–690 with XMM-Newton. A&A **407**, 1079 (2003). https://doi.org/10.1051/0004-6361:20030975. arXiv: astro-ph/0306526
19. L. Boirin, A.N. Parmar, D. Barret, S. Paltani, J.E. Grindlay, Discovery of X-ray absorption features from the dipping low-mass X-ray binary XB 1916–053 with XMM-Newton. A&A **418**, 1061 (2004). https://doi.org/10.1051/0004-6361:20034550. arXiv: astro-ph/0402294
20. W.N. Brandt, N.S. Schulz, The discovery of broad P Cygni X-ray lines from circinus X-1 with the Chandra high-energy transmission grating spectrometer. ApJL **544**, L123 (2000). https://doi.org/10.1086/317313. arXiv: astro-ph/0007406
21. A.C. Brinkman, C.J.T. Gunsing, J.S. Kaastra, R.L.J. van der Meer, R. Mewe, F. Paerels et al., First light measurements of capella with the low-energy transmission grating spectrometer aboard the Chandra X-ray observatory. ApJL **530**, L111 (2000). https://doi.org/10.1086/312504. arXiv: astro-ph/0001034
22. W. Brinkmann, N. Kawai, M. Matsuoka, H.H. Fink, The X-ray emission of SS 433. A&A **241**, 112 (1991)
23. D.J.K. Buisson, D. Altamirano, M. Díaz Trigo, M. Mendez, M. Armas Padilla, N. Castro Segura et al., Soft X-ray emission lines in the X-ray binary Swift J1858.6-0814 observed with XMM-newton reflection grating spectrometer: disc atmosphere or wind? MNRAS **498**, 68 (2020). https://doi.org/10.1093/mnras/staa2258. arXiv: 2007.14407
24. M.J. Burke, M. Gilfanov, R. Sunyaev, A dichotomy between the hard state spectral properties of black hole and neutron star X-ray binaries. MNRAS **466**, 194 (2017). https://doi.org/10.1093/mnras/stw2514. arXiv: 1609.09511
25. E.M. Cackett, J.M. Miller, D.R. Ballantyne, D. Barret, S. Bhattacharyya, M. Boutelier et al., Relativistic lines and reflection from the inner accretion disks around neutron stars. ApJ **720**, 205 (2010). https://doi.org/10.1088/0004-637X/720/1/205. arXiv: 0908.1098
26. C.R. Canizares, J.E. Davis, D. Dewey, K.A. Flanagan, E.B. Galton, D.P. Huenemoerder et al., The chandra high-energy transmission grating: design, fabrication, ground calibration, and 5 years in flight. PASP **117**, 1144 (2005). https://doi.org/10.1086/432898. arXiv: astro-ph/0507035
27. J.I. Castor, D.C. Abbott, R.I. Klein, Radiation-driven winds in Of stars. ApJ **195**, 157 (1975). https://doi.org/10.1086/153315
28. N. Castro Segura, C. Knigge, K.S. Long, D. Altamirano, M. Armas Padilla, C. Bailyn et al., A persistent ultraviolet outflow from an accreting neutron star binary transient. Nature **603**, 52 (2022). https://doi.org/10.1038/s41586-021-04324-2. arXiv: 2203.01372

29. S. Chakravorty, J.C. Lee, J. Neilsen, The effects of thermodynamic stability on wind properties in different low-mass black hole binary states. MNRAS **436**, 560 (2013). https://doi.org/10.1093/mnras/stt1593. arXiv: 1308.4574
30. S. Chakravorty, P.-O. Petrucci, S.R. Datta, J. Ferreira, J. Wilms, J. Jacquemin-Ide et al., Absorption lines from magnetically driven winds in X-ray binaries - II. High resolution observational signatures expected from future X-ray observatories. MNRAS **518**, 1335 (2023). https://doi.org/10.1093/mnras/stac2835. arXiv: 2209.15127
31. S. Chakravorty, P.O. Petrucci, J. Ferreira, G. Henri, R. Belmont, M. Clavel et al., Absorption lines from magnetically driven winds in X-ray binaries. A&A **589**, A119 (2016). https://doi.org/10.1051/0004-6361/201527163. arXiv: 1512.09149
32. C.-Y. Chiang, E.M. Cackett, J.M. Miller, D. Barret, A.C. Fabian, A. D'Aì et al., A test of the nature of the Fe K line in the neutron star low-mass X-ray binary serpens X-1. ApJ **821**, 105 (2016). https://doi.org/10.3847/0004-637X/821/2/105. arXiv: 1509.02969
33. M.J. Church, M. Bałucińska-Church, Measurements of accretion disc corona size in LMXB: consequences for comptonization and LMXB models. MNRAS **348**, 955 (2004). https://doi.org/10.1111/j.1365-2966.2004.07162.x. arXiv: astro-ph/0309212
34. M.J. Church, D. Reed, T. Dotani, M. Bałucińska-Church, A.P. Smale, Discovery of absorption features of the accretion disc corona and systematic acceleration of the X-ray burst rate in XB1323-619. MNRAS **359**, 1336 (2005). https://doi.org/10.1111/j.1365-2966.2005.08728.x. arXiv: astro-ph/0412228
35. J. Contopoulos, R.V.E. Lovelace, Magnetically driven jets and winds: exact solutions. ApJ **429**, 139 (1994). https://doi.org/10.1086/174307
36. S. Corbel, R.P. Fender, A.K. Tzioumis, J.A. Tomsick, J.A. Orosz, J.M. Miller et al., Large-scale, decelerating, relativistic X-ray jets from the Microquasar XTE J1550–564. Science **298**, 196 (2002). https://doi.org/10.1126/science.1075857. arXiv: astro-ph/0210224
37. S. Corbel, M.A. Nowak, R.P. Fender, A.K. Tzioumis, S. Markoff, Radio/X-ray correlation in the low/hard state of GX 339–4. A&A **400**, 1007 (2003). https://doi.org/10.1051/0004-6361:20030090. arXiv: astro-ph/0301436
38. L.R. Corrales, J. García, J. Wilms, F. Baganoff, The dust-scattering component of X-ray extinction: effects on continuum fitting and high-resolution absorption edge structure. MNRAS **458**, 1345 (2016). https://doi.org/10.1093/mnras/stw376. arXiv: 1602.01100
39. E. Costantini, L. Corrales, Interstellar absorption and dust scattering, in *Handbook of X-ray and Gamma-ray Astrophysics*, ed. by C. Bambi, A. Santangelo, (Singapore) (Springer Nature, Singapore, 2022), pp. 1–39. https://doi.org/10.1007/978-981-16-4544-0_93-1
40. E. Costantini, C. Pinto, J.S. Kaastra, J.J.M. in't Zand, M.J. Freyberg, L. Kuiper et al., XMM-Newton observation of 4U 1820-30. Broad band spectrum and the contribution of the cold interstellar medium. A&A **539**, A32 (2012). https://doi.org/10.1051/0004-6361/201117818. arXiv: 1112.4349
41. J. Cottam, S.M. Kahn, A.C. Brinkman, J.W. den Herder, C. Erd, High-resolution spectroscopy of the low-mass X-ray binary ASTROBJ EXO 0748–67 ASTROBJ. A&A **365**, L277 (2001). https://doi.org/10.1051/0004-6361:20000053. arXiv: astro-ph/0010659
42. J. Cottam, M. Sako, S.M. Kahn, F. Paerels, D.A. Liedahl, High-resolution X-ray spectroscopy of the accretion disk corona source 4U 1822-37. ApJL **557**, L101 (2001). https://doi.org/10.1086/323343. arXiv: astro-ph/0107247
43. V.A. Cúneo, T. Muñoz-Darias, J. Sánchez-Sierras, F. Jiménez-Ibarra, M. Armas Padilla, D.A.H. Buckley et al., Discovery of optical outflows and inflows in the black hole candidate GRS 1716-249. MNRAS **498**, 25 (2020). https://doi.org/10.1093/mnras/staa2241. arXiv: 2007.13775
44. A. D'Aí, R. Iaria, T. Di Salvo, G. Lavagetto, N.R. Robba, A complex environment around circinus X-1. ApJ **671**, 2006 (2007). https://doi.org/10.1086/522565
45. A. D'Aì, R. Iaria, T. Di Salvo, A. Riggio, L. Burderi, N.R. Robba, Chandra X-ray spectroscopy of a clear dip in GX 13+1. A&A **564**, A62 (2014). https://doi.org/10.1051/0004-6361/201322044. arXiv: 1403.0071

46. P. Das, O. Porth, A.L. Watts, GRMHD simulations of accreting neutron stars with non-dipole fields. MNRAS **515**, 3144 (2022). https://doi.org/10.1093/mnras/stac1817. arXiv: 2204.00249
47. C. Dashwood Brown, P. Gandhi, P.A. Charles, Measuring interacting binary mass functions with X-ray fluorescence. MNRAS **517**, 2426 (2022). https://doi.org/10.1093/mnras/stac2725. arXiv: 2209.09920
48. N. Degenaar, D. Altamirano, M. Parker, J.C.A. Miller-Jones, J.M. Miller, C.O. Heinke et al., Disc reflection and a possible disc wind during a soft X-ray state in the neutron star low-mass X-ray binary 1RXS J180408.9-342058. MNRAS **461**, 4049 (2016). https://doi.org/10.1093/mnras/stw1593. arXiv: 1607.01780
49. N. Degenaar, D.R. Ballantyne, T. Belloni, M. Chakraborty, Y.-P. Chen, L. Ji et al., Accretion disks and coronae in the X-ray flashlight. Space Sci. Rev. **214**, 15 (2018). https://doi.org/10.1007/s11214-017-0448-3. arXiv: 1711.06272
50. N. Degenaar, J.M. Miller, D. Chakrabarty, F.A. Harrison, E. Kara, A.C. Fabian, A NuSTAR observation of disc reflection from close to the neutron star in 4U 1608–52. MNRAS **451**, L85 (2015). https://doi.org/10.1093/mnrasl/slv072. arXiv: 1505.07112
51. N. Degenaar, J.M. Miller, F.A. Harrison, J.A. Kennea, C. Kouveliotou, G. Younes, High-resolution X-ray spectroscopy of the bursting pulsar GRO J1744–28. ApJL **796**, L9 (2014). https://doi.org/10.1088/2041-8205/796/1/L9. arXiv: 1410.4841
52. N. Degenaar, J.M. Miller, R. Wijnands, D. Altamirano, A.C. Fabian, X-ray emission and absorption features during an energetic thermonuclear X-ray burst from IGR J17062–6143. ApJL **767**, L37 (2013). https://doi.org/10.1088/2041-8205/767/2/L37. arXiv: 1212.4869
53. N. Degenaar, C. Pinto, J.M. Miller, R. Wijnands, D. Altamirano, F. Paerels et al., An in-depth study of a neutron star accreting at low Eddington rate: on the possibility of a truncated disc and an outflow. MNRAS **464**, 398 (2017). https://doi.org/10.1093/mnras/stw2355. arXiv: 1609.04816
54. J.W. den Herder, A.C. Brinkman, S.M. Kahn, G. Branduardi-Raymont, K. Thomsen, H. Aarts et al., The reflection grating spectrometer on board XMM-newton. A&A **365**, L7 (2001). https://doi.org/10.1051/0004-6361:20000058
55. T. Di Salvo, A. Sanna, Accretion powered X-ray millisecond pulsars, in *Astrophysics and Space Science Library*, ed. by S. Bhattacharyya, A. Papitto, D. Bhattacharya. Astrophysics and Space Science Library, vol. 465 (2022), pp. 87–124. https://doi.org/10.1007/978-3-030-85198-9_4
56. M. Díaz Trigo, L. Boirin, Disc atmospheres and winds in X-ray binaries. Acta Polytech. **53**, 659 (2013). arXiv: 1210.0318
57. M. Díaz Trigo, L. Boirin, Accretion disc atmospheres and winds in low-mass X-ray binaries. Astronomische Nachrichten **337**, 368 (2016). https://doi.org/10.1002/asna.201612315. arXiv: 1510.03576
58. M. Díaz Trigo, S. Migliari, J.C.A. Miller-Jones, M. Guainazzi, XMM-Newton observations reveal the disappearance of the wind in 4U 1630–47. A&A **571**, A76 (2014). https://doi.org/10.1051/0004-6361/201424554. arXiv: 1409.3406
59. M. Díaz Trigo, J.C.A. Miller-Jones, S. Migliari, J.W. Broderick, T. Tzioumis, Baryons in the relativistic jets of the stellar-mass black-hole candidate 4U1630-47. Nature **504**, 260 (2013). https://doi.org/10.1038/nature12672. arXiv: 1311.5080
60. M. Díaz Trigo, A.N. Parmar, L. Boirin, M. Méndez, J.S. Kaastra, Spectral changes during dipping in low-mass X-ray binaries due to highly-ionized absorbers. A&A **445**, 179 (2006). https://doi.org/10.1051/0004-6361:20053586. arXiv: astro-ph/0509342
61. M. Díaz Trigo, A.N. Parmar, L. Boirin, C. Motch, A. Talavera, S. Balman, Variations in the dip properties of the low-mass X-ray binary XB 1254–690 observed with XMM-Newton and INTEGRAL. A&A **493**, 145 (2009). https://doi.org/10.1051/0004-6361:200810154. arXiv: 0810.0432
62. M. Díaz Trigo, A.N. Parmar, J. Miller, E. Kuulkers and M.D. Caballero-García, XMM-Newton and INTEGRAL spectroscopy of the microquasar GRO J1655–40 during its, outburst. A&A **462**(2007), 657 (2005). https://doi.org/10.1051/0004-6361:20065406. arXiv: astro-ph/0610873

63. M. Díaz Trigo, L. Sidoli, L. Boirin, A.N. Parmar, XMM-Newton observations of GX 13 + 1: correlation between photoionised absorption and broad line emission. A&A **543**, A50 (2012). https://doi.org/10.1051/0004-6361/201219049. arXiv: 1204.5904
64. C. Done, R. Tomaru, T. Takahashi, Thermal winds in stellar mass black hole and neutron star binary systems. MNRAS **473**, 838 (2018). https://doi.org/10.1093/mnras/stx2400. arXiv: 1612.09377
65. B.T. Draine, *Physics of the Interstellar and Intergalactic Medium* (2011)
66. G. Dubus, C. Done, B.E. Tetarenko, J.-M. Hameury, The impact of thermal winds on the outburst lightcurves of black hole X-ray binaries. A&A **632**, A40 (2019). https://doi.org/10.1051/0004-6361/201936333. arXiv: 1909.13601
67. S. Dyda, R. Dannen, T. Waters, D. Proga, Irradiation of astrophysical objects - SED and flux effects on thermally driven winds. MNRAS **467**, 4161 (2017). https://doi.org/10.1093/mnras/stx406. arXiv: 1610.04292
68. K. Ebisawa, X-ray spectroscopy of binary sources with ASCA - black hole candidates and super-soft sources, in *X-Ray Imaging and Spectroscopy of Cosmic Hot Plasmas*, ed. by F. Makino, K. Mitsuda (1997), p. 427
69. I. El Mellah, F. Casse, A numerical investigation of wind accretion in persistent supergiant X-ray binaries - I. Structure of the flow at the orbital scale. MNRAS **467**, 2585 (2017). https://doi.org/10.1093/mnras/stx225. arXiv: 1609.01532
70. I. El Mellah, V. Grinberg, J.O. Sundqvist, F.A. Driessen, M.A. Leutenegger, Radiography in high mass X-ray binaries. Micro-structure of the stellar wind through variability of the column density. A&A **643**, A9 (2020). https://doi.org/10.1051/0004-6361/202038791. arXiv: 2006.16216
71. I. El Mellah, J.O. Sundqvist, R. Keppens, Accretion from a clumpy massive-star wind in supergiant X-ray binaries. MNRAS **475**, 3240 (2018). https://doi.org/10.1093/mnras/stx3211. arXiv: 1711.08709
72. I. El Mellah, J.O. Sundqvist, R. Keppens, Wind Roche lobe overflow in high-mass X-ray binaries. A possible mass-transfer mechanism for ultraluminous X-ray sources. A&A **622**, L3 (2019). https://doi.org/10.1051/0004-6361/201834543. arXiv: 1810.12937
73. A.A. Esin, J.E. McClintock, R. Narayan, Advection-dominated accretion and the spectral states of black hole X-ray binaries: application to nova muscae 1991. ApJ **489**, 865 (1997). https://doi.org/10.1086/304829. arXiv: astro-ph/9705237
74. M. Espinasse, S. Corbel, P. Kaaret, E. Tremou, G. Migliori, R.M. Plotkin et al., Relativistic X-ray jets from the black hole X-ray binary MAXI J1820+070. ApJL **895**, L31 (2020). https://doi.org/10.3847/2041-8213/ab88b6. arXiv: 2004.06416
75. A.C. Fabian, M.J. Rees, SS 433: a double jet in action? MNRAS **187**, 13P (1979). https://doi.org/10.1093/mnras/187.1.13P
76. R. Fender, T. Belloni, GRS 1915+105 and the disc-jet coupling in accreting black hole systems. ARA&A **42**, 317 (2004). https://doi.org/10.1146/annurev.astro.42.053102.134031. arXiv: astro-ph/0406483
77. R. Fender, S. Corbel, T. Tzioumis, V. McIntyre, D. Campbell-Wilson, M. Nowak et al., Quenching of the Radio Jet during the X-ray high state of GX 339–4. ApJL **519**, L165 (1999). https://doi.org/10.1086/312128. arXiv: astro-ph/9905121
78. R. Fender, T. Muñoz-Darias, The balance of power: accretion and feedback in stellar mass black holes, in *Lecture Notes in Physics*, ed. by F. Haardt, V. Gorini, U. Moschella, A. Treves, M. Colpi, vol. 905 (Springer, Berlin, 2016), p. 65. https://doi.org/10.1007/978-3-319-19416-5_3DOI
79. R.P. Fender, T.M. Belloni, E. Gallo, Towards a unified model for black hole X-ray binary jets. MNRAS **355**, 1105 (2004). https://doi.org/10.1111/j.1365-2966.2004.08384.x. arXiv: astro-ph/0409360
80. R.P. Fender, J. Homan, T.M. Belloni, Jets from black hole X-ray binaries: testing, refining and extending empirical models for the coupling to X-rays. MNRAS **396**, 1370 (2009). https://doi.org/10.1111/j.1365-2966.2009.14841.x. arXiv: 0903.5166

81. Y.X. Feng, A.F. Tennant, S.N. Zhang, Probing the inflow/outflow and accretion disk of cygnus X-1 in the high state with the Chandra high energy transmission grating. ApJ **597**, 1017 (2003). https://doi.org/10.1086/378495. arXiv: astro-ph/0308254
82. G.J. Ferland, K.T. Korista, D.A. Verner, J.W. Ferguson, J.B. Kingdon, E.M. Verner, CLOUDY 90: numerical simulation of plasmas and their spectra. PASP **110**, 761 (1998). https://doi.org/10.1086/316190
83. J. Ferreira, Magnetically-driven jets from Keplerian accretion discs. A&A **319**, 340 (1997). arXiv: astro-ph/9607057
84. D.B. Friend, J.I. Castor, Radiation driven winds in X-ray binaries. ApJ **261**, 293 (1982). https://doi.org/10.1086/160340
85. K. Fukumura, D. Kazanas, I. Contopoulos, E. Behar, Magnetohydrodynamic accretion disk winds as X-ray absorbers in active galactic nuclei. ApJ **715**, 636 (2010). https://doi.org/10.1088/0004-637X/715/1/636. arXiv: 0910.3001
86. K. Fukumura, D. Kazanas, I. Contopoulos, E. Behar, Modeling high-velocity QSO absorbers with photoionized magnetohydrodynamic disk winds. ApJL **723**, L228 (2010). https://doi.org/10.1088/2041-8205/723/2/L228. arXiv: 1009.5644
87. K. Fukumura, D. Kazanas, C. Shrader, E. Behar, F. Tombesi, I. Contopoulos, Magnetic origin of black hole winds across the mass scale. Nature Astron. **1**, 0062 (2017). https://doi.org/10.1038/s41550-017-0062. arXiv: 1702.02197
88. K. Fukumura, D. Kazanas, C. Shrader, F. Tombesi, C. Kalapotharakos, E. Behar, Modeling magnetic disk wind state transitions in black hole X-Ray binaries. ApJ **912**, 86 (2021). https://doi.org/10.3847/1538-4357/abedaf. arXiv: 2103.05891
89. K. Fukumura, F. Tombesi, D. Kazanas, C. Shrader, E. Behar, I. Contopoulos, Magnetically driven accretion disk winds and ultra-fast outflows in PG 1211+143. ApJ **805**, 17 (2015). https://doi.org/10.1088/0004-637X/805/1/17. arXiv: 1503.04074
90. F. Fürst, I. Kreykenbohm, K. Pottschmidt, J. Wilms, M. Hanke, R.E. Rothschild et al., X-ray variation statistics and wind clumping in Vela X-1. A&A **519**, A37 (2010). https://doi.org/10.1051/0004-6361/200913981. arXiv: 1005.5243
91. E. Gallo, R.P. Fender, G.G. Pooley, A universal radio-X-ray correlation in low/hard state black hole binaries. MNRAS **344**, 60 (2003). https://doi.org/10.1046/j.1365-8711.2003.06791.x. arXiv: astro-ph/0305231
92. D.K. Galloway, M.P. Muno, J.M. Hartman, D. Psaltis, D. Chakrabarty, Thermonuclear (Type I) X-ray bursts observed by the rossi X-ray timing explorer. ApJS **179**, 360 (2008). https://doi.org/10.1086/592044. arXiv: astro-ph/0608259
93. E. Gatuzz, E. Churazov, Probing the structure of the gas in the Milky Way through X-ray high-resolution spectroscopy. MNRAS **474**, 696 (2018). https://doi.org/10.1093/mnras/stx2776. arXiv: 1710.08429
94. E. Gatuzz, M. Díaz Trigo, J.C.A. Miller-Jones, S. Migliari, Chandra high-resolution spectra of 4U 1630–47: the disappearance of the wind. MNRAS **482**, 2597 (2019). https://doi.org/10.1093/mnras/sty2850. arXiv: 1810.09464
95. F.P. Gavriil, T.E. Strohmayer, S. Bhattacharyya, An Fe XXIV absorption line in the persistent spectrum of the dipping low-mass X-ray binary 1A 1744–361. ApJ **753**, 2 (2012). https://doi.org/10.1088/0004-637X/753/1/2. arXiv: 0909.1607
96. K.G. Gayley, An improved line-strength parameterization in hot-star winds. ApJ **454**, 410 (1995). https://doi.org/10.1086/176492
97. I.M. George, A.C. Fabian, X-ray reflection from cold matter in active galactic nuclei and X-ray binaries. MNRAS **249**, 352 (1991). https://doi.org/10.1093/mnras/249.2.352
98. D.R. Gies, C.T. Bolton, The optical spectrum of HDE 226868 = cygnus X-1. III. A focused stellar wind model for He II lambda 4686 emission. ApJ **304**, 389 (1986). https://doi.org/10.1086/164172
99. A. Giménez-García, T. Shenar, J.M. Torrejón, L. Oskinova, S. Martínez-Núñez, W.R. Hamann et al., Measuring the stellar wind parameters in IGR J17544–2619 and Vela X-1 constrains the accretion physics in supergiant fast X-ray transient and classical supergiant X-ray binaries. A&A **591**, A26 (2016). https://doi.org/10.1051/0004-6361/201527551. arXiv: 1603.00925

171. A. Marino, T. Di Salvo, L. Burderi, A. Sanna, A. Riggio, A. Papitto et al., Indications of non-conservative mass transfer in AMXPs. A&A **627**, A125 (2019). https://doi.org/10.1051/0004-6361/201834460. arXiv: 1906.01949
172. H.L. Marshall, C.R. Canizares, T. Hillwig, A. Mioduszewski, M. Rupen, N.S. Schulz et al., Multiwavelength observations of the SS 433 Jets. ApJ **775**, 75 (2013). https://doi.org/10.1088/0004-637X/775/1/75. arXiv: 1307.8427
173. H.L. Marshall, C.R. Canizares, N.S. Schulz, The high-resolution X-ray spectrum of SS 433 using the Chandra HETGS. ApJ **564**, 941 (2002). https://doi.org/10.1086/324398. arXiv: astro-ph/0108206
174. H.L. Marshall, R. Rutledge, D.W. Fox, J.M. Miller, R. Guerriero, E. Morgan et al., X-ray spectrum of the rapid burster using the Chandra HETGS. AJ **122**, 21 (2001) https://doi.org/10.1086/321142. arXiv: astro-ph/0103203
175. A. Martocchia, G. Matt, T. Belloni, M. Feroci, V. Karas, G. Ponti, The XMM-newton view of GRS 1915+105. A&A **448**, 677 (2006). https://doi.org/10.1051/0004-6361:20053446. arXiv: astro-ph/0510651
176. D. Mata Sánchez, T. Muñoz-Darias, V.A. Cúneo, M. Armas Padilla, J. Sánchez-Sierras, G. Panizo-Espinar et al., Hard-state optical wind during the discovery outburst of the black hole X-ray dipper MAXI J1803-298. ApJL **926**, L10 (2022). https://doi.org/10.3847/2041-8213/ac502f. arXiv: 2201.09896
177. G. Matt, The iron Kα Compton shoulder in transmitted and reflected spectra. MNRAS **337**, 147 (2002). https://doi.org/10.1046/j.1365-8711.2002.05890.x. arXiv: astro-ph/0207615
178. J.H. Matthews, C. Knigge, K.S. Long, S.A. Sim, N. Higginbottom, The impact of accretion disc winds on the optical spectra of cataclysmic variables. MNRAS **450**, 3331 (2015). https://doi.org/10.1093/mnras/stv867. arXiv: 1504.05590
179. C.W. Mauche, Chandra low energy transmission grating spectrum of SS cygni in outburst. ApJ **610**, 422 (2004). https://doi.org/10.1086/421438. arXiv: astro-ph/0403691
180. C.W. Mauche, D.A. Liedahl, K.B. Fournier, The Fe XXII I(11.92 Å)/I(11.77 Å) density diagnostic applied to the Chandra high energy transmission grating spectrum of EX hydrae. ApJL **588**, L101 (2003). https://doi.org/10.1086/375684. arXiv: astro-ph/0304080
181. C.W. Mauche, D.A. Liedahl, K.B. Fournier, The Fe XXII I(11. 92 Å)/I(11.77 Å) Density Diagnostic, in *IAU Colloq. 190: Magnetic Cataclysmic Variables*, ed. by S. Vrielmann, M. Cropper. Astronomical Society of the Pacific Conference Series, vol. 315 (2004), p. 124. arXiv: astro-ph/0301633
182. J.E. McClintock, R. Shafee, R. Narayan, R.A. Remillard, S.W. Davis, L.-X. Li, The spin of the near-extreme Kerr black hole GRS 1915+105. ApJ **652**, 518 (2006). https://doi.org/10.1086/508457. arXiv: astro-ph/0606076
183. M.L. McCollough, R.K. Smith, L.A. Valencic, Cygnus X-3's little friend. ApJ **762**, 2 (2013). https://doi.org/10.1088/0004-637X/762/1/2. arXiv: 1211.2252
184. I.M. McHardy, E. Koerding, C. Knigge, P. Uttley, R.P. Fender, Active galactic nuclei as scaled-up galactic black holes. Nature **444**, 730 (2006). https://doi.org/10.1038/nature05389. arXiv: astro-ph/0612273
185. P.S. Medvedev, I.I. Khabibullin, S.Y. Sazonov, Diagnostics of parameters for the X-ray Jets of SS 433 from high-resolution Chandra spectroscopy. Astron. Lett. **45**, 299 (2019). https://doi.org/10.1134/S1063773719050049. arXiv: 2005.12416
186. M. Mehdipour, J.S. Kaastra, G.A. Kriss, M. Cappi, P.O. Petrucci, K.C. Steenbrugge et al., Anatomy of the AGN in NGC 5548. I. A global model for the broadband spectral energy distribution. A&A **575**, A22 (2015). https://doi.org/10.1051/0004-6361/201425373. arXiv: 1501.01188
187. M.J. Middleton, D.J. Walton, W. Alston, T. Dauser, S. Eikenberry, Y.F. Jiang et al., NuSTAR reveals the hidden nature of SS433. MNRAS **506**, 1045 (2021). https://doi.org/10.1093/mnras/stab1280. arXiv: 1810.10518
188. S. Migliari, R. Fender, M. Méndez, Iron emission lines from extended X-ray jets in SS 433: reheating of atomic nuclei. Science **297**, 1673 (2002). https://doi.org/10.1126/science.1073660. arXiv: astro-ph/0209131

189. M. Milgrom, On the interpretation of the large variations in the line positions in SS433. A&A **76**, L3 (1979)
190. J.M. Miller, Relativistic X-ray lines from the inner accretion disks around black holes. ARA&A **45**, 441 (2007). https://doi.org/10.1146/annurev.astro.45.051806.110555. arXiv: 0705.0540
191. J.M. Miller, A.C. Fabian, J. Kaastra, T. Kallman, A.L. King, D. Proga et al., Powerful, rotating disk winds from stellar-mass black holes. ApJ **814**, 87 (2015). https://doi.org/10.1088/0004-637X/814/2/87. arXiv: 1510.01177
192. J.M. Miller, A.C. Fabian, R. Wijnands, R.A. Remillard, P. Wojdowski, N.S. Schulz et al., Resolving the composite Fe Kα emission line in the galactic black hole cygnus X-1 with Chandra. ApJ **578**, 348 (2002). https://doi.org/10.1086/342466. arXiv: astro-ph/0202083
193. J.M. Miller, D. Maitra, E.M. Cackett, S. Bhattacharyya, T.E. Strohmayer, A fast X-ray disk wind in the transient pulsar IGR J17480–2446 in Terzan 5. ApJL **731**, L7 (2011). https://doi.org/10.1088/2041-8205/731/1/L7. arXiv: 1101.2377
194. J.M. Miller, S. Mineshige, A. Kubota, S. Yamada, F. Aharonian, C. Done et al., ASTRO-H White Paper - Stellar-Mass Black Holes (2014). arXiv:1412.1173
195. J.M. Miller, J. Raymond, E. Cackett, V. Grinberg, M. Nowak, An ultra-fast X-ray disk wind in the neutron star binary GX 340+0. ApJL **822**, L18 (2016). https://doi.org/10.3847/2041-8205/822/1/L18. arXiv: 1604.03329
196. J.M. Miller, J. Raymond, A. Fabian, D. Steeghs, J. Homan, C. Reynolds et al., The magnetic nature of disk accretion onto black holes. Nature **441**, 953 (2006). https://doi.org/10.1038/nature04912. arXiv: astro-ph/0605390
197. J.M. Miller, J. Raymond, A.C. Fabian, E. Gallo, J. Kaastra, T. Kallman et al., The accretion disk wind in the black hole GRS 1915+105. ApJL **821**, L9 (2016). https://doi.org/10.3847/2041-8205/821/1/L9. arXiv: 1603.01474
198. J.M. Miller, J. Raymond, A.C. Fabian, J. Homan, M.A. Nowak, R. Wijnands et al., Chandra/High energy transmission grating spectrometer spectroscopy of the galactic black hole GX 339–4: A relativistic iron emission line and evidence for a seyfert-like warm absorber. ApJ **601**, 450 (2004). https://doi.org/10.1086/380196. arXiv: astro-ph/0307394
199. J.M. Miller, J. Raymond, A.C. Fabian, C.S. Reynolds, A.L. King, T.R. Kallman et al., The disk-wind-jet connection in the black hole H 1743–322. ApJL **759**, L6 (2012). https://doi.org/10.1088/2041-8205/759/1/L6. arXiv: 1208.4514
200. J.M. Miller, J. Raymond, J. Homan, A.C. Fabian, D. Steeghs, R. Wijnands et al., Simultaneous Chandra and RXTE spectroscopy of the microquasar H1743–322: clues to disk wind and jet formation from a variable ionized outflow. ApJ **646**, 394 (2006). https://doi.org/10.1086/504673. arXiv: astro-ph/0406272
201. J.M. Miller, J. Raymond, T.R. Kallman, D. Maitra, A.C. Fabian, D. Proga et al., Chandra spectroscopy of MAXI J1305–704: detection of an infalling black hole disk wind?. ApJ **788**, 53 (2014). https://doi.org/10.1088/0004-637X/788/1/53. arXiv: 1306.2915
202. J.M. Miller, J. Raymond, C.S. Reynolds, A.C. Fabian, T.R. Kallman, J. Homan, The accretion disk wind in the black hole GRO J1655–40. ApJ **680**, 1359 (2008). https://doi.org/10.1086/588521. arXiv: 0802.2026
203. J.M. Miller, M.T. Reynolds, A disk wind in the black hole candidate MAXI J1803–298. Astron. Telegram **14650**, 1 (2021)
204. J.M. Miller, R. Wijnands, P.M. Rodriguez-Pascual, P. Ferrando, B.M. Gaensler, A. Goldwurm et al., XMM-Newton spectroscopy of the galactic microquasar GRS 1758–258 in the peculiar off/soft state. ApJ **566**, 358 (2002). https://doi.org/10.1086/338038. arXiv: astro-ph/0107514
205. J.M. Miller, P. Wojdowski, N.S. Schulz, H.L. Marshall, A.C. Fabian, R.A. Remillard et al., Revealing the focused companion wind in cygnus X-1 with Chandra. ApJ **620**, 398 (2005). https://doi.org/10.1086/426701. arXiv: astro-ph/0208463
206. I. Miškovičová, N. Hell, M. Hanke, M.A. Nowak, K. Pottschmidt, N.S. Schulz et al., Chandra X-ray spectroscopy of focused wind in the Cygnus X-1 system. II. The non-dip spectrum in the low/hard state - modulations with orbital phase. A&A **590**, A114 (2016). https://doi.org/10.1051/0004-6361/201322490. arXiv: 1604.00364

207. T. Muñoz-Darias, M. Armas Padilla, F. Jiménez-Ibarra, G. Panizo-Espinar, J. Casares, D. Altamirano et al., The changing-look optical wind of the flaring X-ray transient swift J1858.6-0814. ApJL **893**, L19 (2020). https://doi.org/10.3847/2041-8213/ab8381. arXiv: 2003.12073
208. T. Muñoz-Darias, J. Casares, D. Mata Sánchez, R.P. Fender, M. Armas Padilla, M. Linares et al., Regulation of black-hole accretion by a disk wind during a violent outburst of V404 Cygni. Nature **534**, 75 (2016). https://doi.org/10.1038/nature17446. arXiv: 1605.02358
209. T. Muñoz-Darias, F. Jiménez-Ibarra, G. Panizo-Espinar, J. Casares, D. Mata Sánchez, G. Ponti et al., Hard-state accretion disk winds from black holes: the revealing case of MAXI J1820+070. ApJL **879**, L4 (2019). https://doi.org/10.3847/2041-8213/ab2768. arXiv: 1906.04835
210. T. Muñoz-Darias, G. Ponti, Simultaneous X-ray and optical spectroscopy of V404 Cygni supports the multi-phase nature of X-ray binary accretion disc winds. A&A **664**, A104 (2022). https://doi.org/10.1051/0004-6361/202243769. arXiv: 2205.14162
211. T. Muñoz-Darias, M.A.P. Torres, M.R. Garcia, The low-luminosity accretion disc wind of the black hole transient V4641 Sagittarii. MNRAS **479**, 3987 (2018). https://doi.org/10.1093/mnras/sty1711. arXiv: 1806.10881
212. K. Mukai, X-ray emissions from accreting white dwarfs: a review. PASP **129**, 062001 (2017). https://doi.org/10.1088/1538-3873/aa6736. arXiv: 1703.06171
213. N. Murray, J. Chiang, S.A. Grossman, G.M. Voit, Accretion disk winds from active galactic nuclei. ApJ **451**, 498 (1995). https://doi.org/10.1086/176238
214. F. Nagase, S. Hayakawa, N. Sato, K. Masai, H. Inoue, Circumstellar matter in the VELA X-1/HD 77581 system. PASJ **38**, 547 (1986)
215. C.J. Neijssel, S. Vinciguerra, A. Vigna-Gómez, R. Hirai, J.C.A. Miller-Jones, A. Bahramian et al., Wind mass-loss rates of stripped stars inferred from cygnus X-1. ApJ **908**, 118 (2021). https://doi.org/10.3847/1538-4357/abde4a. arXiv: 2102.09092
216. J. Neilsen, The case for massive, evolving winds in black hole X-ray binaries. Adv. Space Res. **52**, 732 (2013). https://doi.org/10.1016/j.asr.2013.04.021. arXiv: 1304.6091
217. J. Neilsen, E. Cackett, R.A. Remillard, J. Homan, J.F. Steiner, K. Gendreau et al., A persistent disk wind in GRS 1915+105 with NICER. ApJL **860**, L19 (2018). https://doi.org/10.3847/2041-8213/aaca96. arXiv: 1806.02342
218. J. Neilsen, M. Coriat, R. Fender, J.C. Lee, G. Ponti, A.K. Tzioumis et al., A link between X-ray emission lines and radio jets in 4U 1630–47 ApJL **784**, L5 (2014). https://doi.org/10.1088/2041-8205/784/1/L5. arXiv: 1402.5140
219. J. Neilsen, J. Homan, A hybrid magnetically/thermally driven wind in the black hole GRO J1655–40 ApJ **750**, 27 (2012). https://doi.org/10.1088/0004-637X/750/1/27. arXiv: 1202.6053
220. J. Neilsen, J.C. Lee, Accretion disk winds as the jet suppression mechanism in the microquasar GRS 1915+105. Nature **458**, 481 (2009). https://doi.org/10.1038/nature07680
221. J. Neilsen, J.C. Lee, M.A. Nowak, K. Dennerl, S.D. Vrtilek, Spectroscopic signatures of the superorbital period in the neutron star binary LMC X-4. ApJ **696**, 182 (2009). https://doi.org/10.1088/0004-637X/696/1/182. arXiv: 0902.0786
222. J. Neilsen, A.J. Petschek, J.C. Lee, Accretion disc wind variability in the states of the microquasar GRS 1915+105. MNRAS **421**, 502 (2012). https://doi.org/10.1111/j.1365-2966.2011.20329.x. arXiv: 1112.1066
223. J. Neilsen, F. Rahoui, J. Homan, M. Buxton, A. Super-Eddington, Compton-thick Wind in GRO J1655–40 ApJ **822**, 20 (2016). https://doi.org/10.3847/0004-637X/822/1/20. arXiv: 1603.04070
224. J. Neilsen, R.A. Remillard, J.C. Lee, The physics of the "Heartbeat" state of GRS 1915+105. ApJ **737**, 69 (2011). https://doi.org/10.1088/0004-637X/737/2/69. arXiv: 1106.0298
225. H. Netzer, A thermal wind model for the X-ray outflow in GRO J1655–40. ApJL **652**, L117 (2006). https://doi.org/10.1086/510067. arXiv: astro-ph/0610231
226. M.A. Nowak, Leveraging high-resolution spectra to understand black hole spectra. Astronomische Nachrichten **338**, 227 (2017). https://doi.org/10.1002/asna.201713335. arXiv: 1612.01998

227. M.A. Nowak, M. Hanke, S.N. Trowbridge, S.B. Markoff, J. Wilms, K. Pottschmidt et al., Corona, jet, and relativistic line models for Suzaku/RXTE/Chandra-HETG observations of the cygnus X-1 hard state. ApJ **728**, 13 (2011). https://doi.org/10.1088/0004-637X/728/1/13. arXiv: 1012.4801
228. M.A. Nowak, A. Paizis, G.K. Jaisawal, J. Chenevez, S. Chaty, F. Fortin et al., Chandra-HETGS characterization of an outflowing wind in the accreting millisecond pulsar IGR J17591–2342. ApJ **874**, 69 (2019). https://doi.org/10.3847/1538-4357/ab0a71. arXiv: 1902.09577
229. L.M. Oskinova, A. Feldmeier, P. Kretschmar, Clumped stellar winds in supergiant high-mass X-ray binaries: X-ray variability and photoionization. MNRAS **421**, 2820 (2012). https://doi.org/10.1111/j.1365-2966.2012.20507.x. arXiv: 1201.1915
230. S.P. Owocki, J.I. Castor, G.B. Rybicki, Time-dependent models of radiatively driven stellar winds. I. Nonlinear evolution of instabilities for a pure absorption model. ApJ **335**, 914 (1988). https://doi.org/10.1086/166977
231. F. Özel, P. Freire, Masses, radii, and the equation of state of neutron stars. ARA&A **54**, 401 (2016). https://doi.org/10.1146/annurev-astro-081915-023322. arXiv: 1603.02698
232. F. Paerels, A.C. Brinkman, R.L.J. van der Meer, J.S. Kaastra, E. Kuulkers, A.J.F. den Boggende et al., Interstellar X-ray absorption spectroscopy of oxygen, neon, and iron with the CHANDRA LETGS spectrum of X0614+091. ApJ **546**, 338 (2001). https://doi.org/10.1086/318251. arXiv: astro-ph/0008349
233. F. Paerels, J. Cottam, M. Sako, D.A. Liedahl, A.C. Brinkman, R.L.J. van der Meer et al., High-resolution spectroscopy of the X-ray-photoionized wind in cygnus X-3 with the chandra high-energy transmission grating spectrometer. ApJL **533**, L135 (2000). https://doi.org/10.1086/312608. arXiv: astro-ph/0002324
234. D. Pandel, F.A. Córdova, K.O. Mason, W.C. Priedhorsky, X-ray observations of the boundary layer in dwarf novae at low accretion rates. ApJ **626**, 396 (2005). https://doi.org/10.1086/429983. arXiv: astro-ph/0503114
235. G. Panizo-Espinar, M. Armas Padilla, T. Muñoz-Darias, K.I.I. Koljonen, V.A. Cúneo, J. Sánchez-Sierras et al., Discovery of optical and infrared accretion disc wind signatures in the black hole candidate MAXI J1348-630. A&A **664**, A100 (2022). https://doi.org/10.1051/0004-6361/202243426. arXiv: 2205.09128
236. K. Parfrey, A. Tchekhovskoy, General-relativistic simulations of four states of accretion onto millisecond pulsars. ApJL **851**, L34 (2017). https://doi.org/10.3847/2041-8213/aa9c85. arXiv: 1708.06362
237. A.N. Parmar, T. Oosterbroek, L. Boirin, D. Lumb, Discovery of narrow X-ray absorption features from the dipping low-mass X-ray binary X 1624–490 with XMM-Newton. A&A **386**, 910 (2002). https://doi.org/10.1051/0004-6361:20020281. arXiv: astro-ph/0202452
238. G. Pelletier, R.E. Pudritz, Hydromagnetic disk winds in young stellar objects and active galactic nuclei. ApJ **394**, 117 (1992). https://doi.org/10.1086/171565
239. N.A. Pereyra, S.P. Owocki, D.J. Hillier, D.A. Turnshek, On the steady nature of line-driven disk winds. ApJ **608**, 454 (2004). https://doi.org/10.1086/386320. arXiv: astro-ph/0311268
240. R. Perna, J. McDowell, K. Menou, J. Raymond, M.V. Medvedev, Chandra observations of the dwarf nova WX Hydri in quiescence. ApJ **598**, 545 (2003). https://doi.org/10.1086/378855. arXiv: astro-ph/0308081
241. P.O. Petrucci, S. Bianchi, G. Ponti, J. Ferreira, G. Marcel, F. Cangemi et al., Expected evolution of disk wind properties along an X-ray binary outburst. A&A **649**, A128 (2021). https://doi.org/10.1051/0004-6361/202039524. arXiv: 2103.04601
242. C. Pinto, E. Costantini, A.C. Fabian, J.S. Kaastra, J.J.M. in't Zand, Unveiling the environment surrounding low-mass X-ray binary SAX J1808.4–3658. A&A **563**, A115 (2014). https://doi.org/10.1051/0004-6361/201323283. arXiv: 1402.0379
243. C. Pinto, J.S. Kaastra, E. Costantini, C. de Vries, Interstellar medium composition through X-ray spectroscopy of low-mass X-ray binaries. A&A **551**, A25 (2013). https://doi.org/10.1051/0004-6361/201220481. arXiv: 1301.1612
244. C. Pinto, J.S. Kaastra, E. Costantini, F. Verbunt, High-resolution X-ray spectroscopy of the interstellar medium. XMM-Newton observation of the LMXB GS 1826–238. A&A **521**, A79 (2010). https://doi.org/10.1051/0004-6361/201014836. arXiv: 1007.2796

245. G. Ponti, S. Bianchi, B. De Marco, A. Bahramian, N. Degenaar, C.O. Heinke, Evolution of the disc atmosphere in the X-ray binary MXB 1659–298, during its 2015–17 outburst. MNRAS **487**, 858 (2019). https://doi.org/10.1093/mnras/stz1245. arXiv: 1905.01308
246. G. Ponti, S. Bianchi, T. Muñoz-Darias, B. De Marco, T. Dwelly, R.P. Fender et al., On the Fe K absorption - accretion state connection in the Galactic Centre neutron star X-ray binary AX J1745.6–2901. MNRAS **446**, 1536 (2015). https://doi.org/10.1093/mnras/stu1853. arXiv: 1409.3224
247. G. Ponti, S. Bianchi, T. Muñoz-Darias, K. Nandra, Measuring masses in low mass X-ray binaries via X-ray spectroscopy: the case of MXB 1659–298. MNRAS **481**, L94 (2018). https://doi.org/10.1093/mnrasl/sly120. arXiv: 1807.04757
248. G. Ponti, K. De, T. Muñoz-Darias, L. Stella, K. Nandra, The puzzling orbital period evolution of the LMXB AX J1745.6–2901. MNRAS **464**, 840 (2017). https://doi.org/10.1093/mnras/stw2317. arXiv: 1511.02855
249. G. Ponti, R.P. Fender, M.C. Begelman, R.J.H. Dunn, J. Neilsen, M. Coriat, Ubiquitous equatorial accretion disc winds in black hole soft states. MNRAS **422**, L11 (2012). https://doi.org/10.1111/j.1745-3933.2012.01224.x. arXiv: 1201.4172
250. G. Ponti, T. Muñoz-Darias, R.P. Fender, A connection between accretion state and Fe K absorption in an accreting neutron star: black hole-like soft-state winds? MNRAS **444**, 1829 (2014). https://doi.org/10.1093/mnras/stu1742. arXiv: 1407.4468
251. G. Prabhakar, S. Mandal, B.G.R., A. Nandi, Wideband study of the brightest black hole X-ray binary 4U 1543-47 in the 2021 outburst: signature of disk-wind regulated accretion. MNRAS (2023). https://doi.org/10.1093/mnras/stad080. arXiv: 2301.13711
252. D. Proga, Winds from accretion disks driven by radiation and magnetocentrifugal force. ApJ **538**, 684 (2000). https://doi.org/10.1086/309154. arXiv: astro-ph/0002441
253. D. Proga, Numerical simulations of mass outflows driven from accretion disks by radiation and magnetic forces. ApJ **585**, 406 (2003). https://doi.org/10.1086/345897. arXiv: astro-ph/0210642
254. D. Proga, Theory of outflows in cataclysmic variables, in *The Astrophysics of Cataclysmic Variables and Related Objects*, ed. by J.M. Hameury, J.P. Lasota. Astronomical Society of the Pacific Conference Series, vol. 330 (2005), p. 103. https://doi.org/10.48550/arXiv.astro-ph/0411200. arXiv: astro-ph/0411200
255. D. Proga, T.R. Kallman, On the role of the ultraviolet and X-ray radiation in driving a disk wind in X-ray binaries. ApJ **565**, 455 (2002). https://doi.org/10.1086/324534. arXiv: astro-ph/0109064
256. D. Proga, T. Waters, S. Dyda, Z. Zhu, Thermal instability in radiation hydrodynamics: instability mechanisms, position-dependent s-curves, and attenuation curves. ApJL **935**, L37 (2022). https://doi.org/10.3847/2041-8213/ac87b0. arXiv: 2207.04032
257. I. Psaradaki, E. Costantini, M. Mehdipour, M. Díaz Trigo, Modelling the disc atmosphere of the low mass X-ray binary EXO 0748–676. A&A **620**, A129 (2018). https://doi.org/10.1051/0004-6361/201834000. arXiv: 1809.06864
258. R.E. Pudritz, R. Ouyed, C. Fendt, A. Brandenburg, Disk winds, jets, and outflows: theoretical and computational foundations, in *Protostars and Planets V*, ed. by B. Reipurth, D. Jewitt, K. Keil (2007), p. 277. https://doi.org/10.48550/arXiv.astro-ph/0603592. arXiv: astro-ph/0603592
259. F. Rahoui, S. Chaty, J. Rodriguez, Y. Fuchs, I.F. Mirabel, G.G. Pooley, Long-term Multi-wavelength Studies of GRS 1915+105. I. A high-energy and mid-infrared focus with RXTE/INTEGRAL and spitzer. ApJ **715**, 1191 (2010). https://doi.org/10.1088/0004-637X/715/2/1191. arXiv: 1004.3032
260. F. Rahoui, M. Coriat, J.C. Lee, Optical and near-infrared spectroscopy of the black hole GX 339–4 - II. The spectroscopic content in the low/hard and high/soft states. MNRAS **442**, 1610 (2014). https://doi.org/10.1093/mnras/stu977. arXiv: 1405.3457
261. G. Raman, C. Maitra, B. Paul, Observation of variable pre-eclipse dips and disc windsin the eclipsing LMXB XTE J1710–281. MNRAS **477**, 5358 (2018). https://doi.org/10.1093/mnras/sty918. arXiv: 1804.06073

262. G. Ramsay, F. Córdova, J. Cottam, K. Mason, R. Much, J. Osborne et al., First XMM-newton observations of a cataclysmic variable II: the X-ray spectrum of OY Car. A&A **365**, L294 (2001). https://doi.org/10.1051/0004-6361:20000042. arXiv: astro-ph/0010416
263. V.R. Rana, K.P. Singh, E.M. Schlegel, P.E. Barrett, Study of Fe Kα lines in nonmagnetic cataclysmic variables using Chandra HEG data. ApJ **642**, 1042 (2006). https://doi.org/10.1086/500316. arXiv: astro-ph/0512274
264. R.A. Remillard, J.E. McClintock, X-Ray properties of black-hole binaries. ARA&A **44**, 49 (2006). https://doi.org/10.1146/annurev.astro.44.051905.092532. arXiv: astro-ph/0606352
265. C.S. Reynolds, Constraints on compton-thick winds from black hole accretion disks: can we see the inner disk? ApJL **759**, L15 (2012). https://doi.org/10.1088/2041-8205/759/1/L15. arXiv: 1210.3029
266. C.S. Reynolds, M.A. Nowak, Fluorescent iron lines as a probe of astrophysical black hole systems. Phys. Rep. **377**, 389 (2003). https://doi.org/10.1016/S0370-1573(02)00584-7. arXiv: astro-ph/0212065
267. M.T. Reynolds, J.M. Miller, Chandra grating spectroscopy of the Be/X-ray binary 1A 0535+262. ApJ **723**, 1799 (2010). https://doi.org/10.1088/0004-637X/723/2/1799. arXiv: 1009.2240
268. T.E. Riley, A.L. Watts, S. Bogdanov, P.S. Ray, R.M. Ludlam, S. Guillot et al., A NICER view of PSR J0030+0451: millisecond pulsar parameter estimation. ApJL **887**, L21 (2019). https://doi.org/10.3847/2041-8213/ab481c. arXiv: 1912.05702
269. M.M. Romanova, G.V. Ustyugova, A.V. Koldoba, R.V.E. Lovelace, Launching of conical winds and axial jets from the disc-magnetosphere boundary: axisymmetric and 3D simulations. MNRAS **399**, 1802 (2009). https://doi.org/10.1111/j.1365-2966.2009.15413.x. arXiv: 0907.3394
270. M.M. Romanova, G.V. Ustyugova, A.V. Koldoba, R.V.E. Lovelace, MRI-driven accretion on to magnetized stars: global 3D MHD simulations of magnetospheric and boundary layer regimes. MNRAS **421**, 63 (2012). https://doi.org/10.1111/j.1365-2966.2011.20055.x. arXiv: 1111.3068
271. T.D. Russell, N. Degenaar, R. Wijnands, J. van den Eijnden, N.V. Gusinskaia, J.W.T. Hessels et al., The radio-bright accreting millisecond X-ray pulsar IGR J17591–2342. ApJL **869**, L16 (2018). https://doi.org/10.3847/2041-8213/aaf4f9. arXiv: 1811.00085
272. M. Sako, D.A. Liedahl, S.M. Kahn, F. Paerels, The X-ray spectrum and global structure of the stellar wind in VELA X-1. ApJ **525**, 921 (1999). https://doi.org/10.1086/307924. arXiv: astro-ph/9907291
273. G. Sala, J. Greiner, J. Vink, F. Haberl, E. Kendziorra, X.L. Zhang, The highly ionized disk wind of ASTROBJ GRO J1655–40 ASTROBJ. A & A **461**, 1049 (2007). https://doi.org/10.1051/0004-6361:20065168. arXiv: astro-ph/0606272
274. J. Sánchez-Sierras, T. Muñoz-Darias, Near-infrared emission lines trace the state-independent accretion disc wind of the black hole transient MAXI J1820+070. A&A **640**, L3 (2020). https://doi.org/10.1051/0004-6361/202038406. arXiv: 2007.07257
275. N.S. Schulz, W.N. Brandt, Variability of the X-Ray P cygni line profiles from circinus X-1 near zero phase. ApJ **572**, 971 (2002). https://doi.org/10.1086/340369. arXiv: astro-ph/0112483
276. N.S. Schulz, C.R. Canizares, J.C. Lee, M. Sako, The ionized stellar wind in vela X-1 during eclipse. ApJL **564**, L21 (2002). https://doi.org/10.1086/338862. arXiv: astro-ph/0111489
277. N.S. Schulz, W. Cui, C.R. Canizares, H.L. Marshall, J.C. Lee, J.M. Miller et al., The first high-resolution X-ray spectrum of cygnus X-1: soft X-ray ionization and absorption. ApJ **565**, 1141 (2002). https://doi.org/10.1086/324482. arXiv: astro-ph/0109236
278. N.S. Schulz, D.P. Huenemoerder, L. Ji, M. Nowak, Y. Yao, C.R. Canizares, Heating in an extended accretion disk corona along the Z-pattern in Cyg X-2. ApJL **692**, L80 (2009). https://doi.org/10.1088/0004-637X/692/2/L80. arXiv: 0812.2264
279. N.S. Schulz, T.E. Kallman, D.K. Galloway, W.N. Brandt, The variable warm absorber in circinus X-1. ApJ **672**, 1091 (2008). https://doi.org/10.1086/523809. arXiv: 0709.3336
280. V. Sguera, A. Tiengo, L. Sidoli, A.J. Bird, Dust-scattering halo and giant hard X-ray flare from the supergiant fast X-ray transient IGR J16479–4514 investigated with XMM-newton

and INTEGRAL. ApJ **900**, 22 (2020). https://doi.org/10.3847/1538-4357/abaa3c. arXiv: 2007.15329
281. M. Shidatsu, C. Done, Application of the thermal wind model to absorption features in the black hole X-ray binary H1743–322. ApJ **885**, 112 (2019). https://doi.org/10.3847/1538-4357/ab46b3. arXiv: 1906.02469
282. M. Shidatsu, C. Done, Y. Ueda, An optically thick disk wind in GRO J1655–40 ApJ **823**, 159 (2016). https://doi.org/10.3847/0004-637X/823/2/159. arXiv: 1604.04346
283. L. Sidoli, T. Oosterbroek, A.N. Parmar, D. Lumb, C. Erd, An XMM-newton study of the X-ray binary MXB 1659–298 and the discovery of narrow X-ray absorption lines. A&A **379**, 540 (2001). https://doi.org/10.1051/0004-6361:20011322. arXiv: astro-ph/0109417
284. L. Sidoli, A.N. Parmar, T. Oosterbroek, D. Lumb, Discovery of complex narrow X-ray absorption features from the low-mass X-ray binary GX 13+1 with XMM-Newton. A&A **385**, 940 (2002). https://doi.org/10.1051/0004-6361:20020192. arXiv: astro-ph/0202077
285. J.F. Steiner, J.E. McClintock, Modeling the jet kinematics of the black hole microquasar XTE J1550–564: a constraint on spin-orbit alignment. ApJ **745**, 136 (2012). https://doi.org/10.1088/0004-637X/745/2/136. arXiv: 1110.6849
286. T.E. Strohmayer, D. Altamirano, Z. Arzoumanian, P.M. Bult, D. Chakrabarty, J. Chenevez et al., NICER discovers spectral lines during photospheric radius expansion bursts from 4U 1820–30: evidence for burst-driven winds. ApJL **878**, L27 (2019). https://doi.org/10.3847/2041-8213/ab25eb. arXiv: 1906.00974
287. T.E. Strohmayer, Z. Arzoumanian, S. Bogdanov, P.M. Bult, D. Chakrabarty, T. Enoto et al., NICER discovers the ultracompact orbit of the accreting millisecond pulsar IGR J17062–6143. ApJL **858**, L13 (2018). https://doi.org/10.3847/2041-8213/aabf44. arXiv: 1808.04392
288. J.O. Sundqvist, S.P. Owocki, Clumping in the inner winds of hot, massive stars from hydrodynamical line-driven instability simulations. MNRAS **428**, 1837 (2013). https://doi.org/10.1093/mnras/sts165. arXiv: 1210.1861
289. A. Suryanarayanan, F. Paerels, M. Leutenegger, The high resolution Fe K spectrum of cygnus X-3 (2022). https://doi.org/10.48550/arXiv.2212.04165. arXiv:2212.04165
290. P. Szkody, K. Nishikida, J.C. Raymond, A. Seth, D.W. Hoard, K.S. Long et al., Chandra spectra of the prototype dwarf nova U Geminorum at Quiescence. ApJ **574**, 942 (2002). https://doi.org/10.1086/341006
291. C.B. Tarter, W.H. Tucker, E.E. Salpeter, The interaction of X-ray sources with optically thin environments. ApJ **156**, 943 (1969). https://doi.org/10.1086/150026
292. M. Tashiro, H. Maejima, K. Toda, R. Kelley, L. Reichenthal, L. Hartz et al., Status of x-ray imaging and spectroscopy mission (XRISM), in *Society of Photo-Optical Instrumentation Engineers (SPIE) Conference Series*, vol. 11444 (2020), p. 1144422. https://doi.org/10.1117/12.2565812
293. B.E. Tetarenko, J.P. Lasota, C.O. Heinke, G. Dubus, G.R. Sivakoff, Strong disk winds traced throughout outbursts in black-hole X-ray binaries. Nature **554**, 69 (2018). https://doi.org/10.1038/nature25159. arXiv: 1801.07203
294. A. Tiengo, G. Vianello, P. Esposito, S. Mereghetti, A. Giuliani, E. Costantini et al., The dust-scattering X-ray rings of the anomalous X-ray pulsar 1E 1547. 0–5408. ApJ **710**, 227 (2010). https://doi.org/10.1088/0004-637X/710/1/227. arXiv: 0911.3064
295. R. Tomaru, C. Done, J. Mao, What powers the wind from the black hole accretion disc in GRO J1655-40 MNRAS (2022). https://doi.org/10.1093/mnras/stac3210. arXiv: 2204.08802
296. R. Tomaru, C. Done, H. Odaka, S. Watanabe, T. Takahashi, Monte Carlo simulations of the detailed iron absorption line profiles from thermal winds in X-ray binaries. MNRAS **476**, 1776 (2018). https://doi.org/10.1093/mnras/sty336. arXiv: 1802.07019
297. R. Tomaru, C. Done, K. Ohsuga, H. Odaka, T. Takahashi, The thermal-radiative wind in low-mass X-ray binary H1743–322 - II. Iron line predictions from Monte Carlo radiation transfer. MNRAS **494**, 3413 (2020). https://doi.org/10.1093/mnras/staa961. arXiv: 1911.01660
298. R. Tomaru, C. Done, K. Ohsuga, H. Odaka, T. Takahashi, The thermal-radiative wind in the neutron star low-mass X-ray binary GX 13 + 1. MNRAS **497**, 4970 (2020). https://doi.org/10.1093/mnras/staa2254. arXiv: 2007.14607

299. J.M. Torrejón, N.S. Schulz, M.A. Nowak, T.R. Kallman, A Chandra survey of fluorescence fe lines in X-ray binaries at high resolution. ApJ **715**, 947 (2010). https://doi.org/10.1088/0004-637X/715/2/947. arXiv: 1004.2345
300. D.F. Torres, G.E. Romero, X. Barcons, Y. Lu, Probing the precession of the inner accretion disk in cygnus X-1. ApJ **626**, 1015 (2005). https://doi.org/10.1086/430125. arXiv: astro-ph/0503186
301. N. Trueba, J.M. Miller, A.C. Fabian, J. Kaastra, T. Kallman, A. Lohfink et al., A spectroscopic angle on central engine size scales in accreting neutron stars. ApJ **925**, 113 (2022). https://doi.org/10.3847/1538-4357/ac3766. arXiv: 2111.04764
302. N. Trueba, J.M. Miller, A.C. Fabian, J. Kaastra, T. Kallman, A. Lohfink et al., A redshifted inner disk atmosphere and transient absorbers in the ultracompact neutron star X-ray binary 4U 1916–053. ApJL **899**, L16 (2020). https://doi.org/10.3847/2041-8213/aba9de. arXiv: 2008.01083
303. N. Trueba, J.M. Miller, J. Kaastra, A. Zoghbi, A.C. Fabian, T. Kallman et al., A comprehensive chandra study of the disk wind in the black hole candidate 4U 1630–472. ApJ **886**, 104 (2019). https://doi.org/10.3847/1538-4357/ab4f70. arXiv: 1910.11243
304. P. Tzanavaris, T. Yaqoob, New constraints on the geometry and kinematics of matter surrounding the accretion flow in X-ray binaries from Chandra high-energy transmission grating X-ray spectroscopy. ApJ **855**, 25 (2018). https://doi.org/10.3847/1538-4357/aaaab6. arXiv: 1801.08544
305. Y. Ueda, K. Asai, K. Yamaoka, T. Dotani, H. Inoue, Discovery of an iron K absorption line in the low-mass X-ray binary GX 13+1. ApJL **556**, L87 (2001). https://doi.org/10.1086/323007
306. Y. Ueda, K. Honda, H. Takahashi, C. Done, H. Shirai, Y. Fukazawa et al., Suzaku observation of GRS 1915+105: evolution of accretion disk structure during limit-cycle oscillation. ApJ **713**, 257 (2010). https://doi.org/10.1088/0004-637X/713/1/257. arXiv: 1003.0317
307. Y. Ueda, H. Inoue, Y. Tanaka, K. Ebisawa, F. Nagase, T. Kotani et al., Detection of absorption-line features in the X-ray spectra of the galactic superluminal source GRO J1655–40. ApJ **492**, 782 (1998). https://doi.org/10.1086/305063
308. Y. Ueda, H. Murakami, K. Yamaoka, T. Dotani, K. Ebisawa, Chandra high-resolution spectroscopy of the absorption-line features in the low-mass X-ray binary GX 13+1. ApJ **609**, 325 (2004). https://doi.org/10.1086/420973
309. Y. Ueda, K. Yamaoka, R. Remillard, GRS 1915+105 in "Soft State": nature of accretion disk wind and origin of X-ray emission. ApJ **695**, 888 (2009). https://doi.org/10.1088/0004-637X/695/2/888. arXiv: 0901.1982
310. G.V. Ustyugova, A.V. Koldoba, M.M. Romanova, R.V.E. Lovelace, "Propeller" regime of disk accretion to rapidly rotating stars. ApJ **646**, 304 (2006). https://doi.org/10.1086/503379. arXiv: astro-ph/0603249
311. L.A. Valencic, R.K. Smith, Interstellar dust properties from a survey of X-ray halos. ApJ **809**, 66 (2015). https://doi.org/10.1088/0004-637X/809/1/66
312. J. van den Eijnden, N. Degenaar, C. Pinto, A. Patruno, K. Wette, C. Messenger et al., The very faint X-ray binary IGR J17062–6143: a truncated disc, no pulsations, and a possible outflow. MNRAS **475**, 2027 (2018). https://doi.org/10.1093/mnras/stx3224
313. J. van den Eijnden, N. Degenaar, T.D. Russell, R. Wijnands, A. Bahramian, J.C.A. Miller-Jones et al., A new radio census of neutron star X-ray binaries. MNRAS **507**, 3899 (2021). https://doi.org/10.1093/mnras/stab1995. arXiv: 2107.05286
314. J. van den Eijnden, N. Degenaar, N.S. Schulz, M.A. Nowak, R. Wijnands, T.D. Russell et al., Chandra reveals a possible ultrafast outflow in the super-Eddington Be/X-ray binary Swift J0243.6+6124. MNRAS **487**, 4355 (2019). https://doi.org/10.1093/mnras/stz1548. arXiv: 1906.01597
315. J.C.A. van Peet, E. Costantini, M. Méndez, F.B.S. Paerels, J. Cottam, Properties of the ionised plasma in the vicinity of the neutron-star X-ray binary EXO 0748–676. A&A **497**, 805 (2009). https://doi.org/10.1051/0004-6361/200811181. arXiv: 0902.4470
316. P.A.J. Vitello, I. Shlosman, Line-driven winds from accretion disks. I. Effects of the ionization structure. ApJ **327**, 680 (1988). https://doi.org/10.1086/166225

temperatures which were in between the UV-dominated (~0.01 keV) spectra of supermassive black holes (SMBHs) powering active galactic nuclei (AGN) and the X-ray peaked (~1 keV) spectra in the stellar-mass black holes (BHs) powering X-ray binaries. An accretion disc with $kT \sim 0.1$ keV and a bolometric luminosity of 10^{40} erg s^{-1} would correspond to BHs of several hundred Solar masses accreting at relatively low Eddington-normalised rates, which are expected to produce hard X-ray spectra through comparison with Galactic XRBs [63, 108]. Amongst the best candidates for IMBHs there is the hyper-luminous source (HLX-1) in ESO 243–49 [38]. Its outburst and spectral states are similar to the sub-Eddington Galactic BHs. With a peak luminosity of 10^{42} erg s^{-1}, it likely has a mass of $10^{4-5} M_{\odot}$, and could be a failed tidal disruption event owing to a recent delay in the outbursts [92].

Two alternative scenarios invoked either extreme geometrical beaming in Eddington limited stellar-remnant black holes [74] or super-Eddington accretion onto such objects (black holes were typically assumed in this scenario), similarly to the Galactic microquasar SS 433 [14] but potentially viewed at a more favourable angle (our view of SS433 is almost perfectly edge-on).

The ULX field was revolutionised after the discoveries of X-ray pulsations from the extreme ULX M82 X-2 with *NuSTAR* (peak luminosity of $L_X \sim 2 \times 10^{40}$ erg s^{-1} [8]), clearly demonstrating this source is powered by a highly super-Eddington (and strongly magnetised) neutron star. Since this discovery, a handful of other ULXs have been confirmed as neutron star accretors via the detection of pulsations with either *XMM-Newton* or *NuSTAR* [24, 41, 59, 60, 150, 152]. More detail on the properties of ULX variability at short- and long-term time scales, including fast-coherent and slow-(super)orbital modulations, is reported in Sect. 12.2.

Deep observations with *XMM-Newton* and, later on, *NuSTAR* have shown that most ULXs actually show X-ray spectra that differ from the sub-Eddington accretion states seen in Galactic black hole XRBs, further supporting the super-Eddington accretion scenario. ULX spectra typically show a strong curvature in the ~2–10 keV band before breaking to very steep spectra above ~10 keV, with spectral slopes of $\Gamma \sim 3$ [7, 184], and the brightest sources are now often interpreted as exhibiting two thermal components that dominate below 10 keV (with typical temperatures ~0.2–0.5 and ~2–4 keV). It has therefore become largely obsolete to relate ULX spectra to the standard XRB accretion states. Instead, ULX spectra are now more typically discussed in the context of the so-called '*ultra-luminous state*' [46], a distinct accretion state related to super-Eddington accretion. Observationally, the ultra-luminous state is further classified according to three main regimes depending on the effective spectral slope below 10 keV: soft ultra-luminous (SUL, in which $\Gamma > 2$, or equivalently the cooler thermal component dominates) or hard ultra-luminous (HUL, in which $\Gamma < 2$, where the hotter thermal component dominates). In the latter, if the X-ray spectrum has a single peak and a blackbody-like shape, it is called broadened disc regime (BD, [162]). In addition to these classifications, there is also a rare population of extremely soft ULXs which have spectra that are primarily dominated by even cooler thermal emission ($kT \sim 0.1$ keV) and are often referred to as 'ultra-luminous supersoft sources' (ULSs). More detail on X-ray broadband spectral properties of ULXs is given in Sect. 12.3.

Atomic features provide key information on accretion flow geometry/dynamics and on the presence of any outflows (which are naturally predicted for super-Eddington accretion), but have been challenging to detect in ULX spectra. Nevertheless, early studies spotted some spectral features around 1 keV where the resolving power of CCD spectrometers is rather low [160], and more recent studies have suggested that the shape and variability of these features imply that they are associated with the ULX itself rather than to the hot interstellar medium (ISM) of the host galaxy [105, 163]. ULXs are generally difficult targets for high-resolution X-ray spectroscopy. Despite their high X-ray luminosities, the Mpc distances normally limit their flux in the canonical 0.3–10 keV band to less than a few 10^{-12} erg s^{-1} cm^{-2}. This makes it difficult to fill the thousands of energy channels in high-resolution grating spectrometers with sufficient photons. However, the more recent availability of deep observations (300–500 ks) with moderate effective area detectors, mainly the Reflection Grating Spectrometers (RGS) aboard *XMM-Newton*, finally enabled the detection and identification of emission and absorption lines in ULX X-ray spectra [127]. The emission lines are commonly found at their laboratory wavelengths, with the exception of NGC 5204 (UL)X-1 where they are blueshifted by $\sim$0.3c [83]. Multiple observations show that the emission lines vary over time and have huge X-ray luminosities ($L_X \sim 10^{38}$ erg s^{-1}; [131]) that are orders of magnitude brighter than in Eddington-limited Galactic XRBs [5, 141].

The absorption lines are instead almost always highly blueshifted (0.1–0.3 c) and vary with the ULX regime and amongst ULXs with different spectral slope [84, 128, 131]. *Chandra* gratings confirmed similar outflows in a Galactic transient ULX ([173]). Further work on moderate-resolution detectors confirmed the presence of blueshifted absorption and emission features [181, 189]. These discoveries, particularly the relativistically blueshifted absorption lines, likely reveal the long-sought powerful winds predicted by the theoretical simulations of radiation pressure in super-Eddington accretion discs [119]. Detail on the methods used in high-resolution X-ray spectroscopy and their applications onto ULX spectra is provided in Sect. 12.4.

12.1.3 Brief Summary of ULX Multi-wavelength Observations

Observations of Galactic XRBs in other energy bands improve our knowledge, especially on the nature of the donor, the mass of the compact object, the presence of collimated outflows (jets) and the local environment. For ULXs, this is complicated by their larger distances (at least two orders of magnitude than Galactic XRBs).

Starting in the early 2000s, it was found that many bright ULXs are spatially associated with an optical [121] and/or radio nebula [64]. Originally they were thought to be powered by radiation and being mainly photoionised owing to their strong He II (4686 Å), [Ne V] (3426 Å) and [O IV] (25.89 μm) line emission. The He II line requires photons with energies in excess of 54.4 eV, which implies strong X-ray illumination of the ISM by the ULX and rather low beaming, even accounting for precession. For Holmberg II X-1 a lower limit on the luminosity required to illuminate its surrounding nebula of 1.1×10^{40} erg s^{-1} was obtained [16], which—given the long recombination time of He III—indicates that the average luminosity has been at least 10^{40} erg s^{-1} for several thousand years [66].

The discovery of large velocity broadening (up to 150 km s^{-1}), i.e. outflows, and the presence of shocks and lines from collisional excitation (e.g. [O I] at 6300 Å) indicated that in many cases mechanical power due, e.g., to winds or jets is actually inflating them. A clear example is the optical bubble, previously thought to be a nebula, around NGC 1313 (UL)X-1. Recent, accurate work with the *Very Large Telescope (VLT)*/MUSE has confirmed the collisional nature of this interstellar cavity [51]. The bubbles are young (10^{5-6} yr) and their Hα luminosities require a mechanical power of 10^{39-40} erg s^{-1} [121], which agrees with early estimates of the mechanical power associated with the X-ray winds in ULXs [130]. More detail on the comparison between the properties of ULX X-ray winds and multi-wavelength observations is provided in Sect. 12.5.

High-quality optical spectra of ULXs also revealed the presence of winds in the form of broad (500–1500 km s^{-1}) emission lines from the Balmer series and from He II [36]. This implies that caution is needed when attempting to use the emission lines to measure radial velocity curves and the mass of the compact object. The supersoft ULX M81 ULS-1 shows a blueshifted Hα line, with the blueshift varying with time and corresponding to a range of projected velocities between 0.14–0.17 c [94]. This indicates that it originates in either a relativistic baryonic jet like in SS 433 (see below), or a conical wind similar to those observed in X-rays [127].

In some cases, jets could play a role in powering ULX bubbles. There is indeed evidence of jets through radio observations of a few nearby ULXs such as Holmberg II X-1, M31 ULX, and NGC 7793 S26 [29, 102, 122]. The latter is particularly interesting as the jet mechanical power of a few 10^{40} erg s^{-1} is orders of magnitude higher than the X-ray luminosity, such that this system strongly resembles the case of the Galactic microquasar SS 433 [35, 98]. There is a large consensus that SS 433 is a compact object accreting at a super-Eddington rate of $\sim 10^{-4} M_{\odot}$ yr^{-1} via Roche lobe overflow from a massive donor, likely an evolved A supergiant, and it is X-ray weak due to strong circumstellar obscuration owing to an edge-on viewing angle. Should it be observed face-on, i.e. closer to the accretion disc axis, it would appear as a ULX [106, 177]. The M31 ULX was detected in the radio whilst exhibiting a thermal spectrum, similar to the XRB soft X-ray states. The radio emission was likely due to discrete ejecta similar to the Galactic BH GRS 1915+105. They also have a comparable X-ray peak luminosity of 1.3×10^{39} erg s^{-1}, suggesting that M31 ULX is a BH accreting near its Eddington limit. For ESO 243–49 HLX-1 and NGC 2276–3c, instead, the detection of flaring radio emission and hard X-ray spectral states would suggest the presence of an Eddington-limited IMBH as the accretor.

Substantial work has been done in the last decade in order to search for the compact optical counterparts of ULXs, with a view to identifying their donor stars and, in turn, obtaining dynamical mass constraints for the accretors via radial velocity (RV) studies. ULX distances imply searches with current facilities are only sensitive to optical counterparts brighter than ∼20th magnitude, even for massive donors, which makes their detection difficult. Moreover, the high probability of ULXs being located in star-forming regions, i.e. crowded fields, complicates the situation even further. Here the best results have been obtained through the simultaneous use of the high-spatial-resolution images (<0.3″) from *Chandra* and the *Hubble Space Telescope (HST)*. For about 20 ULXs unique optical counterparts were found [47, 166]. The magnitudes and B-V colours suggest OB supergiant companion stars if these counterparts are really dominated by the stellar companion, but the optical emission could be significantly affected by X-ray reprocessing in the outer disc. Indeed, this is confirmed in some ULXs by the detection of strong optical variability [166].

These efforts have resulted in two initial claims for dynamical mass estimates that support the presence of stellar-remnant accretors (as opposed to IMBHs). The first is M 101 ULX-1, where evidence for a Wolf-Rayet companion is seen, and a probable mass of 20–30 $M_\odot$ for the accretor is proposed (with a lower limit of 5 $M_\odot$, which would require the source to be powered by a black hole). However, caution is needed here, as the sampling of the RV data is very sparse, and the stellar velocity shifts are inferred from emission lines, which may be problematic in the presence of strong outflows from the ULX (as noted above). The second is NGC 7793 P13, where a B9Ia supergiant companion is seen, and an upper limit to the accretor mass of 15 $M_\odot$ for the accretor is claimed based on the stellar companion type and the detection of a ∼64-day optical photometric period (interpreted to be the binary orbit [114]). As it turns out, NGC 7793 P13 is the second source confirmed to host a neutron star through the detection of X-ray pulsations [41], consistent with the dynamical mass constraint.

VLT/X-shooter spectra of the pulsating NS NGC 300 ULX-1, covering the wavelength range 3,500–23,000 Å clearly showed the presence of a red supergiant (RSG) donor star that is best matched by a stellar atmosphere with $T_{\rm eff}$ = 3,650–3,900 K and $\log(L_{\rm bol}/L_\odot) = 4.25 \pm 0.10$, which yields a stellar radius $R/R_\odot = 310 \pm 70$ [55]. Given the large donor-to-compact object mass ratio, orbital modulations of the radial velocity of the RSG are likely undetectable. Other ULXs have an RSG as a donor [54] with NGC 253 ULX favouring a massive BH ($\gtrsim$ 50 $M_\odot$) as the compact object [53]. In these cases, long-term monitoring may be used to search for line modulations and allow for further dynamical mass estimates in the future.

The community of experts in stellar population synthesis has been actively working on the predictions of ULX binary systems properties. For instance, it was shown that extremely high, super-Eddington accretion rates, of $10^{-3}\ M_\odot\ {\rm yr}^{-1}$ or above can be easily sustained for $\sim 10^4$ yr in BHs and NSs with companion stars of similar masses [144, 193]. Exploring the different formation channels of merging double compact objects (DCOs: BH-BH/BH-NS/NS-NS) using the STARTRACK code, it was found that in the local Universe typically 50% of merging BH-BH progenitor binaries would have evolved through a phase of super-Eddington accretion (when only one of the binary members had evolved into a compact object) and would have potentially appeared as a ULX [112]. This indicates that ULXs can be used to study the origin of gravitational wave (GW) sources. The fraction of observed ULXs that will form merging DCOs in future varies between 5 and 40%. More detail on the accretion rates and the comparison with observations is provided in Sect. 12.5.

12.2 X-Ray Timing Properties

One of the key observational aspects of accretion-powered sources is temporal variability, and ULXs are no different. These variability properties are often characterised via Fourier techniques and have revealed a wide range of phenomenology in Galactic XRBs, including broadband noise, quasi-periodic oscillations and coherent pulsations (which can reveal the nature of the compact object) on short timescales, as well as potentially orbital or super-orbital variations on (typically) longer timescales (e.g. [146]). In addition to these flux variations, time lags between different energy bands may unveil photon reprocessing or disc perturbations, and help to reconstruct the accretion geometry. Here we provide a brief overview on ULX variability properties focusing first on short-term and then long-term variability timescales.

12.2.1 Pulsations

Undeniably the most important recent discovery on X-ray timing properties of ULXs is the coherent pulsations in a growing sample of ULXs. To date, there are 6 'traditional' ULXs that are known to be pulsars which show sustained luminosities comfortably in excess of 10^{39} erg s^{-1}, although there is plenty of speculation that more of the ULX population will reveal themselves to be pulsars over time. In addition to these sources, there are a similar number of transient Be X-ray binaries with neutron star accretors in the local group that occasionally peak just above 10^{39} erg s^{-1} whilst in outburst, and so could also formally be considered ULX pulsars (e.g. [155, 175, 194]).

M 82 X-2 was the first ULX to show pulsations [8] with a spin period of 1.37 s. The evolution of the pulsations also revealed a 2.5 day orbital modulation and a continuous spin-up of $\dot{P} = -2 \times 10^{-10}$ s s^{-1} (see Fig. 12.1, left panel). Coherent pulsations are the robust signature of a rotating neutron star, which indicates that the observed luminosity surpasses the Eddington limit by orders of magnitude. Long-term monitoring of the pulse period has shown that the neutron star exhibits periods of both secular spin-up and spin-down, potentially indicating the neutron star is close to spin equilibrium, and also that the orbital period is evolving owing to the extreme mass transfer, confirming accretion rates that significantly exceed the Eddington limit [9, 10].

After M82 X-2, the best-studied ULX pulsars are NGC 7793 P13 and NGC 5907 ULX-1. NGC 7793 P13 is the second pulsating ULX (PULX) discovered [41, 60]. It shows pulsations at 0.43 s with a secular spin-up of $\dot{P} = -3 \times 10^{-11}$ s s^{-1}. In this system, the X-ray pulsations reveal a much longer orbital period of 67 days, similar to

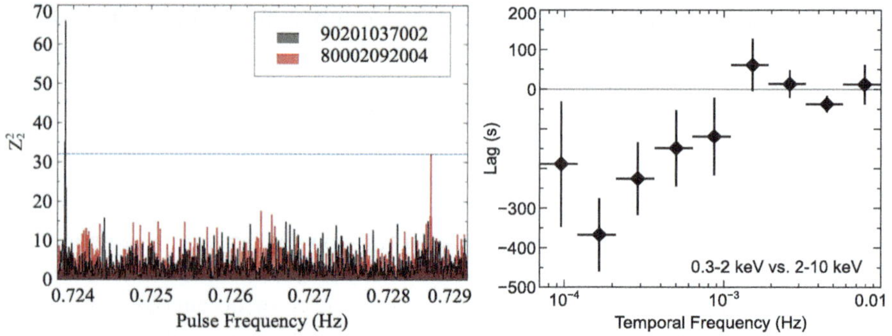

Fig. 12.1 Left: Detection of pulsations in M 82 X-2 with *NUSTAR* in 2014 (red) and 2016 (black). The horizontal line indicates the 3σ level for the given number of trials [9]. Right: soft (0.3–2 keV) versus hard (2–10 keV) lags as a function of temporal frequency for NGC 1313 X-1 (*XMM-Newton*/EPIC-pn data). Negative lags are soft band lags. The variability on long timescales show a soft band lag of ~150 s [72]

but formally slightly longer than the optical photometric period (which now appears to be a *sub*-orbital periodicity [43]). The most extreme PULX is NGC 5907 (UL)X-1, which reaches a peak luminosity $\sim 10^{41}$ erg s^{-1}, corresponding to ~500 times the Eddington limit of a neutron star [42, 59]. This sources also exhibit pulsations on timescales of ~1 s and a strong $\dot{P} = -8 \times 10^{-10}$ s s^{-1}. The orbital solution is not as well determined in this case, but the best estimates from the evolution of the X-ray pulsations put the orbital period in the range ~5–30 days [59].

PULXs typically show nearly sinusoidal pulse profiles which would argue against the presence of strong beaming or pencil-beam geometry, in broad agreement with the large emitting region and moderate beaming implied by the presence of the photoionised components of ULX bubbles (Sect. 12.1.3). The pulsed fraction is typically ~10% in the *XMM-Newton* band (0.3–10 keV), but rises with the energy such that typical pulsed fractions above 10 keV are ~20–30% (although the most extreme case, NGC 300 ULX-1, shows an extreme pulsed fraction of ~80% at these energies [24]). PULX spectra are discussed in Sect. 12.3, but tend to fit in the hard end of ULX spectra. Whilst the traditional PULXs do spend extended periods at their extreme luminosities, almost all also show very strong long-term flux variations, and the most extreme cases can also drop down to fluxes below 10^{38} erg s^{-1} [15]. Section 12.5 reports some theoretical interpretations.

12.2.2 *Quasi-periodic Oscillations*

High-frequency quasi-periodic oscillations (HF-QPOs) can be identified with their main harmonics above 100 Hz thanks to their 3:2 frequency ratio [146]. They are often used to probe the inner edge of the accretion disc owing to their high coherence and dependence on the BH mass and spin. They tend to be found mainly during BH hard states. *RXTE* observations of M 82 X-1 provided evidence for a pair of QPOs at 3.32 ± 0.06 and 5.07 ± 0.06 Hz, i.e. with a harmonic ratio compatible with HF-QPOs seen in Galactic BHs. Each QPO has a rather low significance (3.7σ and 2.7σ), which combined yields $4.7\,\sigma$. Assuming that the QPOs are of the same nature, scaling according to the frequency-mass relation, leads to a mass estimate of $428 \pm 105 M_\odot$ for the compact object mass in M 82 X-1, i.e. an IMBH [123].

In Galactic BHs, low frequency (LF-) QPOs are divided into three types, A, B, and C [25]. Type-C frequency is correlated with the spectral index and the disk flux. The QPOs detected in M82 X-1 and NGC 5408 X-1 are similar to BH type-C QPOs. In M82 X-1, however, the QPO frequency is uncorrelated with the spectral index, which cautions against its use as a BH mass indicator. This was confirmed in NGC 5408 X-1 where the QPO frequencies are not correlated with the continuum noise break frequencies and the spectral parameters [101].

Quasi-periodicities can also be produced by other phenomena such as dips. The *XMM-Newton* light curve of the supersoft source in NGC 247 shows strong flux dips on time scales of 200 s or longer. These cannot be explained just by an increase of photoelectric absorption [39]. Power spectra and coherence reveal the dipping preferentially occurs on 5 ks and 10 ks timescales [4]. A likely explanation is therefore occultation of the central X-ray source by an optically thick structure, such as warping of the accretion disc, or obscuration by a wind launched from the accretion disc, or both. This supports the idea that supersoft ULXs are viewed close to edge-on (see Sect. 12.5). An alternative explanation invokes the occurrence of the propeller effect. Assuming this as an origin for the jump in flux and, thereby, in luminosity from 2.5×10^{39} erg s^{-1} to 2.3×10^{38} erg s^{-1}, a magnetic field of 2×10^{11} G was constrained [31] which is in line with estimates in other ULXs [77].

12.2.3 Time Delays Between Energy Bands

An advanced technique to decompose, in a model-independent way, the spectrum of a variable source and, in particular, the structure of an accretion disc consists of searching for correlated variability and time delays between time series of different energy bands. The low count rates seen from ULXs combined with the lower levels of short-timescale variability they exhibit in comparison to Galactic XRBs, make it difficult with the current instruments. In the last decade, dedicated work has found evidence for delays in the soft X-ray band at various frequencies: from $\sim$10 mHz in NGC 5408 X-1 [56]) down to much lower frequencies, $f \sim 0.1-1$ mHz in other ULXs [72, 128, 134] (Fig. 12.1, right panel). Given the two orders of magnitude difference in their time lags, it is not obvious that these are due to the same mechanism. The covariance spectra suggest that emission contributing to the lags is largely associated with the hotter of the two thermal-like components, likely originating from the inner accretion flow (see Sect. 12.3). All these soft lags are however $\sim$5–20%

of the corresponding characteristic variability timescales. If these soft lags can be understood in the context of a unified picture of ULXs, then lag timescales may provide constraints on the density and extent of the accretion flow (see Sect. 12.5).

12.2.4 Long-Term Modulations

Since the launch of the *Rossi X-ray Timing Explorer (RXTE)* and, later on, the *Neil Gehrels Swift Observatory* it was possible to obtain long and regularly sampled light curves of X-ray sources. The brightest, persistent, ULXs with fluence $\gtrsim 10^{-12}$ erg s^{-1} cm^{-2} make very good *Swift* targets. This enables the search for long-term (days to months) periodicities which could be associated with orbital or super-orbital (e.g. precession) variability. This is common practice in Galactic XRBs where periodic cycles can identify disc instabilities, enhanced accretion rates at specific orbital phases (e.g. periastron) and warps in the accretion disc [138, 192].

The first evidence for a ULX long-term (62-day) periodicity in a ULX was a 62-day period reported for the M82 galaxy based on data from *RXTE* [65]. Although initially assumed to be associated with M82 X-1, as this is typically the brightest X-ray source in the M82 galaxy, it is now understood that this period is actually associated with the ULX pulsar M82 X-2 [21, 142]. As such, given that the orbital period for this system is known to be ~2.5 days, the 62-day variations must be super-orbital in nature. *Swift* monitoring significantly enlarged the detections of long-term periodicities, including a longer period of 115 days found in NGC 5408 X-1 [161], which then shifted to 136 days [6] suggesting that the modulation is super-orbital.

Long-term X-ray periods are relatively common amongst the PULX population, although they appear to have a range of origins. [182] report a 78-day X-ray period in NGC5907 ULX-1, which is likely super-orbital in nature given the current best estimates for the orbital period (between 5–30 days [59]). Similarly, M 51 ULX-7 shows super-orbital X-ray variations with a period that drifts from ~38–44 days [22, 174]; here the orbital period is known to be ~2 days based again on the evolution of the X-ray pulsations [150]. Although NGC7793 P13 shows a long-term X-ray periodicity with a 67-day timescale, the best current solution for the binary orbit based on the X-ray pulsations suggests that the variations are orbital in nature.

At the other end of the mass scale, the IMBH candidate ESO 243–49 HLX-1 regularly goes into outburst (as mentioned in Sect. 12.1.2). Initially it appeared as though these outbursts had a periodic recurrence time of ~300 days [49], but more recently the period between outbursts has been increasing dramatically [92]. Amongst the several interpretations here, one potential option is an IMBH that is fed by winds from a giant star with a tidally stripped envelope [110].

12.3 Broadband X-Ray Spectroscopy

Historically speaking, ULX spectra have typically been compared to the accretion states seen in Galactic black hole XRBs, which are broadly characterised by the relative contributions of the accretion disc (which provides thermal blackbody emission at lower energies, often characterised by temperatures $kT \sim 1$ keV) and the up-scattering 'corona' (which provides non-thermal Comptonised emission at higher energies, often characterised by electron temperatures $kT_e \sim 50-100$ keV). Most of these sources are transient LMXBs that accrete via Roche lobe overflow. When they go through an outburst, during which they can reach luminosities comparable to the Eddington limit, in addition to the strong flux variability they also exhibit strong spectral variability (i.e. strong changes in their spectral 'hardness'), with most sources tending to follow a relatively well-defined pattern of behaviour. Sources generally rise through the 'hard' state (in which the corona dominates the observed X-ray spectrum, and persistent radio jets are seen), before transitioning to the 'soft' state (in which the accretion disc dominates, and winds are often observed) when they reach higher luminosities. As the sources fade, they transition back to the hard state and eventually return to quiescence (note that there is typically some hysteresis seen, such that the hard-to-soft and soft-to-hard transitions during the rise and decay, respectively, do not occur at the same luminosities (see Fig. 12.2, left panel [151]). There are also a small number of more persistent black hole HMXBs (most notably Cygnus X-1) which broadly show similar states to their LMXB cousins but do not fade to quiescence on observable timescales (as they primarily accrete from the stellar wind of their companion).

Early observations of ULXs with ASCA yielded the first high-counts spectra in the 0.5–10 keV band [27]. Some spectra appeared thermal in nature, albeit with a slightly broader profile and higher disc temperatures than typically seen in classic Galactic XRB soft states, whilst others appeared power-law-like and showed evidence for short-timescale variability analogous to the hard state (e.g. [89, 97]). There was therefore speculation that ULXs were showing the same accretion states as seen in Galactic black hole XRBs, but at higher luminosities, thereby implying the presence of higher-mass black holes.

12.3.1 The Golden Age of ULX X-Ray Spectroscopy

The launch of *Chandra* [191] and *XMM-Newton* [61] brought X-ray detectors with improved effective area, facilitating more detailed studies of ULX spectra. Early observations with these facilities revealed spectra that required two continuum components. These could be well explained by a disc–corona model, which initially seemed to further support the connection to sub-Eddington accretion states. The disc temperatures obtained were relatively cool ($kT \sim 0.3$ keV), which would correspond to an IMBH with mass $\sim$1,000 $M_\odot$ [108], as the inner disc temperature should scale with the mass as $T_{\rm in} \propto M^{-0.25}$.

However, sub-Eddington thin discs are expected to show a relationship between their bolometric luminosity and inner temperature $L_{\rm disc} \propto T_{\rm in}^4$. As multi-epoch spectra were available it became clear that ULXs exhibit spectral variability that disagreed with this trend, instead showing flatter relations [12, 137]. Moreover, the discovery of an inverted $L_{\rm disc} \propto T_{\rm in}^{-4}$ trend in the most luminous ($L_{\rm X} > 10^{40}$ erg s^{-1}) ULXs with multiple high-quality spectra [68, 147] has been interpreted as due to beaming caused by a large-scale cone inflated by strong radiation in a super-Eddington accretion regime [75]. More discussion on super-Eddington accretion and geometrical beaming is provided in Sect. 12.5.2.

In addition, as deep observations became available, evidence started to emerge that the higher energy component was not power-law-like (e.g. [46]), instead seeming to show some subtle curvature, casting doubt on the standard disc-corona models that had been considered previously. A key development in broadband X-ray spectroscopy came with the launch of *NuSTAR* in 2012 [52], covering a higher energy bandpass (3–78 keV) than *XMM-Newton* and *Chandra*, and carrying the first imaging X-ray optics to extend above 10 keV. *NuSTAR* observations (often in coordination with *XMM-Newton* to provide genuine broadband observations) have provided spectra of ULXs up to $\sim$30–40 keV (in the best cases), and unambiguously confirmed that the higher energy emission seen by *XMM-Newton* and *Chandra* is not power-law-like, and the curvature indicated by these missions represents a genuine spectral cutoff (e.g. [7, 179]; see Fig. 12.2). The broadband spectra revealed were not consistent with any of the standard sub-Eddington accretion states, strongly arguing that ULXs are primarily a population of super-Eddington accretors (instead of sub-Eddington

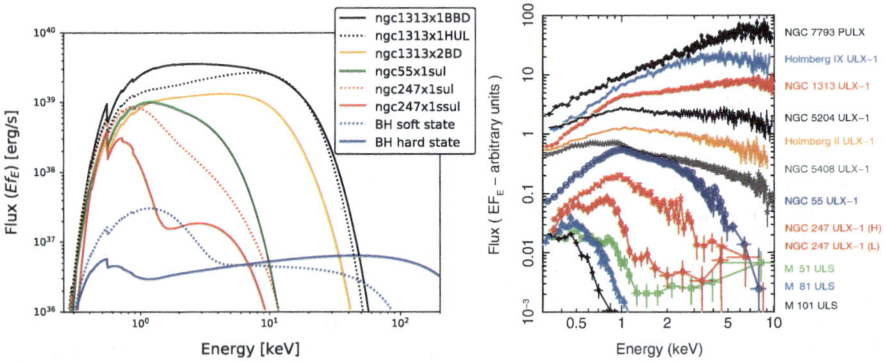

Fig. 12.2 Left: Comparison of typical spectra of ULXs and Galactic BH binaries. ULX spectra are classified as supersoft (SSUL or ULS), soft (SUL) and hard (HUL) ultra-luminous and broadened disc (BD), and the brighter broadened disc regime (BBD). Right: X-ray spectra of some amongst the brightest ULXs with hardness increasing from bottom to top taken with *XMM-Newton* (mainly) and *Chandra* (ULS). Y-axis units are $E \times F_E$ in an arbitrary scale for displaying purposes [12, 130]

IMBHs). The existence of ULX pulsars was also discovered by *NuSTAR* around this time, further confirming these conclusions.

Owing to their distinct spectral appearance, comparisons between ULX spectra and the standard sub-Eddington accretion states have become largely obsolete. Instead, ULXs are now mostly interpreted as representing a new, super-Eddington accretion state, dubbed the *ultra-luminous state* [46]. However, a rich diversity of spectra are seen from the ULX population (see Fig. 12.2), and so the ultra-luminous state has itself been empirically divided into 4 main sub-classifications, depending on the different spectral properties that ULXs exhibit in the 0.3–10 keV band [162]:

- **Hard ultra-luminous** (HUL)—the 0.3–10 keV spectrum shows two distinct continuum components. If modelled with a cool accretion disc and a higher energy power-law, the latter is brighter and yields slopes $\Gamma < 2$. Several ULXs (and all pulsating ULXs) show HUL spectra [133, 184]. Some notable examples are: NGC 7793 (p)ULX, Holmberg IX X-1 and NGC 1313 X-1 in Fig. 12.2 (right).
- **Soft ultra-luminous** (SUL)—the spectrum is similar to the HUL in the sense that two continuum components are required, but here the thermal component is brighter and $\Gamma > 2$. SUL spectra yield $\mathrm{kT_{BB}} \sim 0.1 - 0.3$ keV, whilst the value can be slightly higher in the HUL regime. Notable examples of SUL sources are: NGC 55 ULX-1, NGC 5408 X-1, NGC 247 X-1 (H) and Holmberg II X-1.
- **Supersoft ultra-luminous** (SSUL or ULS)—these spectra are almost entirely dominated by a cool blackbody-like component, with $kT \sim 0.1$ keV. They have a much fainter (a factor ten) hard tail with $\Gamma = 2 - 4$. SSUL spectra are broadly similar to those during the supersoft (SS) phases of novae, although ULSs are persistent and up to 100 times brighter. Some notable examples are shown in Fig. 12.2 including a low-flux epoch of M 101 ULX-1 and NGC 247 X-1 (L).
- **Broadened disc** (BD)—these spectra also seem to be dominated by a single thermal component, but it is hot and has a broader profile than expected from a standard thin accretion disc [153]. In the 0.3–10 keV band, they can be described with an advection-dominated disc model with $p \sim 0.5 - 0.6$, where $T(R) \propto R^{-p}$ ($p = 0.75$ for a standard thin disc), and $kT_{\mathrm{in}} \sim 1 - 2$ keV. Notable examples are NGC 1313 X-1 (high-BD or BBD) and X-2 (low-BD), and M 33 X-8.

The diversity of these spectra is proposed to be primarily related to the fact that at super-Eddington accretion rates, the inner regions of the accretion disc are expected to be both geometrically and optically thick, and to launch powerful outflows [3, 137, 153]. For such a geometry there is a natural expectation that the observed properties of a given source will depend on the angle it is viewed at [104]. The general picture often now invoked is that the innermost (and hottest) regions of the accretion disc are the primary origin of the hotter of the two emission components that dominate below 10 keV, but are most readily viewed when the disc is observed close to face-on owing to their geometrically thick nature, resulting in a HUL source. As the viewing angle is increased these hotter regions are eventually blocked from view, meaning that cooler thermal emission from larger radii and possibly the outflow (which may itself be optically thick) dominate the observed spectra, resulting first in a SUL source, and at even higher inclinations a SSUL source. Broadened disc sources stand out as something of an exception as they are not necessarily related to this viewing angle dependence, but are often thought to represent the transitional regime around the Eddington threshold in which the disc has started to deviate from a normal thin disc. There are exceptions to this, where very luminous examples of BD spectra have been seen. Therefore, the full geometric effects associated with strongly super-Eddington accretion have not yet been fully established. It is important to note that many ULXs switch between different regimes as they vary with time. For instance, NGC 1313 X-1 is known to show SUL, BD and HUL spectra at different epochs with some BD spectra fainter and others brighter than its HUL spectrum (see Fig. 12.2). NGC 247 X-1 instead shows notable transitions between SUL and SSUL spectra. This could be due to the precession of the accretion flow, or to changes in the scale height of the funnel formed by the inner disc/wind.

12.3.2 Moving Beyond Simple Spectral Models

Whilst the discovery of ULX pulsars and the non-standard broadband spectra observed have demonstrated that the majority of ULXs are super-Eddington accretors, their broadband X-ray spectra are not necessarily solely described by thermal emission from a super-Eddington accretion disc. In addition to the lower-energy thermal components, a further relevant ingredient is a discovery in *XMM-Newton* and *NuSTAR* spectra of a third, harder continuum component that extends beyond the Wien tail from these thermal components. This third component seems to be present in all ULXs observed by *NuSTAR* to date with sufficient statistics above 10 keV [184]. The nature of this component is not well understood, although it is thought to be produced either by scattering in an optically-thin corona or in the accretion column onto a magnetised neutron star (phase-resolved spectroscopy of the known ULX pulsars has found that the pulsed emission from the accretion columns dominates the highest energy emission in those sources [19, 184, 185]).

Detailed spectral modelling of ULXs is also often complicated by the presence of spectral residuals such as those around 1 keV (most likely from winds, see Sect. 12.4) and at lower energies possibly due to variability in the line-of-sight neutral column density. The latter is known to occur for some Galactic XRBs, particularly in HMXBs. Some absorption is certainly produced by the Galactic ISM with abundances close to Solar (albeit with a gradient towards the centre of the Milky Way [125]). In many cases, there is, however, excess absorption in ULX spectra from a few 10^{21} cm^{-2} due to either the host galaxy ISM, the circumstellar medium (CSM) around the ULX [195], or even the ULX wind [105]. Further modelling complications might arise from non-Solar metallicity in the host galaxy (which is often sub-Solar for ULX hosts) or in the ULX CSM [195].

12.3.3 Insights on the $L - T$ Relation and Spectral Evolution

The last decade has seen significant interest in attempts to understand how ULX spectra evolve, and to determine the physical origin of their spectral components. This was enabled by long-term monitoring primarily with *Swift* and multiple deep observations with *XMM-Newton*, *NuSTAR* and, in a few cases, *Chandra* and *Suzaku*. First attempts of accounting for the wind effects and the dependence on both mass accretion rate and line of sight showed encouraging results in the description of the trends between spectral hardness and variability power. The covariance spectra provided evidence on correlated variability mainly in the hard X-ray band with a shape similar to that of the hot (~1–3 keV) component [104].

In 2017, the deepest broadband view to date of a bright ULX, NGC 1313 X-1, was performed, combining substantial investments from *XMM-Newton* (750 ks), *NUSTAR* (500 ks) and *Chandra* (500 ks). This allowed a detailed study of the evolution of each individual spectral component over a range of timescales spanning weeks to years. As noted above, the spectrum below 10 keV was modelled with two thermal components (i.e. a DISKBB+DISKPBB combination in XSPEC; both are multi-colour blackbody accretion disc models, but the latter also includes the radial temperature index as a free parameter). The softer thermal component (0.1–0.3 keV) component and the high-energy emission above 10 keV did not show strong variability. In contrast, the hotter (and dominant) thermal component (~1–3 keV) showed a surprising variability pattern which could not be described with a simple trend. Instead the data seemed to exhibit two correlated $L - T$ relationships which split above $\sim 10^{40}$ erg s^{-1}, or alternatively at temperatures above ~1.9 keV [187]. In Fig. 12.3 (left panel) two different modelling approaches for the high-energy continuum are shown: one assumes the source is a non-magnetised accretor (e.g. a BH) and that the highest energy emission arises in an up-scattering corona (with a power-law spectral profile), whilst the other assumes a magnetised accretor and that the highest energy emission arises in a ULX-pulsar-like accretion column.[2] Qualitatively similar results are seen, but the cause of this behaviour remains poorly understood.

The campaign on NGC 1313 was particularly productive since it also enabled the discovery of time lags (see Sect. 12.2.3) and variable winds (see Sect. 12.4.3) in X-1, as well as pulsations in X-2 [152]. A similar, follow-up, spectroscopic work on NGC 1313 X-2 focused on the full set of available observations with *XMM-Newton*—spanning a 17-year baseline—in order to extend the variability range and constrain the $L - T$ relationship (NGC 1313 X-2 is typically both softer and fainter than X-1, making it harder to study with other facilities). As X-2 is now known to be a pulsating ULX, a magnetic accretor model was then adopted (formally a DISKBB+DISKBB+CUTOFFPL model, with the parameters of the latter component fixed again to the values highlighted above). Trends of $L \propto T^{-4}$ and $L \propto T^3$ were obtained for the cooler and hotter thermal components, respectively, suggesting that the inner

[2] Here the emission from the accretion columns is treated empirically, assuming a CUTOFFPL model with parameters $\Gamma = 0.59$ and $E_{\rm cut} = 7.9$ keV; this is based on the average parameters found for the pulsed emission from the known ULX pulsars via phase-resolved spectroscopy.

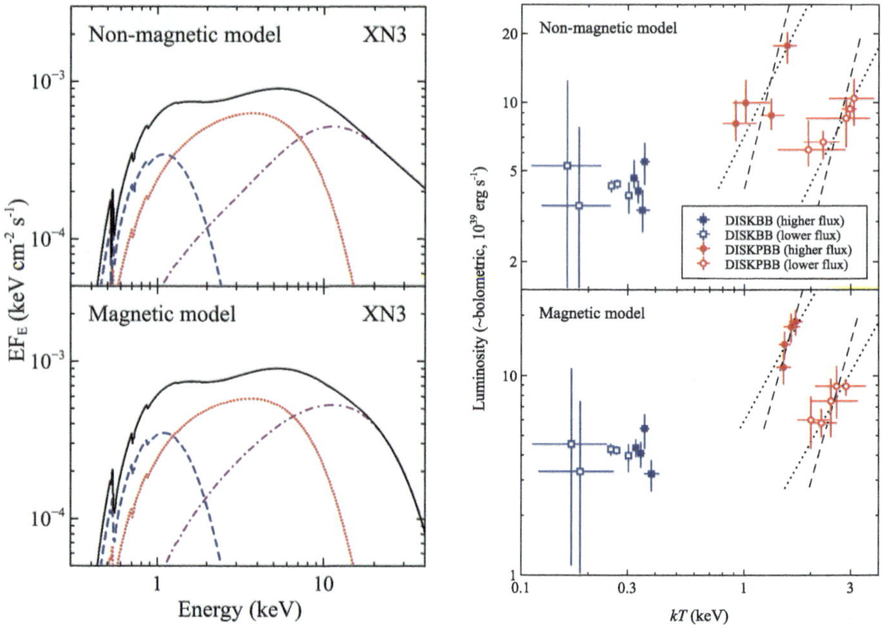

Fig. 12.3 Left: Example model fits to the broadband spectra of NGC 1313 X-1 (a low-flux epoch) for two alternative accretor models. The total model is shown in solid black, the cooler DISKBB and the hotter DISKPBB components are shown in dashed blue and dotted red, and the highest energy component (either SIMPL or CUTOFFPL) is shown in dash-dotted magenta. Right: corresponding $L - T$ trends for 9 epochs. Epochs of low-flux (HUL) and high-flux (BBD) regimes are labelled. The DISKPBB component shows distinct tracks at high and low luminosities. Each track broadly agrees with the $L \propto T^2$ (thick disc) and $L \propto T^4$ (thin disc) predictions [187]

disc is not geometrically thin and that the cooler component is perhaps tracing an expanding disc photosphere or soft X-ray emission from the wind [147].

Another deep X-ray campaign was performed on the supersoft/SUL source NGC 247 ULX-1 with *XMM-Newton* (800 ks). Starting in December 2019 and consisting of 8 uninterrupted exposures, it yielded the clearest view of an extremely soft ULX. These data revealed clear evidence of winds (see Sect. 12.4.4.2) and quasi-periodicities (see Sect. 12.2.2). A detailed X-ray spectroscopic analysis was performed to study the spectral evolution over time and within various ranges of spectral hardness. The results show a clear anti-correlation between the temperature and radius of the cool blackbody component [31] in agreement with an outflowing photosphere model [159], as well as with early results with different facilities [39] and results for other ULSs [172].

More recent work performed on a sample of 17 ULXs observed with a range of X-ray facilities has suggested that the hardest sources might be powered by strongly-magnetised neutron stars with the high-energy emission likely dominated by the accretion column. This is supported by the strong variability of such components

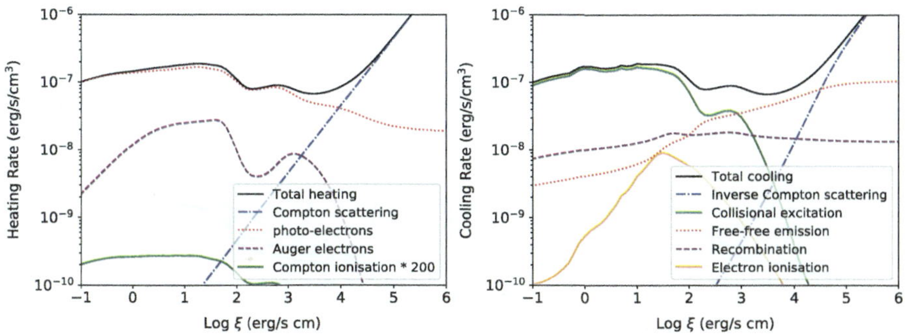

Fig. 12.5 Heating rates (left panel) and cooling rates (right panel) calculated for NGC 1313 X-1 (HUL) with the PION code available in SPEX [100, 130]

The ionisation balance of photoionised plasmas is usually described with the relationship between the temperature and the ionisation parameter (see Fig. 12.6, left panel). These curves correspond to the computation for the SEDs shown in Fig. 12.4 (left panel) and the hard/soft XRB state in Fig. 12.2 (left panel); the branches where the curves flatten indicate where Compton heating is important and show the temperature expected for a thermal wind. SEDs of sources with harder X-ray spectra, e.g. Sy1 NGC 5548 and the XRB hard state, are characterised by excess heating and high Compton temperature. A better visualisation of the ionisation balance is given by the stability curves (or 'S curves'). These show the temperature or the ionisation parameter as a function of the ratio between the radiation pressure (F/c) and thermal pressure ($n_{\mathrm{H}}kT$), which is given by $\Xi = F/n_{\mathrm{H}}ckT = 19222\,\xi/T$, with $F = L/4\pi R^2$ [88]. Along these curves heating equals cooling and, therefore, the gas is in thermal balance. On the left side of the S curve, cooling dominates overheating, whilst on the right side, heating dominates over cooling. Where the S curve has a positive gradient, the photoionised gas is thermally stable, i.e. small perturbations upwards (downwards) will be balanced by an increase of cooling (heating). Negative gradient branches are characterised by plasma that is thermally unstable to temperature perturbations. The stability curves computed for our representative SEDs are shown in Fig. 12.6 (right panel) and underline how hard X-ray spectra correspond to S curves with large regions of instability. Photoionisation modelling of ULX X-ray spectra does indeed show that the wind is found preferentially along thermally stable branches [130]. This occurs also for thermal winds in XRBs, which appear mainly in soft states with highly stable SEDs [135], and AGN warm absorbers [70]. For more detail on ULX wind detections and properties see Sects. 12.4.3 and 12.5.2.

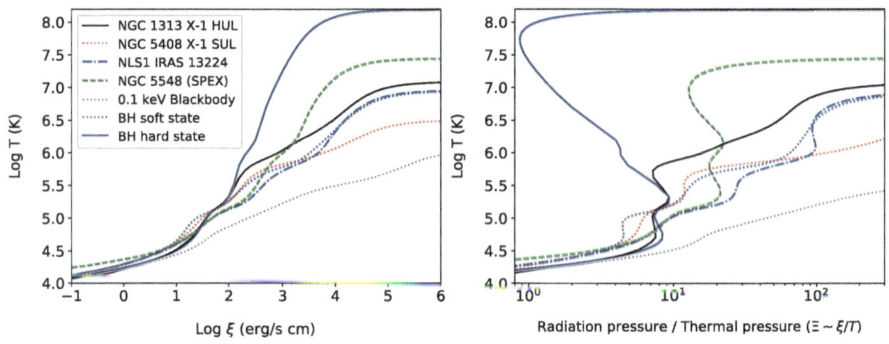

Fig. 12.6 Left panel: curves of ionisation balance ($kT - \xi$) for a variety of SEDs from Galactic XRBs, ULXs and AGN (see Figs. 12.2 and 12.4). Right panel: corresponding stability curves ($kT - \Xi$) with negative slopes indicating unstable branches [130]

12.4.3 The Quest for Spectral Lines in X-Ray Spectra

The first well-exposed ($\sim$100 ks) RGS spectrum of Holmberg II X-1, one of the 5 brightest ULXs (in flux), showed an excess of counts at 569 ± 9 eV. With an equivalent width of 28 ± 12 eV, it was consistent with emission from a blend of lines from the He-like O VII triplet (at 561–574 eV or 21.6–22.1 Å), which would correspond to a luminosity of $\sim 10^{38}$ erg s^{-1} [48]. However, given the low significance of the feature, it was interpreted as being due to incorrect modelling of the neutral O edge in the ISM model caused by the assumption of Solar metallicity.

The first major step in this field was achieved through the stacking of several observations totalling over 300 ks with the high spectral resolution RGS aboard *XMM-Newton*. Such stacking was possible in two sources, NGC 1313 X-1 and NGC 5408 X-1, and enabled the 1 keV features in the CCD-resolution spectra of ULXs to be resolved into a forest of emission and absorption lines [127]. The emission lines are found near their laboratory wavelengths but their strength and position slightly vary with the source luminosity and spectral hardness [131], which indeed confirms they are associated with the ULX itself (see Fig. 12.7, left panel). When modelled with line-emitting plasma in either photoionisation or collisionally-ionisation equilibrium they yield X-ray luminosities of $L_X \sim 10^{38}$ erg s^{-1}, orders of magnitude brighter than the emission lines detected in the sub-Eddington Galactic XRBs [5]. The strong absorption features above (below) 1 keV in NGC 1313 X-1 (NGC 5408 X-1) could not be identified with the laboratory energies of the dominant transitions. In order to find the exact line centroids and compute possible Doppler shifts, a spectral scan was performed using a moving gaussian line with a step of 250 km s^{-1}, about 1/3 of the RGS spectral resolution. Figure 12.7 (right panel) shows the results of this scan for three flux-resolved spectra of NGC 1313 X-1 adopting three different line widths. Apart from the O VII He-like triplet, O VIII and Ne X Ly-α emission lines, there is a strong absorption feature bluewards of the Ne X line ($\gtrsim$1 keV) and other

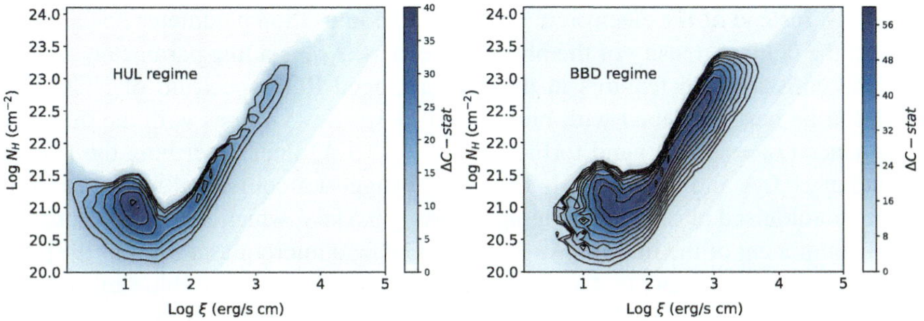

Fig. 12.9 Photoionisation emission model scan of the 2017 HUL (L-I, left) and BBD spectra (I-B, right panel) of NGC 1313 X-1 (RGS $\sim$ 0.3–2 keV, EPIC $\sim$ 2–10 keV). The color is coded according to the ΔC-stat fit improvement to the continuum model. The black contours refer to the 2.5, 3.0, 3.5, 4.0σ, etc. confidence levels estimated through Monte Carlo simulations (see Sect. 12.4.4.3). Negative values of ΔC-stat for high $N_{\rm H}$ at low-to-mild ξ are set to zero for display purposes [131]

plasmas. The RGS data available for NGC 1313 X-1 is so deep that the emission from the He-like O VII triplet (resonance line at 21.6 Å, intercombination line at 21.8 Å and forbidden line at 22.1 Å) can be resolved. Accounting for uncertainties in the line widths, the continuum and the RGS calibration, the following ratios were estimated: $r = 0.6\pm^{1.0}_{0.4}$ and $g = 1.3\pm^{1.6}_{0.8}$ [131]. Despite the large uncertainties, it is clear that they are compatible with a high-density hybrid gas ($n_{\rm H} \sim 10^{10-12}$ cm^{-3}) where both recombination and collisional processes occur [136], at least for the cool O VII emitting plasma.

Using the PION code available in SPEX, a scan of the $N_{\rm H} - \xi$ parameter space was performed for NGC 1313 X-1 RGS spectra adopting Solar metallicity, plasma density $n_{\rm H} = 10^{10}$ cm^{-3}, a full solid angle $\Omega = 4\pi$, velocity dispersion of 1,000 km s^{-1} and no Doppler shift ($v_{\rm LOS} = 0$). Figure 12.9 shows the ΔC-stat spectral improvements to a simple continuum model in the $N_{\rm H} - \xi$ space. The significance of the detections was constrained by running MC simulations (see Sect. 12.4.4.3). The plots show that the plasma is multi-phase, with a higher ionisation component that becomes dominant when the source is brighter (i.e. when the spectrum is in the bright broadened disc regime; right panel). This may indicate either a connection between the line emission and the accretion rate or that our view of the inner regions of the accretion disc changes with time (perhaps via disc precession [131]).

For a few ULXs the RGS spectra show very strong emission lines with wavelengths that are incompatible with well-known transitions. A notable example are those from the intermediate spectral hardness HUL NGC 5204 X-1 and the soft-regime spectra of NGC 55 ULX-1 and NGC 247 ULX-1. In these cases, the scanning physical models require a deep exploration of a parameter space which includes at least the LOS velocity and a parameter describing the state of the plasma. For model grids of CIE line-emission, one can search through a grid of line of sight velocity ($v_{\rm LOS}$), velocity dispersion (v_σ) and electron temperature (k$T_{\rm e}$). The normalisation or emission measure ($n_{\rm H} n_{\rm e} V$) is a free parameter. In the case of line emission from PIE plasmas, instead of the electron temperature, the ionisation parameter ξ is adopted, whilst the column density of the photoionised gas, N_H, is a free parameter.

The emission-like features in the time-averaged RGS spectrum of NGC 5204 X-1 can be best described with blueshifted (0.3 c) Fe XVII lines with the dominant resonance ($\lambda_0 = 15.0$ Å) and forbidden ($\lambda_0 = 17.1$ Å) lines describing the features found at 10.0 Å and 11.3 Å [83]. This would suggest a collisionally-ionised rather than photoionised plasma with a temperature $\gtrsim 0.5$ keV, which is very similar to the cool component of the multiphase jet in the Galactic microquasar SS 433 [98].

It is important to note that similar outflows producing strongly blueshifted emission lines around 1 keV have also been found, albeit at lower significance and only during particular epochs, in a few Galactic transients. For instance, the *Chandra*/LETGS spectrum extracted during a very long burst of the NS LMXB SAX J1808.4–3658 exhibited a strange system of emission lines between 10–14 Å (mainly Ne X) that can be best described with CIE emission blueshifted by about 0.1 c [126]. Another example is the very faint X-ray binary IGR J17062-6143 which showed a variable pattern of emission lines from the same ion but a lower blueshift ($\sim$0.05 c) in a *Chandra*/HETGS spectrum [32]. In the former case, the outflow might have been driven by the powerful burst (amongst the brightest ever detected), whilst in the latter case the very low accretion rate might suggest an origin due to the propeller effect near the NS magnetosphere. Both scenarios might be relevant in ULXs since many of them are likely powered by super-Eddington magnetised neutron stars.

The emission lines detected in the SUL RGS spectra of NGC 55 ULX-1 [128] and NGC 247 ULX-1 [132] can be better described with emission from photoionised plasma as expected for a photoionised wind. The centroids of the lines are more blueshifted at higher fluxes which might indicate an enhanced radiation pressure at higher accretion rates. Further discussion is presented in Sect. 12.5.2.

12.4.4.2 Absorption-Line Gas

As shown in Fig. 12.8, most absorption lines are not consistent with the laboratory wavelengths of the dominant transitions expected in the soft X-ray band (0.3–2 keV). Therefore the scan of physical models must include v_{LOS} as a main parameter to account for the Doppler shift. The normalisation is represented by the column density for both CIE and PIE plasma models here. The ionisation balance is computed identically to the line-emitting gas but with the difference that here it can be done more quickly since the number of dominant transitions is smaller, which is mainly characterised by ground-state (often 1s–2p level) transitions. It is also well known that forbidden lines in absorption have typically a smaller oscillator strength than resonance lines. In SPEX other than the aforementioned PION there is the faster (and more stable) XABS component which adopts a pre-calculated ionisation balance.

The first application of the SPEX/XABS component to model the absorption lines in the RGS spectra of two ULXs, the HUL NGC 1313 X-1 and the SUL NGC 5408 X-1, revealed that they were compatible with absorption from mildly to highly ionised ionic species, the same as those seen in emission (such as O VII-VIII, Ne IX-X

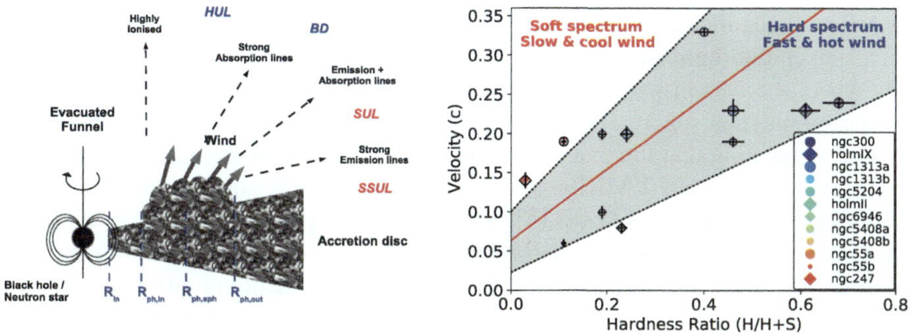

Fig. 12.11 Left panel: schematic cartoon of a super-Eddington accretion disc. The accretion disc is geometrically thick around the spherisation radius, $R_{\rm sph}$, where an optically-thick clumpy outflow is launched by radiation pressure. The wind becomes optically thin at $R_{\rm ph,out}$. Along LOS closer to the disc axis the spectra will appear hard. At moderate inclinations, the spectra will appear soft with strong absorption lines from the wind. At high inclinations the continuum is significantly absorbed and strong emission lines appear in the spectra with the latter becoming supersoft. Right panel: correlation between LOS velocity, ionisation parameter (point size) and hardness ratio (1–10 keV flux/0.3–10 keV flux) for ULX wind detections (combining different sources and, where available, multiple epochs [130])

The properties of the winds, e.g. velocity, outflow rate and temperature, depend on the accretion rate and the mass of the compact object. Magneto-hydrodynamic simulations of super-Eddington accretion onto black holes [119] have shown that above the supercritical accretion rate ($\dot{M}_{\rm cr} = 9/4\,\dot{M}_{\rm Edd}$ [137]) the radiation pressure inflates the inner accretion flow, which becomes geometrically thick, and launches fast winds; the escape velocity exceeds 0.1 c in the innermost regions ($R \lesssim 100\,R_{\rm G}$). These super-Eddington winds are expected to be launched from inside the spherisation radius, $R_{\rm sph} = (\dot{M}/\dot{M}_{\rm Edd})R_{\rm in}$, where $R_{\rm in}$ is the inner disc radius, which in absence of disc truncation is considered to be about 6 $R_{\rm G}$ (see Fig. 12.11, left panel). The winds are also expected to have a mild ionisation state ($\log \xi \sim 2-4$) owing to the relatively soft SEDs of super-Eddington accretion discs. Such ionisation parameters are fully consistent with the properties of the extreme outflows found in ULXs. The discovery of these outflows, together with their unusual broadband spectra and the discovery of pulsating ULXs, has helped establish the ULX population as the best local examples of super-Eddington accretors.

The presence of transient neutron stars that can reach luminosities in excess of the Eddington limit ($L_{\rm Edd,NS} \sim 2 \times 10^{38}$ erg s^{-1}) was already established 40 years ago [155]. However, the discovery of the first pulsating ULX in M 82 [8] and, especially, the HLX NGC 5907 X-1 [59], showed that these objects can reach highly super-Eddington luminosities; the latter reaches 10^{41} erg s^{-1}, i.e. 500 times $L_{\rm Edd}$. In principle, extreme magnetic field strengths (magnetar-level, $B \gg 10^{13}$ G) can suppress the Thomson electron scattering cross-section. This would reduce the impact of radiation pressure and, therefore, raise the effective Eddington limit, which may allow the NS to reach extreme rates [117]. However, the presence of ultrafast winds in many ULXs, including those with pulsations [84], would rather suggest that 'classical' super-Eddington accretion flows do exist (i.e. such that the geometrically thick regions of the disc in which radiation pressure exceeds gravity do form) outside of the pulsar magnetosphere; this would correspond to the scenario in which $R_{sp} > R_M$, in turn suggesting that the magnetic field strengths are more moderate.

Some transient BH LMXBs, such as GRS 1915+105, sometimes reach mildly super-Eddington luminosities [115]. SMBHs powering NLSy1 AGN can also shine up to a few times L_{Edd} [129]. However, stellar-mass compact objects in the most extreme ULXs would require accretion rates (relative to Eddington) orders of magnitude higher. This can be achieved, for instance, through thermal-timescale mass transfer in high mass binaries [74] when the companion star evolves, filling its Roche lobe. Binary evolution models have shown that the mass transfer can be very high during this phase, peaking at $10^{-3}\ M_\odot\ \mathrm{yr}^{-1}$ for $\sim 10^4$ yr [144, 193]. This can occur either for a BH with a companion star heavier than $\sim 10\ M_\odot$ or a NS accreting from a Helium-burning secondary with mass $\sim$1–2 $M_\odot$. These systems can often reach 1,000 $\dot{M}_{Edd}$ during the 10^5 yr thermal-timescale phase and are prime candidates for explaining the presence of extreme ULXs.

12.5.2 Super-Eddington Disc-Wind Structure

At $\dot{M} > \dot{M}_{Edd}$, the accretion becomes radiatively inefficient due to strong advection, photon trapping and heat transfer within the disc. At higher accretion rates, the luminosity no longer increases linearly with $\dot{M}$, and should instead scale logarithmically. However, at the same time the disc is expected to become geometrically thick, assuming a funnel shape which, via scattering, can collimate photons produced in the inner regions into the evacuated cones away from the disc plane (geometrical beaming). This can help sources appear to reach high luminosities if spherically-symmetric emission is assumed. Accounting for all the effects, the observed or apparent luminosity calculated assuming spherical symmetry can be expressed as $L \simeq L_{Edd}[1 + \ln(\dot{M}/\dot{M}_{Edd})/b]$, where b is the beaming factor. This is proposed to scale with the accretion rate as $b = 73(\dot{M}/\dot{M}_{Edd})^{-2}$ [75].

The current scenario [137] suggests the presence of three main zones that characterise the structure of a super-Eddington accretion disc (see Fig. 12.11). Zone 1 or the innermost region ($R < R_{ph,in}$), where the wind is optically thin, the disc can be approximated with a hot accretion flow and produce the component with kT ~few keV observed in ULXs. Zone 2 between the inner edge of the photosphere and the spherisation radius ($R_{ph,in} < R < R_{sph}$) where the wind is opaque and forms a continuation of the accretion disc. As an expanding photosphere, this will have a $T \propto R^{-1/2}$ profile with typical temperatures of a few times 0.1 keV. Zone 3 with $R_{sph} < R < R_{ph,out}$ where the optical depth of the wind falls as $1/R$. The luminosity of this region is Eddington limited and the temperature profile follows the standard $T \propto R^{-3/4}$ trend.

The funnel shape of the disc-wind structure implies that along the disc axis, the wind is either absent or highly ionised and the X-ray spectra would therefore appear harder (HUL) as the observed emission is dominated by the hot inner accretion flow. At larger viewing angles, the optically-thick clumps of the wind absorb the X-ray photons from the innermost hot regions and the spectra appear softer with strong absorption lines. At very high inclinations strong emission lines become visible against a highly-suppressed continuum. If the LOS is crossed by large wind clumps, or even by the thick disc itself, the spectra will appear supersoft. These general predictions are confirmed by the correlation between the LOS velocity and the ionisation parameter of the wind with the observed spectral hardness (characterised as a hardness ratio, defined as 1–10 keV flux/0.3–10 keV flux) measured for a small sample of ULXs with high-quality RGS spectra (see Fig. 12.11, right panel [130]). They are also supported by the anti-correlation between the spectral hardness and the number of lines detected in the RGS spectra of a sample of 19 ULXs [86]. The transition from an intermediate-luminosity HUL ($\Gamma < 2$) spectrum to a bright broadened disc (BBD) is likely due to a higher accretion rate as confirmed by the presence of a cooler and slower wind in the BBD spectrum of NGC 1313 X-1, which is expected for an expansion of $R_{\rm sph}$, i.e. a larger launching radius [131]. Here we assume that the wind speed is associated to the escape velocity at the wind launching radius $v_{\rm esc} = \sqrt{2GM/R}$.

Although there is not yet a large consensus, it is plausible that the ubiquitous low-velocity emission lines are produced by slow-moving thermal winds (see Sect. 12.4.1) as these are unlikely to obscure the inner accretion flow but will emit as the gas cools and recombines [107, 132]. This would occur near the Compton radius $R_{IC} = GM\mu m_{\rm p}/(kT_{IC})$. For a $10\,M_{\odot}$ black hole and a Compton temperature of $T_{IC} \sim 10^{6-7}$ K (i.e. the temperature where the $T - \xi$ curve flattens for the ULXs in Fig. 12.6, left panel), this would correspond to a radius of a few times $10^5\,R_{\rm G}$. Alternative solutions would be reprocessing in the outer disc (especially for the cool O VII component), or emission from a bulge near the magnetosphere [81] (particularly for the hotter O VIII-Ne X component). In principle wind collisions with the CSM could contribute but they would vary on much longer timescales than what is observed, but internal shocks of the ULX wind (if heated above 1 keV) or collisions with the stellar wind of the companion star might also be a possible explanation.

12.5.2.1 The Case of Magnetised Neutron Stars

As previously mentioned, some ULXs host neutron stars as shown by the discovery of pulsations in a handful of sources. M 82 X-2 [8] and NGC 5907 X-1 [59] are very bright and could potentially host NS with high magnetic fields, i.e. $B \sim 10^{12-14}$ G. Here, the Thomson electron scattering opacity can be significantly suppressed and facilitate high luminosities [117] with a significant truncation of the inner accretion disc. This might prevent the formation of the spherisation radius with the matter channeled along the magnetic field lines and forming of an accretion column. The presence of relativistic winds in many ULXs, including some with pulsations [84, 86, 173], would suggest small truncation and low-to-moderate magnetic fields.

GR-MHD simulations of super-Eddington accreting NS with low magnetic fields have shown properties similar to black holes [120]. The net mass-accretion rate onto the NS exceeds the critical rate albeit being 20–30% that of BHs for the same input $\dot{M}$. This means that the mass-outflow rate is a few times larger in flows around non-magnetic NSs than in flows around BHs.

At intermediate magnetic fields ($B \sim 10^{10}$ G) the supercritical disc is truncated at a few NS radii, where the magnetic pressure associated with the magnetic field of the central NS balances with the radiation pressure in the disc. The transport of angular momentum spins up the NS of about -10^{-11} s s^{-1} or even more. The ejecta consists of collimated outflows (jets moving at $0.4\,c$) and winds with typical velocities of $0.1\,c$, the latter with a much higher density [165]. Similar work on NS with a magnetic field $B \sim 10^{12}$ G has shown that copious photons are generated at the shock near the NS surface and escape from the sidewall of the accretion columns but high-density inflow and low-density outflows appear within the column. The matter significantly slows down and much of it does not reach the surface due to a strong radiation force [73]. At high accretion rates with the aid of advection, the matter can broaden the column, mass-load the field lines, and produce radiation-driven, mildly relativistic ejecta with properties similar to those seen in ULX winds [2]. Here, beaming effects are expected to be qualitatively similar to those of outflows in black hole ULXs [76].

The predicted spin up rates and outflows are remarkably consistent with the observations [8, 127] which would provide strong evidence for ULXs to be mainly powered by neutron stars with magnetic fields $B \sim 10^{10-13}$ G, $\dot{M} \sim 10 - 100\,\dot{M}_{\rm Edd}$, and magnetospheric radii $R_M \lesssim R_{\rm sph} \sim 10^{7-8}$ cm or $10 - 100$ NS radii [77, 118]. We adopted the standard definition $R_M \sim 7 \times 10^7\,\Lambda m^{1/7} R_6^{10/7} B_{12}^{4/7} L_{39}^{-2/7}$ cm, where m = $M_{\rm NS}/M_\odot$, $B_{12} = B/10^{12}$ G, $L_{39} = L/10^{39}$ erg s^{-1}, $R_6 = R_{\rm NS}/10^6$ cm and $\Lambda = 0.5$ is a constant corresponding to a specific accretion disc geometry [117]. The discovery of pulsations in ULXs characterised by harder (typically HUL) spectra would confirm the funnel shape of the system although this could also be a result of the accretion column structure. Moreover, given that the magnetic field might be buried by accretion over time, a large fraction of the ULX population likely consists of NS that do not exhibit pulsations. This is suggested by the remarkable similarity between the spectra of ULX with/without pulsations [184].

12.5.3 Feedback and Growth Rate

ULX winds have extreme velocities and are therefore expected to have a significant impact on their surrounding environments. Many ULXs, including NGC 1313 X-1, are indeed surrounded by huge interstellar cavities. Some of them have a supersonic expansion rate (80–250 km s^{-1}) and optical line ratios which suggest a mechanical expansion driven by outflows [51]. In these cases, they are called bubbles and show typical ages of 10^{5-6} yr and Hα luminosities that require a mechanical power of 10^{39-40} erg s^{-1} [121]. In other cases the optical lines seem to be more consistent with photoionisation [16], which could be caused by the ULX radiation field (in this case they are known as nebulae). Regardless of the exact driving mechanism, it is clear that ULXs can deposit huge amounts of energy into the surrounding ISM.

It is possible to estimate the mechanical power of ULX winds and the amount of matter that is carried by the wind. The mass outflow rate is taken from the mass conservation law: $\dot{M}_w = 4\,\pi\,R^2\,\rho\,v_w\,\Omega\,C$, where Ω and C are the solid angle and the volume filling factor (or *clumpiness*), respectively, $\rho = n_{\rm H}\,m_p\,\mu$ is the gas density, and R is the distance from the ionising source. From this, an expression for the kinetic luminosity can be derived by substituting the $R^2\rho$ factor from the definition of the ionisation parameter ($\xi = L_{\rm ion}/n_{\rm H}R^2$), giving $L_w = 0.5\,\dot{M}_w\,v_w^2 = 2\,\pi\,m_p\,\mu\,\Omega\,C\,L_{\rm ion}\,v_w^3\,/\,\xi$.

According to GR-MHD simulations [80], radiatively-driven winds in super-Eddington systems are highly porous ($C \sim 0.01 - 0.1$) and have solid angles of about $\Omega/4\pi \gtrsim 0.3$. Inserting the typical values of ionisation parameters found in ULX winds ($\log \xi \sim$ 2–4) in Eq. (23) of [80] we estimate comparable values of clumpiness. The fraction of ULXs with detection of emission lines is very high ($>60\%$ [86]), whilst absorption lines (i.e. fast winds) are detected in a smaller fraction of the sample (e.g. ~30% [130]). Assuming the detection fraction as a proxy for the solid angle implies values broadly consistent with the simulations [80].

The Ω and C parameters along with ξ and $v_{\rm LOS}$ estimated for the wind through the photoionisation models (see Sect. 12.4.4.2) yield a kinetic power $L_w = 10^{39-40}$ erg s^{-1}, which is comparable to the ULXs X-ray luminosity. This means that about 50% of the total energy budget is expended powering the winds observed, and their mechanical power is sufficiently high to inflate the bubbles. Using the estimates of L_w, the age (τ) of the bubbles and adopting a typical ISM density $\rho_{\rm ISM} \sim 10^{-24}$ g cm^{-3}, it is possible to predict a bubble size $R_{\rm bubble} = 0.76(\dot{M}_w\,v_w^2/2\,\rho_{\rm ISM})^{1/5}\tau^{3/5} \sim 100$ pc [26], which is in excellent agreement with the observations [121].[4]

Regarding the wind mass outflow rate, assuming a standard accretion efficiency, i.e. $\eta = 0.1$, we estimate $\dot{M}_w^{\rm max}$ ~90% of the total inflow rate through the outer disc. However, if we take into account advection and photon trapping we estimate a more realistic mass loss rate of 10–50% [118]. This means that the compact object would still grow rapidly despite the presence of these powerful outflows.

The total energy output of a ULX integrated over its time of activity can be as high as 10^{53} erg, and released within a region of ~1 kpc^2. The present population of ULXs in nearby galaxies is rather small (1–2 per galaxy). However, in the local Universe the number of ULXs per galaxy increases in galaxies with lower metallicity [91] and, especially, with higher star formation rates [90, 111]. It is therefore possible that at $z = 2 - 3$, near the peak of star formation, ULXs were much more abundant and

[4] Here, a lifetime of 10^6 yr is adopted from the current expansion velocities of ~100 km s^{-1} and the expansion law $R \sim t^{3/5}$. The long recombination time of He III (see Sect. 12.1.3 [16]) and, alternatively, arguments related to the maximum amount of matter that can be accreted onto a NS without becoming a BH [130] agree with such long timescales.

had a significant effect on the evolution of their host galaxies. The effects, perhaps comparable to those of supernovae, may help to explain the inconsistencies between the observed amount of galaxies and the (higher) theoretical predictions in the low-mass end of the mass function of galaxies [154].

12.5.4 ULX as Probes of the Primordial Black Holes

Winds in active galactic nuclei are common features and include a variety of phases from the slow ($v_{\rm LOS} \lesssim 5{,}000\,{\rm km\,s^{-1}}$) warm absorbers to the ultrafast outflows reaching $0.4\,c$ (see Sect. 12.4.1). The fraction of AGN exhibiting UFOs is of about 40% [169] but the fraction might increase at higher accretion rates as shown by the frequent detections in quasars and NLSy1 [85, 129, 145, 186, 197]. This could be an indication that radiation pressure starts to be important and to contribute to the launch of the winds which at low $\dot{M}$ might be driven by magnetic fields instead. The main difference between ULX and AGN UFOs is that in the latter case they exhibit higher values of ionisation parameter (on average) due to the harder SEDs of AGN. However, amongst the AGN UFOs, the highly-accreting NLSy1 show cooler winds with properties more similar to those seen in ULXs.

The discovery of AGN powered by SMBHs with masses $\sim 10^9 M_\odot$ at redshift $\gtrsim 7$ when the Universe was less than a Gyr old [11, 37] indicated that some channels of the fast growth of SMBHs are required. Possible channels for the formation of supermassive black holes are the merging of many smaller black holes, Eddington-limited accretion onto IMBH seeds or super-Eddington accretion. Presently, all these channels provide viable solutions: recent gravitational wave discoveries have shown that merging amongst compact objects does occur [1], although it might be not straightforward to merge a million BHs due to the system instability. Of course, BH-BH mergers may have helped to reach large black hole masses of e.g. $\sim 10^{3-4} M_\odot$ in the very early Universe [176] but then a substantial accretion rate has to be sustained in order to reach the SMBH masses in a few hundred million years [62]. Accretion is a mass-scale invariant, even if in the case of SMBHs they primarily accrete from the ISM (e.g. during galactic mergers) rather than a binary companion (although in rare cases they can still accrete from stars during tidal disruption events; TDEs).

TDEs are transient flares of activity caused by partial or total disruptions of stars that get too close to a SMBH. Typical BH masses are $\lesssim 10^8\,M_\odot$, and the peak accretion rates during these events are widely expected to be highly super-Eddington. They are typically characterised by supersoft X-ray spectra ($kT_{\rm BB} \sim 0.1\,{\rm keV}$) and are believed to reach fallback luminosities $\sim 100\,L_{\rm Edd}$ for a short period [196]. One very bright TDE that occurred in a nearby galaxy provided a high-quality, high-resolution spectrum which showed evidence of radiation-pressure driven winds at low and high velocities [71, 109]. For these reasons, TDEs and NLSy1 (both characterised by SMBHs with moderate masses and high accretion rates) could be considered the supermassive counterparts of ULXs [30]. GR-MHD simulations confirm that the black hole can rapidly grow with a growth time scale given by $\tau_{\rm growth} = M/\dot{M} =$

$4.5 \times 10^6(\dot{m}/100)^{-1}$ yr [119]. This is a fairly short timescale and indicates that SMBHs can be formed by super-Eddington accretion provided that such a high accretion rate is sustained for long enough given that at most about 50% of the matter is lost into the wind [130].

Finally, even though it now seems like they would be in the minority, it is still possible that some ULXs could host IMBHs ($10^{2-5} M_\odot$) and shed light onto another of the potential SMBH formation pathways, with ESO 243–49 HLX-1 ($L^{\rm max} = 2 \times 10^{42}$ erg s^{-1}) and M 82 X-1 ($L^{\rm max} \sim 10^{41}$ erg s^{-1}) amongst the best candidates. They both follow an $L \propto T^4$ relationship which, together with their X-ray variability and radio properties (the latter for HLX-1), would suggest masses of a few 10,000 $M_\odot$ and 400 $M_\odot$, respectively [123, 190]. However, there are large uncertainties in these measurements and some results would instead suggest masses of $\sim 10^5 M_\odot$ for the former [167] and $20 - 118\ M_\odot$ for the latter [18]. This would place ESO 243–49 HLX-1 in the low end of the SMBHs and M82 X-1 in the high end of the stellar-mass BHs. HLX-1 is indeed found in a cluster of stars and which might be the remnant of a tidally stripped dwarf satellite galaxy [158] with the SMBH accreting from a captured, partially-disrupted, companion star [92].

12.6 Future Prospects

Novel techniques in X-ray timing and spectroscopy, combined with theoretical simulations, have provided a remarkable improvement in our knowledge of the astrophysics of ULXs. However, despite the significant recent progress, several important questions remain unanswered. *What is the relative fraction of BH-ULXs and NS-ULXs? Are there any spectroscopic differences between them? Are the spectral transitions triggered by orbital variations in the accretion rate or by stochastic variations in the wind? How does ULXs feedback evolve over time and affect the evolution of their host galaxies?* To solve these issues a combination of accurate plasma modelling, spectral-timing studies and more sensitive observations is required.

12.6.1 Current Limitations

It is difficult to distinguish BHs from NSs in absence of pulsations, cyclotron lines and dynamical mass measurements (see Sects. 12.2.1 and 12.1.3). As mentioned in Sect. 12.5.2.1, some differences are expected to be found in wind properties depending on the accretor. In principle QPOs, sharp and high-amplitude flux drops in the light curves and $L - T$ trends may place further constraints, but these strongly depend on theoretical models and their interpretations. Of course, having two or more methods pointing towards the same solution e.g. a NS or a BH would be encouraging but in many cases we face signal-to-noise limitations. With the current X-ray instruments (such as *XMM-Newton*) only about 20 ULXs have statistics sufficiently good to search for winds through individual lines in grating spectra. Only about 30 sources have time series with sufficient statistics to detect pulsations. CRSFs are difficult to detect as they require high sensitivity across a very broad energy band, which is possible to achieve only for a handful of sources. This means we lack the statistics that are necessary to search for putative ULX sub-samples with different properties, although new deep observations could serve this purpose.

One way to study in detail the size of the accretion disc, and the wind launching region and place constraints on the accretor nature would be using timing arguments such as energy-dependent lags and the response of the wind to continuum variability. However, we lack both the spectral resolution and the sensitivity in ULX X-ray time series. A significant detection of spectral lines requires observations longer than about 100 ks with the grating spectrometers which prevent us from resolving the timescales of continuum variability and the 100–1,000 s delays between different energy bands [4, 72, 80]. This also creates ambiguity in the variability process, the structure and the size of the emitting region.

The emission line plasma is clearly multiphase but the current sensitivity is not sufficient to distinguish between CIE and PIE equilibrium in most observations, which prevents a full understanding of the nature of the X-ray line emission. An accurate estimate of the thermal structure and plasma response to continuum changes (which seems to occur in time-integrated flux-resolved spectra [31, 131, 132]) would allow us to distinguish amongst different interpretations of the absorption component (e.g. from close to the NS accretion column [2] or at $R_{\rm sph}$ [137]) and provide further detail of the ULX effects on the binary system and the surrounding medium.

Another issue is the lack of sensitivity and spectral resolution in the Fe K band (6–10 keV) which limits our capability of detecting the hottest components of the wind launched from the inner disc. These are expected to be detected in absorption primarily in ULXs with hard spectra (HUL), and would fill an observational gap that remains present in the outflow structure. Currently, the only Fe K detections are those for NGC 1313 X-1, NGC 300 ULX-1 and NGC 4045 HLX-1 [22, 84, 181].

12.6.2 Future Missions and Technologies

There are a number of future X-ray missions/mission concepts (both approved and under study) that will enhance our ability to study hot and energetic plasmas, both in the next few years and on longer timescales. Some will carry instruments with a larger collecting area and increased spectral resolution, which will improve our sensitivity to atomic lines, others will extend the X-ray energy band over which sensitive observations can be taken, which will improve our knowledge of the spectral continuum and our ability to detect of broad features such as CRSFs.

12.6.2.1 High Spectral Resolution Detectors

The sensitivity to weak lines scales with the so-called *figure of merit* which is typically defined as the product of the resolving power ($R = E/\Delta E$) and the effective area: $FoM = \sqrt{A_{eff}\, R}$ (but sometimes as $\sqrt{A_{eff}/\Delta E}$ instead). A remarkable improvement with respect to the gratings aboard *XMM-Newton* (Chap. 2) and *Chandra* (Chap. 3) was achieved thanks to the launch of *Hitomi* (previously known as *ASTRO-H*, see Chap. 5) in 2016. This satellite carried a micro-calorimeter detector which measured the energy of an incoming photon through a thermometer cooled down to 50 mK. This device achieved an excellent resolution of 5 eV in the Fe K band and was able to cover the 0.3–12 keV band. The effective area of the calorimeter was better than all of the high-resolution instruments currently operating above 1 keV. Although an observation of NGC 1313 X-1 was approved for the performance verification phase, the satellite was unfortunately lost after 5 weeks, but not before showing its capabilities thanks to a deep observation of the Perseus galaxy cluster [57]. The *X-ray Imaging and Spectroscopy mission (XRISM)*, planned for launch in 2023, will once again provide the high-resolution capabilities of its predecessor through the *Resolve* micro-calorimeter spectrometer. Similarly to *Hitomi*, *XRISM/Resolve* is expected to yield a FoM improvement by a factor 5 around 1 keV and above 10 in the Fe K band with respect to present instruments, albeit with a limited spatial resolution ($PSF \sim 1'$). The *Arcus* mission concept [156] has complementary capabilities. It consists of multiple grating detectors with a high effective area and a spectral resolution that provides an order of magnitude improvement in the FoM below 1 keV. Finally, the largest planned X-ray mission relevant for high-resolution studies of ULXs will be the *Advanced Telescope for High-ENergy Astrophysics (Athena)*, which is currently under ESA study. This facility is planned for launch after 2035 and will bear the most powerful high-resolution X-ray instrument so-far conceived: the *X-ray Integral Field Unit (X-IFU)* [13]. At the time of writing, *X-IFU* is designed to achieve a 2.5 eV spectral resolution coupled with a high effective area (1 m^2) and a good spatial resolution (5–10$''$). With an FoM that is a factor 10 better than any of the current or planned instruments at $\gtrsim 1$ keV, and coverage of the 0.2–12 keV band, *Athena* will revolutionise high-resolution studies of ULXs enabling us to (1) probe shorter variability timescales, (2) resolve ULXs in crowded fields and 3) observe more distant and fainter sources thereby boosting the sample of ULXs with detected spectral lines.

Here we showcase the capabilities of *XRISM* and *Athena* by presenting some simulated observations. In Fig. 12.12 top panels we show the results of photoionisation absorption model scans of the HUL and BBD spectra of NGC 1313 X-1 (see also Fig. 12.7 top and bottom panels). Over 300 ks of exposure time is required by *XMM-Newton/RGS* in order to achieve significant detections (up to 4 σ accounting for the L-E effect with MC simulations [131]). Starting from the best-fitting continuum + wind models, we simulated *Resolve* and *X-IFU* observations for the HUL and BBD spectra of NGC 1313 X-1, then we removed the ionised wind from the spectral model and performed an identical routine (as applied to the *XMM-Newton* datasets) to search for the evidence of outflows in the simulated spectra. The simu-

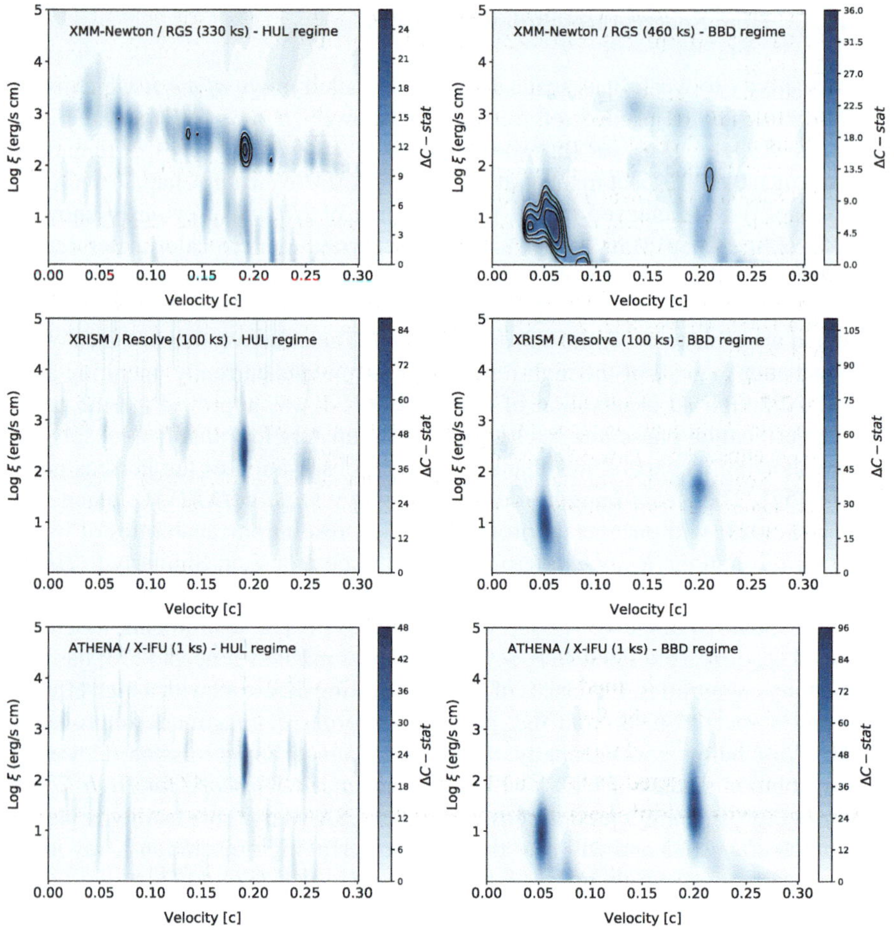

Fig. 12.12 Top panels: photoionisation absorption model scan of the 2006–2012 low-flux HUL spectrum (left, 330 ks exposure) and the 2017 high-flux BBD spectrum (right, 460 ks) of NGC 1313 X-1 taken with *XMM-Newton* (see also Fig. 12.7). Colour scale and labels are same as in Fig. 12.10 [131]. Middle and bottom panels: corresponding simulations with the *XRISM/Resolve* (100 ks) and *Athena/X-IFU* (1 ks) micro-calorimeters adopting the same the best-fit continuum + wind model

lations performed with *Resolve* (middle panels) used a clean exposure of 100 ks and those with *X-IFU* (bottom panel in Fig. 12.7) an exposure of just 1 ks. It is evident that *XRISM* will significantly reduce the present issues with degeneracy between different (ξ, $v_{\rm LOS}$) solutions, and also improve the detection of multiple components (see the case for the more complex outflow in the BBD spectrum). Given the values of $\Delta C - stat$ for a 100 ks *XRISM* exposure, it is obvious that even exposures shorter by a factor 2 would produce excellent results. The results obtained with *X-IFU* will be transformative, given that small snapshots of just 1 ks are required to achieve significant wind detections in the brightest ULXs. This will enable us to probe the timescales of the X-ray variability of the wind, and the nature of the soft X-ray lags [56, 72, 128] and the overall disc structure. Longer exposures will instead allow us to search for winds in much more distant sources, significantly increasing the sample of wind detections in ULXs. Similar conclusions might be derived about our future capabilities to search for further putative narrow proton CRSFs in ULXs.

12.6.2.2 Broadband and Other X-Ray detectors

In Sect. 12.3 we have discussed the importance of an accurate knowledge of the continuum shape of ULX X-ray spectra and the complex multi-component modelling. A correct determination of the continuum has also proved to be mandatory in order to search for narrow and broad spectral features which provided evidence for winds and CRSFs (see Sect. 12.4). Broadband X-ray detectors are also equipped with the highest effective area resulting in the best counting statistics and are, therefore, the ideal instruments to search for pulsations and other timing features.

Amongst the best facilities for ULX-related science there is the *enhanced X-ray Timing and Polarimetry* (*eXTP*) mission (2027-, [198]). It will provide a high effective area in the hard X-ray band (>2 keV) which together with a low background due to the low Earth orbit will dramatically improve the detection of highly-ionised wind components in the Fe K band and cyclotron lines with respect to *XMM-Newton* and *NUSTAR*. We have performed a simple simulation of an 80 ks observation of PULX NGC 300 ULX-1 using the *Spectroscopic Focusing Array (SFA)* aboard *eXTP* and the best-fit continuum + wind model [84]. We have found that the Fe XXV absorption line complex may be detected at 5 σ. The use of physical model scans (see Sect. 12.4.4) will allow us to fit multiple lines at once and achieve such significance at much shorter exposure times, thereby probing variability timescales of a few hours. Moreover, for the brightest sources, *eXTP* might provide the first estimates of polarised flux as an alternative probe of the system geometry.

Although not formally a new mission, a facility that still has to show its potential is *eROSITA*. Currently on hold, the mission scanned the whole sky for four years every six months. It is therefore plausible that, as soon as the data become public, many new transient ULXs will be discovered and followed up with dedicated observations and other facilities. Amongst the proposed X-ray missions, the *High Energy X-ray Probe (HEX-P)* [96] is expected to achieve an effective area 10 times better than the present detector (*NUSTAR*) at energies around 10 keV, along with the first focused coverage up to 200 keV. Unlike *NUSTAR*, this facility will also cover the soft X-ray band down to 0.2 keV. This will be ideal to estimate the high-energy curvature of ULX spectra, study the structure of accretion columns, and detect pulsations and CRSFs in ULXs. We have run another simulation for PULX NGC 300 ULX-1 adopting the CRSF best-fit model, i.e. a broad ($\sigma \sim 3$ keV) feature at 13 keV [184]. A 100 ks observation with *HEX-P* will not only achieve a result comparable to that obtained with 200 ks of observations each with *XMM-Newton* and *NUSTAR*, but the broader band coverage will also break the degeneracy between the line energy and width. Moreover, *HEX-P* will provide the necessary statistics to distinguish between a CRSF model and alternative continuum models that do not require CRSFs [82]. Other concept missions with great potential thanks to their high effective area, spatial and (for the latter) spectral resolution are *AXIS* and *Lynx*. *AXIS* [116] would boost our capabilities to detect ULXs, particularly at high distances and in crowded fields. *Lynx* [45] should also be equipped with both micro-calorimeter and grating spectrometers, providing for the first time high spatial ($0.5''$) and spectral ($R > 2{,}000$) resolution in the whole 0.2–10 keV bandpass, which can boost ULX wind studies at greater distances and significantly enlarge the sample.

Acknowledgements –We acknowledge support from the European Union's Horizon 2020 Programme under the AHEAD2020 project (grant agreement n. 871158). The authors are grateful to Peter Kosec for reading the draft and providing useful insights for the discussion.

References

1. B.P. Abbott, R. Abbott, T.D. Abbott, M.R. Abernathy et al., PhRvL **116**, 061102 (2016)
2. P. Abolmasov, G. Lipunova, Submitted to MNRAS (2022). arXiv:2207.12312
3. M.A. Abramowicz, B. Czerny, J.P. Lasota, E. Szuszkiewicz, ApJ **332**, 646 (1988)
4. W.N. Alston, C. Pinto, D. Barret, A. D'Aì et al., MNRAS **505**, 3722 (2021)
5. R. Amato, V. Grinberg, N. Hell, S. Bianchi, C. Pinto et al., A&A **648**, A105 (2021)
6. T. An, X.L. Lu, J.Y. Wang, A & A **585**, A89 (2016)
7. M. Bachetti, V. Rana, D.J. Walton et al., ApJ **778**, 163 (2013)
8. M. Bachetti, F. Harrison, D.J. Walton et al., Nature **514**, 202 (2014)
9. M. Bachetti, T.J. Maccarone, M. Brightman, M.C. Brumback et al., ApJ **891**, 44 (2020)
10. M. Bachetti, M. Heida, T. Maccarone, D. Huppenkothen et al., ApJ **937**, 125 (2022)
11. E. Bañados, B.P. Venemans, C. Mazzucchelli, E.P. Farina et al., Nature **553**, 473 (2018)
12. F. Barra, C. Pinto, D.J. Walton et al., MNRAS **516**, 3972 (2022)
13. D. Barret, V. Albouys, J.-W. den Herder, et al., Submitted to SPIE (2022). arXiv:2208.14562
14. M.C. Begelman, ApJ **568**, L97 (2002)
15. A. Belfiore, P. Esposito, F. Pintore, G. Novara et al., NatAs **4**, 147 (2020)
16. C.T. Berghea, R.P. Dudik et al., ApJ **708**, 364 (2010)
17. M.C. Bernadich, A.D. Schwope, K. Kovlakas et al., A & A **659**, A188 (2022)
18. M. Brightman, F.A. Harrison, D. Barret et al., ApJ **829**, 2 (2016)
19. M. Brightman, F. Harrison, D.J. Walton et al., ApJ **816**, 60 (2016)
20. M. Brightman, F.A. Harrison, F. Fürst et al., NatAs **2**, 312 (2018)
21. M. Brightman, F.A. Harrison, M. Bachetti, Y. Xu, F. Fürst et al., ApJ **873**, 115 (2019)
22. M. Brightman, P. Kosec, F. Fürst, H. Earnshaw, M. Heida et al., ApJ **929**, 138 (2022)
23. I. Caballero, J. Wilms, Mem. Soc. Astron. Italiana **83**, 230 (2012)
24. S. Carpano, F. Haberl, C. Maitra, G. Vasilopoulos, MNRAS **476**, L45 (2018)
25. P. Casella, T. Belloni, L. Stella, ApJ **629**, 403–407 (2005)
26. J. Castor, R. McCray, R. Weaver, ApJL **200**, L107 (1975)
27. E.J.M. Colbert, R.F. Mushotzky, ApJ **519**, 89–107 (1999)
28. M. Cropper, R. Soria, R.F. Mushotzky, K. Wu et al., MNRAS **349**, 39–51 (2004)
29. D. Cseh, P. Kaaret, S. Corbel, F. Grise, C. Lang et al., MNRAS **439**, L1 (2014)
30. L. Dai, J.C. McKinney, N. Roth et al., ApJL **859**, L20 (2018)
31. A. D'Aì, C. Pinto, M. Del Santo, F. Pintore, R. Soria et al., MNRAS **507**, 5567 (2021)
32. N. Degenaar, C. Pinto, J.M. Miller et al., MNRAS **464**, 398 (2017)
33. H.P. Earnshaw, T.P. Roberts, R. Sathyaprakash, MNRAS **476**, 4272 (2018)
34. G. Fabbiano, ARA & A **27**, 87 (1989)

35. S. Fabrika, ASPRv **12**, 1F (2004)
36. S. Fabrika, Y. Ueda, A. Vinokurov et al., Nat. Phys. **11**(551), 553 (2015)
37. X. Fan, M.A. Strauss, D.P. Schneider, R.H. Becker, et al., AJ **125**, 1649 (2003)
38. S.A. Farrell, N.A. Webb, D. Barret, O. Godet, J.M. Rodrigues, Nature **460**, 73 (2009)
39. H. Feng, L. Tao, P. Kaaret, F. Grisé, ApJ **831**, 117 (2016)
40. G.J. Ferland, K.T. Korista, D.A. Verner et al., PASP **110**, 761 (1998)
41. F. Fürst, D.J. Walton, F.A. Harrison, D. Stern, D. Barret et al., ApJL **831**, L14 (2016)
42. F. Fürst, D.J. Walton, D. Stern, M. Bachetti, D. Barret et al., ApJ **834**, 77 (2017)
43. F. Fürst, D.J. Walton, M. Heida, M. Bachetti, C. Pinto et al., A & A **651**, A75 (2021)
44. K. Fukumura, D. Kazanas, C. Shrader et al., NatAs **1**, 0062 (2017)
45. J.A. Gaskin, D.A. Swartz, A. Vikhlinin et al., JATIS **5**, 021001 (2019)
46. J.C. Gladstone, T.P. Roberts, C. Done, MNRAS **397**, 1836 (2009)
47. J.C. Gladstone, C. Copperwheat, C.O. Heinke, T.P. Roberts et al., ApJs **206**, 14 (2013)
48. M.R. Goad, T.P. Roberts, J.N. Reeves, P. Uttley, MNRAS **365**, 191–198 (2006)
49. O. Godet, J.C. Lombardi, F. Antonini, D. Barret, N.A. Webb et al., ApJ **793**, 105 (2014)
50. A. Gúrpide, O. Godet, F. Koliopanos, N. Webb, J.-F. Olive, A & A **649**, A104 (2021)
51. A. Gúrpide, M. Parra, O. Godet, T. Contini, J.-F. Olive, A & A **666**, A100 (2022)
52. F.A. Harrison, W.W. Craig, F.E. Christensen, C.J. Hailey et al., ApJ **770**, 103 (2013)
53. M. Heida, M.A.P. Torres, P.G. Jonker, M. Servillat et al., MNRAS **453**, 3510 (2015)
54. M. Heida, P.G. Jonker, M.A.P. Torres, T.P. Roberts et al., MNRAS **459**, 771 (2016)
55. M. Heida, R.M. Lau, B. Davies, M. Brightman, F. Fürst et al., ApJL **883**, L34 (2019)
56. L.M. Heil, S. Vaughan, MNRAS **405**, L86–L89 (2010)
57. Hitomi Collaboration, Nature **535**, 117 (2016)
58. Y. Inoue, K. Yabe, Y. Ueda, PASJ **73**, 1315 (2021)
59. G.L. Israel, A. Belfiore, L. Stella, P. Esposito, P. Casella et al., Science **355**, 817 (2017)
60. G.L. Israel, A. Papitto, P. Esposito, L. Stella, L. Zampieri et al., MNRAS **466**, L48 (2017)
61. F. Jansen, D. Lumb, B. Altieri et al., A & A **365**, L1 (2001)
62. C. Jin, C. Done, M. Ward, MNRAS **455**, 691–702 (2016)
63. P. Kaaret, A.H. Prestwich et al., MNRAS **321**, L29 (2001)
64. P. Kaaret, M.J. Ward, A. Zezas, MNRAS **351**, L83 (2004)
65. P. Kaaret, M.G. Simet, C.C. Lang, ApJ **646**, 174–183 (2006)
66. P. Kaaret, H. Feng, T.P. Roberts, ARA & A **55**, 303 (2017)
67. J.S. Kaastra, et al., UV and x-ray spectroscopy of astrophysical and laboratory **411** (1996)
68. J.J.E. Kajava, J. Poutanen, MNRAS **398**, 1450–1460 (2009)
69. T. Kallman, M. Bautista, ApJS **133**, 221 (2001)
70. T. Kallman, A. Dorodnitsyn, ApJ **884**, 111 (2019)
71. E. Kara, L. Dai, C.S. Reynolds, T. Kallman, MNRAS **474**, 3593 (2018)
72. E. Kara, C. Pinto, D.J. Walton, W.N. Alston, M. Bachetti et al., MNRAS **491**, 5172 (2020)
73. T. Kawashima, K. Ohsuga, PASJ **72**, 15 (2020)
74. A.R. King, M.B. Davies, M.J. Ward, G. Fabbiano, M. Elvis, ApJ **552**, L109 (2001)
75. A.R. King, MNRAS **393**, L41–L44 (2009)
76. A. King, J.-P. Lasota, W. Kluźniak, MNRAS **468**, L59 (2017)
77. A. King, J.-P. Lasota, MNRAS **494**, 3611 (2020)
78. A.L. King, J.M. Miller, J. Raymond, A.C. Fabian et al., ApJL **746**, L20 (2012)
79. A.L. King, D.J. Walton, J.M. Miller, D. Barret et al., ApJL **784**, L2 (2014)
80. H. Kobayashi, K. Ohsuga, H.R. Takahashi, T. Kawashima et al., PASJ **70**, 22 (2018)
81. F. Koliopanos, M. Gilfanov, MNRAS **456**, 3535 (2016)
82. F. Koliopanos, G. Vasilopoulos, J. Buchner et al., A & A **621**, A118 (2019)
83. P. Kosec, C. Pinto, A.C. Fabian, D.J. Walton, MNRAS **473**, 5680 (2018)
84. P. Kosec, C. Pinto, D.J. Walton, A.C. Fabian, M. Bachetti et al., MNRAS **479**, 3978 (2018)
85. P. Kosec, D.J.K. Buisson, M.L. Parker, C. Pinto et al., MNRAS **481**, 947 (2018)
86. P. Kosec, C. Pinto, C.S. Reynolds, M. Guainazzi, E. Kara et al., MNRAS **508**, 3569 (2021)
87. K. Kovlakas, A. Zezas, J.J. Andrews, A. Basu-Zych et al., MNRAS **498**, 4790 (2020)
88. J.H. Krolik, C.F. McKee, C.B. Tarter, ApJ **249**, 422 (1981)

89. A. Kubota, C. Done, K. Makishima, MNRAS **337**, L11–L15 (2002)
90. B.D. Lehmer, R.T. Eufrasio, P. Tzanavaris et al., ApJS **243**, 3 (2019)
91. B.D. Lehmer, R.T. Eufrasio, A. Basu-Zych et al., ApJ **907**, 17 (2021)
92. L.C.-C. Lin, C.-P. Hu, K.-L. Li, J. Takata, D.C.-C. Yen et al., MNRAS **491**, 5682 (2020)
93. J.-F. Liu, J.N. Bregman, ApJS **157**, 59 (2005)
94. J.-F. Liu, Y. Bai, S. Wang, Nature **528**, 108 (2015)
95. K.S. Long, S. Dodorico, P.A. Charles, M.A. Dopita, ApJL **246**, L61 (1981)
96. K.K. Madsen, F. Harrison, D. Broadway et al., SPIE **10699**, 106996M (2018)
97. K. Makishima, A. Kubota, T. Mizuno et al., ApJ **535**, 632–643 (2000)
98. H.L. Marshall, C.R. Canizares, N.S. Schulz, ApJ **564**, 941 (2002)
99. G.A. Matzeu, J.N. Reeves, V. Braito, E. Nardini et al., MNRAS **472**, L15 (2017)
100. M. Mehdipour, J.S. Kaastra, T. Kallman, A & A **596**, A65 (2016)
101. M.J. Middleton, T.P. Roberts, C. Done, F.E. Jackson, MNRAS **411**, 644–652 (2011)
102. M.J. Middleton, J.C.A. Miller-Jones et al., Nature **493**, 187 (2013)
103. M.J. Middleton, D.J. Walton, T.P. Roberts, L. Heil, MNRAS **438**, 51 (2014)
104. M.J. Middleton, L. Heil, F. Pintore, D.J. Walton, T.P. Roberts, MNRAS **447**, 3243 (2015)
105. M.J. Middleton, D.J. Walton, A. Fabian, T.P. Roberts et al., MNRAS **454**, 3134 (2015)
106. M.J. Middleton, D.J. Walton, W. Alston, T. Dauser et al., MNRAS **506**, 1045 (2021)
107. M.J. Middleton, N. Higginbottom, C. Knigge et al., MNRAS **509**, 1119 (2022)
108. J.M. Miller, G. Fabbiano, M.C. Miller, A.C. Fabian, ApJ **585**, L37 (2003)
109. J.M. Miller, J.S. Kaastra, M.C. Miller et al., Nature **526**, 542 (2015)
110. M.C. Miller, S.A. Farrell, T.J. Maccarone, ApJ **788**, 116 (2014)
111. S. Mineo, M. Gilfanov, R. Sunyaev, MNRAS **419**, 2095 (2012)
112. S. Mondal, K. Belczyński, G. Wiktorowicz et al., MNRAS **491**, 2747 (2020)
113. S. Mondal, A. Różańska, B. De Marco, A. Markowitz, MNRAS **505**, L106 (2021)
114. C. Motch, M.W. Pakull, R. Soria, F. Grisé, G. Pietrzyński, Nature **514**, 198 (2014)
115. S.E. Motta, J.J.E. Kajava, M. Giustini, D.R.A. Williams et al., MNRAS **503**, 152 (2021)
116. R. Mushotzky, AXIS team. LPICo **2135**, 5025 (2019)
117. A.A. Mushtukov, V.F. Suleimanov, S.S. Tsygankov, J. Poutanen, MNRAS **454**, 2539 (2015)
118. A.A. Mushtukov, A. Ingram, M. Middleton et al., MNRAS **484**, 687 (2019)
119. K. Ohsuga, M. Mori, T. Nakamoto, S. Mineshige, ApJ **628**, 368 (2005)
120. K. Ohsuga, PASJ **59**, 1033 (2007)
121. M.W. Pakull, L. Mirioni, RMxAC **15**, 197 (2003)
122. M.W. Pakull, R. Soria, C. Motch, Nature **466**, 209 (2010)
123. D.R. Pasham, T.E. Strohmayer, R.F. Mushotzky, Nature **513**, 74–76 (2014)
124. C. Pinto, J.-U. Ness, F. Verbunt, J.S. Kaastra et al., A & A **543**, A134 (2012)
125. C. Pinto, J.S. Kaastra, E. Costantini, C. de Vries, A & A **551**, 25 (2013)
126. C. Pinto, E. Costantini, A.C. Fabian, J.S. Kaastra, J.J.M. in't Zand, A & A **563**, A115 (2014)
127. C. Pinto, M.J. Middleton, A.C. Fabian, Nature **533**, 64 (2016)
128. C. Pinto, W. Alston, R. Soria, M.J. Middleton, D.J. Walton et al., MNRAS **468**, 2865 (2017)
129. C. Pinto, W. Alston, M.L. Parker, A.C. Fabian et al., MNRAS **476**, 1021 (2018)
130. C. Pinto, M. Mehdipour, D.J. Walton, M.J. Middleton et al., MNRAS **491**, 5702 (2020)
131. C. Pinto, D.J. Walton, E. Kara, M.L. Parker, R. Soria et al., MNRAS **492**, 4646 (2020)
132. C. Pinto, R. Soria, D.J. Walton, A. D'Aì, F. Pintore et al., MNRAS **505**, 5058 (2021)
133. F. Pintore, L. Zampieri, L. Stella, A. Wolter et al., ApJ **836**, 113 (2017)
134. F. Pintore, S. Motta, C. Pinto, M.G. Bernardini et al., MNRAS **504**, 551 (2021)
135. G. Ponti, R.P. Fender, M.C. Begelman, R.J.H. Dunn et al., MNRAS **422**, L11 (2012)
136. D. Porquet, J. Dubau, A & AS **143**, 495 (2000)
137. J. Poutanen, G. Lipunova, P. Fabrika, et al., MNRAS **377**, 1187 (2007)
138. W.C. Priedhorsky, S.S. Holt, SSRv **45**, 291 (1987)
139. D. Proga, J.M. Stone, T.R. Kallman, ApJ **543**, 686 (2000)
140. R. Protassov, D.A. van Dyk, A. Connors et al., ApJ **571**, 545 (2002)
141. I. Psaradaki, E. Costantini, M. Mehdipour, Trigo M. Díaz, A & A **620**, A129 (2018)
142. Y. Qiu, J. Liu, J. Guo, J. Wang, ApJL **809**, L28 (2015)

Table 13.1 Summary of past high-resolution X-ray spectroscopy of SNRs

Name	Distance[a] (kpc)	Age (yr)	Type	Refs. for high-resolution X-ray spectroscopy			
				Einstein	Chandra	XMM-Newton	Hitomi
Cygnus loop	0.74±0.03	$1–2\times10^4$	CC	[148]	—	[143]	—
RX J1713.7-3946	0.9±0.6	1629	CC	—	—	[141]	—
Puppis A	1.3±0.3	4450±750	CC	[22, 164, 165]	—	[77, 79]	—
SN 1006	∼2	1016	Ia	—	—	[20, 149, 153]	—
RCW86	2.2±0.4	1837	Ia	—	—	[18]	—
Tycho's SNR	3±1	450	Ia	—	[101]	[29, 162]	—
Crab Nebula	$3.37^{+4.04}_{-0.11}$	968	CC	[129, 130]	[156, 157]	[66]	[55]
Cas A	$3.4^{+0.3}_{-0.1}$	342±19	CC	[96]	[90, 124]	[14]	—
G296.1-0.5	4.3±0.8	∼20000	CC	—	—	[25, 139]	—
G21.5-0.9	4.4±0.2	870^{+200}_{-150}	CC	—	—	—	[56]
Kepler's SNR	∼5	418	Ia	—	[100]	[10, 70, 75]	—
G292.0 + 1.8	6.2±0.8	3000±60	CC	—	[152]	[8]	—
SN 1987A	LMC	35	CC	—	[2, 15, 35, 36, 98, 99, 119, 136, 173–175]	[49, 51, 135]	—
SNR 0509-67.5	LMC	310^{+40}_{-30}	Ia	—	—	[84, 85, 159]	—
SNR 0519-69.0	LMC	600±200	Ia	—	—	[83, 84, 159]	—
N103B	LMC	∼800	Ia	—	—	[144, 170]	—
SNR 0540-69.3	LMC	∼1200	CC	—	—	[147]	—
N132D	LMC	∼2500	CC	[61]	[24]	[6, 137]	[54]
SNR 0506-68.0	LMC	∼4000	CC	—	—	[19]	—
DEM L71	LMC	∼4400	Ia	—	—	[146]	—
N49	LMC	∼4800	CC	—	—	[3]	—
SNR 0454-6713	LMC	∼8000	Ia	—	—	[131]	—
SNR 0453.6-6829	LMC	∼13000	CC	—	—	[53, 86]	—
SNR 0453-6655	LMC	∼70000	CC	—	—	[132]	—
1E 0102.2-7219	SMC	∼2000	CC	—	[24, 43, 145]	[112, 118]	—
SNR 0103-72.6	SMC	∼18000	CC	—	—	[145]	—
SN 1996cr	Circinus	26	CC	—	[34, 38, 106]	—	—
SN 1978K	NGC 1313	44	CC	—	—	[28]	—

[a]Distances to LMC, SMC, Circinus, and NGC 1313 are 50 kpc [111], 62 kpc [48], 3.8 Mpc [82], and 4.61 Mpc [114], respectively

References for basic information about each SNR: Cygnus Loop [41], RX J1713.7-3946 [117], Puppis A [5, 117, 121], SN 1006 [71], RCW86 [117], Tycho's SNR [30, 123], Crab Nebula [13, 44], Cas A [81], G296.1-0.5 [62, 133], G21.5-0.9 [9, 116], Kepler's SNR [151], G292.0+1.8 [46, 117], SN 1987A [97], SNR 0509-67.5 [58, 120], SNR 0519-69.0 [120, 161], N103B [120, 158], SNR 0540-69.3 [94], N132D [154], SNR 0506-68.0 [19, 59], DEM L71 [47], N49 [110], SNR 0454-6713 [131], SNR 0453.6-6829 [45], SNR 0453-6655 [132], 1E 0102.2–7219 [42, 168], SNR 0103-72.6 [109], SN 1996cr [4], SN 1978K [125]

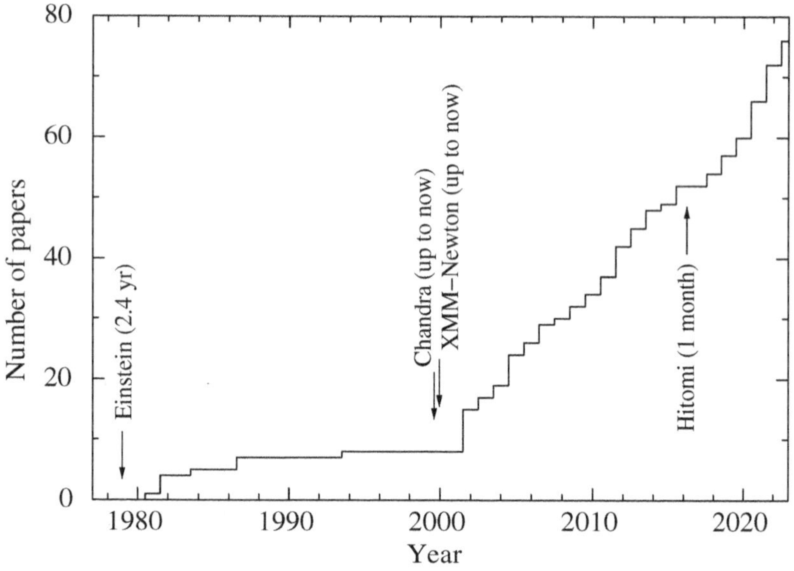

Fig. 13.1 Cumulative number of papers related to high-resolution X-ray spectroscopy of SNRs as a function of year. The launches of the X-ray astronomy satellites, Einstein, Chandra, XMM-Newton, and Hitomi are indicated as arrows

13.2 Plasma Diagnostics

SNR plasmas are often in non-equilibrium (or transient) conditions. This is because the density of the interstellar medium (ISM) is very low (typically 0.3 cc: [7]). Shocks in the ISM are the so-called "collisionless shocks", at which the ordered ion kinetic energy is dissipated into random thermal motions not by Coulomb collisions but through collisionless interactions involving magnetic fields. The collisionless shocks do not immediately lead to equilibrium behind the shock, and it takes a long time for the shock-heated plasmas to be equilibrated.

There are two kinds of non-equilibration of interest, i.e., ionisation and kinetic temperature. In non-equilibrium ionisation (NEI), the ionisation rate is not equal to the recombination rate. If the ionisation rate is higher or lower than the recombination rate, the plasma is referred to as under- or over-ionisation, respectively. Both NEI conditions have been observed in SNRs. The non-equilibrium in kinetic temperature usually means different kinetic temperatures among elements. In this case, lighter elements are naturally expected to have lower temperatures, which has been observed in some SNRs. The opposite situation has not yet been observed in SNRs, although measurements of ion temperatures are still scarce. High-resolution X-ray spectroscopy offers excellent opportunities to study both types of non-equilibrations. Some results on the SNR plasma diagnostics will be reviewed below.

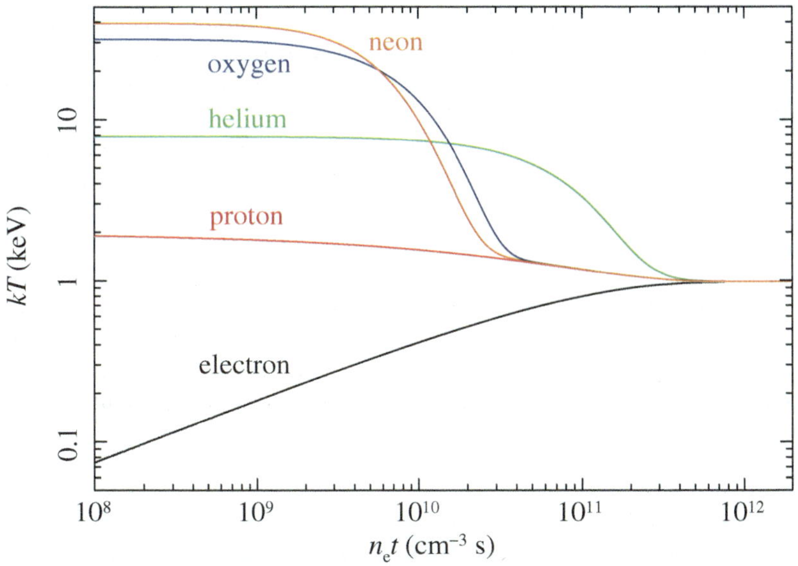

Fig. 13.2 Left: Ion population of Si in an ionising plasma as a function of ionisation timescale. The electron temperature is assumed to be constant at 1.5 keV. This figure is created with SPEX. Right: Same as left, but for a recombining plasma. The initial and final temperatures are assumed to be 10 and 0.1 keV, respectively

13.2.1 Ionization States

Just after shock passages, post-shock plasmas are ionised only weakly, and then ionisation is thought to proceed slowly via Coulomb collisions mainly by free electrons. The density is so low that it takes a long time for the plasma to reach ionisation equilibrium via Coulomb collisions. In this case, the ionisation proceeds with time. Hence we see "ionising plasma" or "under-ionised plasma", in which the plasma is less ionised than what is expected for the equilibrium plasma at the same electron temperature. As an example, Fig. 13.2 left shows the ionisation fraction of Si ions as a function of ionisation timescale, $n_e t$, where n_e is the free electron density and t is the time after the shock heating. The plot is generated with the SPEX code [68], assuming a constant temperature of 1.5 keV. Another type of NEI is the so-called "recombining plasma" or "over-ionised plasma", in which the ionisation state is greater than what is expected for the equilibrium plasma with the same electron temperature. Figure 13.2 right shows how the recombination proceeds with the ionisation timescale for the recombining plasma, where we assume the initial equilibrated plasma with a temperature of 10 keV and the final equilibrated temperature of 0.1 keV.

High-resolution X-ray spectroscopy allows one to measure the ionisation state (or ionisation timescale) of the plasma. Historically, the NEI conditions in SNR

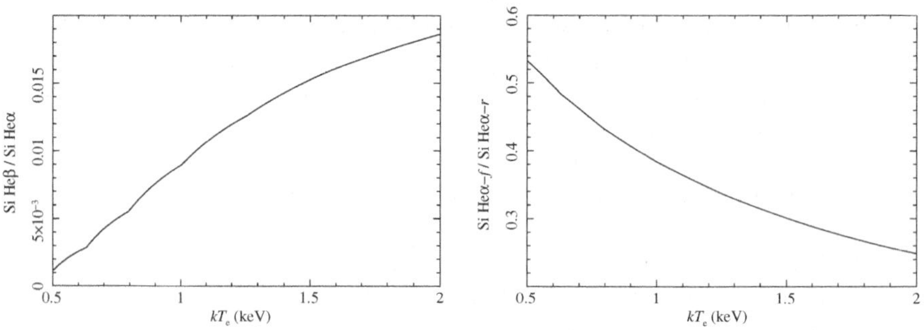

Fig. 13.3 Line intensity ratios of Si ions in an ionizing plasma. **a** Si Lyα/Si Heβ as a function of ionisation timescale at a fixed electron temperature of 1.5 keV. This figure is created with SPEX. **b** Same as **a** but for Si Heα-f/Si Heα-r. **c** Si Heβ/Si Heα as a function of electron temperature at a fixed ionisation timescale of 10^{11} cm^{-3} s. **d** Same as **c** but for Si Heα-f/Si Heα-r

plasmas were first found with the advent of the Einstein FPCS. The ionisation state can be estimated by line intensity ratios of different transitions involving different ionisation states of the same element. Practically, the flux ratio observed at the Earth is the emissivity ratio modified by a factor describing the relative interstellar absorption at the two lines and is given by:

$$\frac{F_1}{F_2} = \frac{n_{\rm Zz}}{n_{\rm Zx}}\frac{\Omega_1}{\Omega_2}\exp[(E_2 - E_1)/kT_{\rm e}]\exp[(\sigma_{\rm E_2} - \sigma_{\rm E_1})N_{\rm H}]), \tag{13.1}$$

where $n_{\rm Zz}$ and $n_{\rm Zx}$ are number densities of ions $\rm Z^{Z+}$ and $\rm Z^{x+}$, Ω is the effective collision strength or oscillator strength, $\sigma_{\rm E}$ gives the absorption cross section at energy E, and $N_{\rm H}$ is the hydrogen column density for the intervening material. The cleanest flux ratio is Lyα ($1s^2 - 1s2p$)/Heβ ($1s^2 - 1s3p$), because these two lines are located so closely that the exponential terms in Eq. 13.1 are of order 1. Oscillator strengths are given in the literature. Therefore, it is straightforward to retrieve the ion fraction between He-like and H-like ions, which leads to the ionisation timescale. Figure 13.3a displays a flux ratio between Si Lyα and Si Heβ as a function of ionisation timescale, which is calculated at an assumed constant temperature of 1.5 keV with the SPEX code. The Lyα/Heβ ratio increases monotonically with increasing ionisation timescale, as expected. By comparing the ion fractions with the electron temperature, the ionisation timescales have been estimated. Figure 13.4 demonstrates such an example; the O Lyα to O Heβ ratio as well as the G ratio (see Sect. 13.2.2) are used to estimate ion fractions and electron temperatures for a young Galactic SNR, RCW 86 [18]. In this way, NEI conditions have been found for many SNRs including Puppis A [164, 165], Cygnus Loop [148], and 1E 0102.2-7219 [118], and so on. It should be noted that NEI conditions have also been suggested from observations with instruments having moderate spectral resolution [102, 142].

Alternatively, flux ratios of the Kα lines in He-like ions can be used to estimate the NEI condition. Detailed descriptions of Heα diagnostics can be found in [113]. We

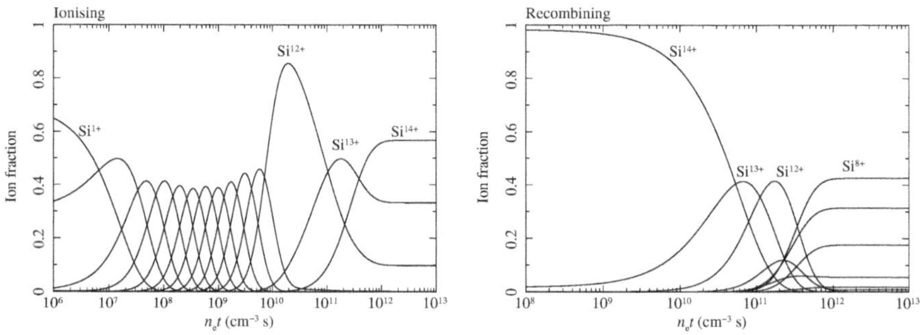

Fig. 13.4 Left: RGS spectrum of the northwestern rim of RCW 86. The solid red line represents the best-fit model consisting of an absorbed, two NEI plus one power-law component. Right: Grid of temperature and ionisation timescale created with SPEX. The line ratios observed at the different four regions of the SNR are plotted by partially-filled circles as indicated in the right-upper corner. These figures are taken from [18]

here focus on one powerful diagnostic parameter, i.e., the forbidden (f) to resonance (r) line ratio as shown in Fig. 13.3. The resonance lines are emitted at the moment of collisional excitations mainly due to electrons. On the other hand, collisional excitations from the ground level to the triplet level ($^3P_{0,1,2}$ and $^3S_1$; see, e.g., Fig. 1 of [113]) are much less efficient according to the selection rule. Thus, the intercombination (i) and forbidden (f) lines are emitted after either the inner-shell ionisations of Li-like ions or recombinations of the H-like ions. These processes favour the forbidden triplet levels because of its high statistical weight. Therefore, at the very low-ionisation condition in which Li-like ions are more populated than He-like and H-like ions, we expect a high f/r ratio. Also, at equilibrium in which H-like ions are the dominant species, the Heα lines are dominated by recombination processes. Because recombination processes (both radiative and di-electronic recombinations) favour the triplet level due to the high statistical weight, we expect an intense Heα-f line (for low-density plasmas), leading to a high f/r ratio. In contrast, when the abundance of He-like ions becomes dominant, with little fractions of Li-like and H-like ions, we expect a low f/r ratio. Such a trend can be readily seen in Fig. 13.3 right showing f/r ratios in He-like Si ions as a function of ionisation timescale. The low f/r ratios observed in many SNRs are considered as such a mild NEI condition: Puppis A [164, 165], Cygnus Loop [148], and 1E 0102.2-7219 [118].

On the other hand, an enhanced O VII f/r ratio was found in middle-aged SNRs, DEM L71 and 0506-68 in the LMC. [19, 146] argued that the high f/r ratio is most naturally explained by contamination of recombining plasmas, given that the X-ray spectrum can be best explained by a cooling (or recombining) plasma model. However, radiative recombination edges, which are strong signatures of rapidly cooling plasmas found in mixed-morphology SNRs like IC 443 [171] and W49B [105], were not observed in the two LMC SNRs. Therefore, other possibilities like resonance line scattering are still left as a viable mechanism to create the high f/r ratio.

13.2.2 Kinetic Temperatures

In addition to the non-equilibrium of ionisation states, kinetic temperatures are also thought to be non-equilibrated in SNR plasmas, namely different elements have different temperatures. This is naturally expected because, in the limit of true collisionless plasma, the particles will have temperatures proportional to their masses. The post-shock temperature T_a for particle species a with mass m_a is given by conservation of mass, momentum, and energy across the shock discontinuity as $kT_a = 3/16 m_a v_s^2$ for shock velocity v_s [150], assuming that the gas pressure is much greater than the magnetic pressure and the cosmic-ray particles' pressure. Then, the distinct temperatures gradually approach equilibration via Coulomb collisions. The timescales for electron-proton and electron-electron equilibrations are given by $t_{ep} \sim 9800\,\mathrm{yr}\ (n_p/1\,\mathrm{cm}^{-3})^{-1}(kT/1\,\mathrm{keV})^{1.5}$ and $t_{ee} \sim 16\,\mathrm{yr}\ (n_e/1\,\mathrm{cm}^{-3})^{-1}(kT/1\,\mathrm{keV})^{1.5}$, respectively.

Figure 13.5 illustrates the effects of non-equilibration of temperatures at the shock front. The equilibration time depends on the density, the mass ratio and the square of the charge number of the particles (see [150] for more details). The ion-ion equilibration, therefore, proceeds faster than electron-ion equilibration. Interestingly, equilibration between heavy elements and protons proceeds faster than He-proton equilibration due to the charge dependence.

High-resolution X-ray spectroscopy is a must to measure the degree of non-equilibration in the kinetic temperature. Electron temperatures can be estimated mainly by intensity ratios of lines from the same ion, such as α/β and β/γ. This would be easily understood from Eq. 13.1, where we can eliminate the ion fraction terms if we focus on the same ion. In fact, Fig. 13.3c clearly demonstrates that the Si Heβ/Si Heα ratio strongly depends on the electron temperature. Another useful clue to inferring the electron temperature is the f/r ratio in He-like ions. We note that the G($\equiv (f+i)/r$) ratio has been often used for this temperature diagnostics as was demonstrated in Fig. 13.4. However, G ratios can be usually approximated to f/r ratios, because the i lines are usually much weaker than f and r lines. The

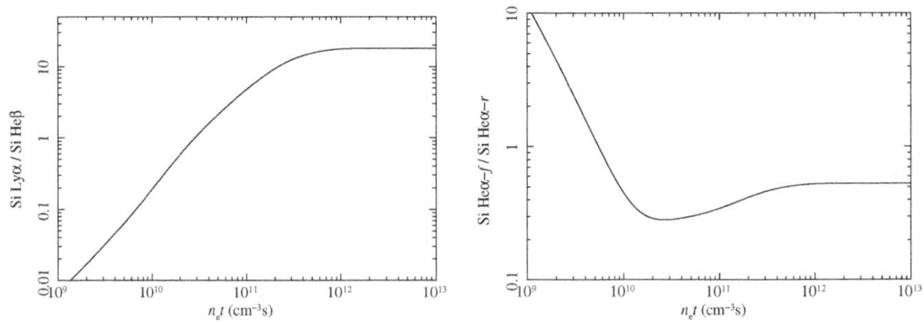

Fig. 13.5 Temperature evolution of shock-heated plasma as a function of ionisation timescale. The shock speed is assumed to be $1000\,\mathrm{km\,s^{-1}}$. The equilibration proceeds with Coulomb collisions

collisional excitation rates have different temperature dependence between f and r lines; the intensity of the r line increases more rapidly with the temperature than the f line [113]. Therefore, the f/r ratio decreases with increasing temperature, as can be seen in Fig. 13.3d. These line ratios (α/β, β/γ, and f/r) have been measured for many SNRs, yielding electron temperatures.

In reality, it is not rare that the temperatures from f/r ratios are inconsistent with those from β/α ratios or global fittings [124, 137]. This could be explained by contaminations of additional emission processes that will be described in the next subsection. Therefore, electron temperatures estimated from f/r ratios (or G ratios) are subject to some systematic uncertainties for the moment.

Ion temperatures can be estimated from line broadenings. The velocity distribution of thermal particles (ions) follows the Maxwellian distribution: $f(v_x) = \sqrt{m/2\pi kT}\exp(-mv_x^2/2kT)$. The lines emitted from ions are Doppler shifted following this velocity distribution so that the line profile can be represented by a Gaussian distribution with $\sigma = E_0/c\sqrt{kT/m}$, where E_0 is the line energy. Combining it with the relation between shock speed and post-shock temperature ($kT_a = 3/16m_a v_s^2$ as described above), we can estimate an expected line width to be $\sigma = 1.4\,\mathrm{eV}(E_0/1\,\mathrm{keV})(v_s/1000\,\mathrm{km\,s^{-1}})$. Because this width is much smaller than the spectral resolution of most past instruments aboard X-ray astronomy satellites, there are only a few successful measurements of ion temperatures. Significant X-ray detections of line broadenings were limited only for SN 1006s northwestern knot [20, 153] and SN 1987A [98]. In both cases, ion temperatures were found to be much higher than the electron temperature, evidencing the non-equilibration of temperatures in collisionless shocks. Moreover, [98] reported that ion temperatures increase with the ion mass from proton to Fe as $T_i/T_p = km_a$ with $k = 0.90 \pm 0.12$. Therefore, the ion post-shock temperature is consistent with being mass-proportional, which is expected for a truly collisionless shock.

Even upper limits on the line widths can give us useful information. The RGS spectrum taken from fast-moving ejecta knots in Puppis A revealed line broadening at O Lyα to be $\sigma \lesssim 0.9\,\mathrm{eV}$, indicating an oxygen temperature of $\lesssim 30\,\mathrm{keV}$ [77]. This oxygen temperature, combined with the electron temperature and ionisation timescale, led to a conclusion that the ejecta knots were heated by a reverse shock with a velocity of 600–1200 $\mathrm{km\,s^{-1}}$.

13.3 New Spectral Features

Recent high-resolution X-ray spectroscopy of SNR plasmas has been revealing that line intensity ratios are often inconsistent with a conventional thermal emission models. There are two main possibilities proposed to explain anomalous line ratios: charge exchange (CX) and resonance scattering (RS). Both effects were predicted to be present in SNRs three decades ago [67, 88, 167], but are not yet established firmly.

The CX process simply transfers (an) electron(s) from one atom to another. No photon is created via this electron transfer itself, but the electron maintains roughly the same binding energy in the process of moving, so the recipient ion is usually left in a highly-excited state (typically $n = 5$ to $n = 10$) that then will be stabilised by radiative cascades (see [134] for detailed explanations on CX processes). From an observational perspective, it is important to note that (1) CX into a He-like triplet state will lead to enhanced f and i lines (because almost any exchange into a triplet state results in such lines), and (2) CX into the He-like singlet state will lead to enhanced (relative to a purely collisional model in the ground state) high-n resonance transitions. These features are similarly expected in the recombining plasmas, because X-rays are emitted via radiative cascade in both processes. One remarkable difference is that radiative recombination edges, which are created when free electrons move to ions, should be seen in recombining plasmas, but not in CX processes.

CX emission in SNRs is thought to play an enhanced role where neutrals are mixed with shocked hot gas. [167] first performed detailed calculations of CX-induced X-ray emission in a SNR. They found that CX emission generally contributes only 10^{-3} to 10^{-5} of the collisionally excited lines in the entire SNR. On the other hand, [88] later examined projected emission profiles for both CX and thermal emission in SNRs, and noted that CX X-ray emission may be comparable with thermal emission in thin layers at the SNR edge. It is also pointed out that the relative importance of CX X-ray emission to thermal emission is proportional to a quantity, $n_c v_r n_e^{-2}$, with n_c being the cloud density, with v_r the relative velocity between neutrals and ions, and n_e the electron density of the hot plasma. Thus, the higher the density contrast (n_c/n_e), the stronger the presence of CX X-ray emission would become. A strong emission feature at ~0.7 keV in parts of outermost rims of the Cygnus Loop was best interpreted as the O He$\gamma + \delta + \epsilon$ arising from CX between neutrals and H-like O ions [78]. Later, [143] discovered a high f/r ratio of O Heα at the southwestern knot of the Cygnus Loop, using the XMM-Newton RGS as shown in Fig. 13.6. This anomalous f/r ratio is best interpreted as a result of a contribution from CX emission. High f/r ratios were also found in some other SNRs including Puppis A [79], G296.1-0.5 [139], and SNR 0453.6-6829 [86], suggesting the presence of CX X-ray emission. However, other possibilities such as resonance scattering or inner-shell ionization are not ruled out.

The importance of RS in SNR plasmas was first pointed out by [67], who argued that the optical depth for resonance lines cannot be assumed to be negligibly small for bright SNRs. If there is a sufficient ion column density along a particular line of sight, resonance line photons can be scattered out of the line of sight to appear at another location. The SNR rims are promising sites where RS effects can be significant because lines of sight (and thus large optical depths) are usually long enough there. Observationally, lines with large oscillator strengths (e.g., Heα-r and Lyα) will be suppressed at SNR rims, whereas those with small oscillator strengths (e.g., Heα-f and Heβ) will not be affected.

Anomalously high f/r ratios of O Heα observed in some SNRs like DEM L71 [146] and N49 [3] in the LMC were best interpreted as the suppression of r lines due to the effect of RS. Using the RGS data, [146] revealed spatial variations in the f/r

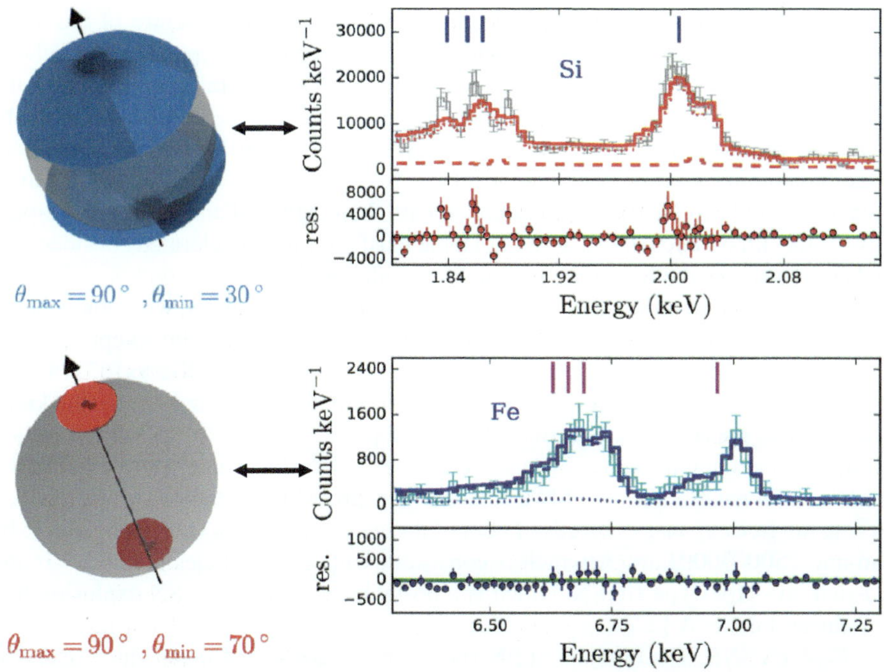

Fig. 13.8 Left: Chandra X-ray image of Cas A with regions for the high-resolution spectroscopy with the Chandra HETG. Right: Example Chandra HETG spectrum of the Si K band taken from the region R1 indicated in the left panel. The best-fit model is plotted with a red line. The nominal positions of the Si lines are marked. These figures are taken from [90]. © AAS. Reproduced with permission

in Fig. 13.8 right, with the best-fit blueshift of 2600 km s^{-1}. Overall, the knots in the southeastern portion mostly show blueshifted velocities up to −2500 km s^{-1}, whereas knots in the northwestern side mostly show redshifted velocities up to +4000 km s^{-1}, which is in general agreement with past observations with lower spatial and spectral resolutions. More recently, [124] investigated temporal evolution using Chandra HETG observations spanning a 10 yr baseline. Due to the low ionisation timescale and the high electron density derived from the X-ray analysis, the small features were expected to show significant amounts of plasma evolution during the 10 yr baseline. However, most of them showed insignificant time evolution, suggesting a much lower electron density for the Si-rich plasmas and thus a longer timescale for evolution of the plasma.

G292.0 + 1.8: Another relatively young core-collapse SNR. The Chandra HETG revealed Doppler shifts in emission lines from metal-rich ejecta, providing 3D structures of clumpy ejecta material in the remnant [8]. The distribution of ejecta knots in velocity versus projected-radius space suggests an expanding ejecta shell with a projected angular thickness of ∼90″ (corresponding to ∼3 pc at d = 6 kpc). Based on this geometrical distribution of the ejecta knots, the location of the reverse shock is estimated approximately at the distance of $\sim$4 pc from the centre of the SNR, putting it in close proximity to the outer boundary of the radio pulsar wind nebula. It should be noted that [152] presented a high-resolution X-ray spectrum with the XMM-Newton RGS for the central belt-like feature across the remnant. No significant line broadening is indicated by the O Lyα with $\sigma < 730\,\mathrm{km\,s^{-1}}$, but significant line broadening seems to be present for the Ne Lyα with $\sigma \sim 1500\,\mathrm{km\,s^{-1}}$. Given that most of O and Ne lines are associated with the ISM/CSM and SN ejecta, respectively, this kinematic result suggests that the blastwave has decelerated considerably, whereas some of the ejecta are still moving with a high velocity.

Puppis A: One of the three "oxygen-rich" SNRs in our Galaxy, with others being Cas A and G292.0+1.8. The X-ray emission is dominated by the swept-up CSM and/or ISM, whereas some ejecta features have been identified in X-rays [63, 74, 76]. [77] observed O-rich ejecta knots and filaments with the RGS, finding that the knots located near the centre of the remnant have a Doppler velocity of $\sim$1500 $\mathrm{km\,s^{-1}}$ blueward and the filament located in the eastern rim has a Doppler velocity of $\sim$650 $\mathrm{km\,s^{-1}}$ redward. Given that they are SN ejecta, they are probably heated by reverse shocks. Then, the free expansion speeds of both of these ejecta features were estimated to be around 2500–3000 $\mathrm{km\,s^{-1}}$, which is consistent with a typical ejecta velocity in the O-rich layer for Type IIb SN, which is a possible subtype of the SN explosion that produced Puppis A [27].

Kepler's SNR: The remnant of SN 1604—the most recent Galactic historical SN. This SNR is a rare class of Type Ia SNRs that show interactions with the CSM. The X-ray emission comes from both the CSM and SN ejecta that is associated with the forward and reverses shocks, respectively. Reference [127] measured proper motions and Doppler velocities for 14 compact X-ray knots, using data from non-dispersive CCDs aboard Chandra. They found high Doppler velocities of up to $\sim 10^4\,\mathrm{km\,s^{-1}}$ for five out of the 14 knots. Such a high speed is comparable to the typical Si velocity seen in SNe Ia near maximum light. Later, [100] measured precise Doppler shifts of Si Heα lines for metal-rich ejecta knots, using high-resolution X-ray spectra with the Chandra HETG. They found that some of the knots seem to be expanding nearly freely. In addition, 8 out of the 15 ejecta knots show a statistically significant (at the 90% confidence level) redshifted spectrum, but only two show blueshifted spectra. This may suggest an asymmetry in the ejecta distribution in Kepler's SNR along the line of sight, although a larger sample size is required to confirm this interpretation. Using the XMM-Newton RGS, [70] measured Doppler velocities of N and O lines which are dominated by the CSM. The lines are overall blueshifted in a range of 0–500 $\mathrm{km\,s^{-1}}$. On the other hand, the central bar structure shows an interesting spatial variation in that the northwestern and southeastern halves are blueshifted and redshifted, respectively. Such velocity structures are consistent with previous optical measurements [12], implying a torus-like shape of the CSM distribution.

Tycho's SNR: The remnant of SN 1572—a confirmed normal Type Ia SN, evidenced by the light-echo spectrum obtained with modern instruments [87]. Reference [126] measured Doppler velocities for 27 compact X-ray knots using non-dispersive CCDs aboard Chandra. The highest-velocity knots are located near the centre, while the low-velocity ones appear near the edge as expected for a generally spherical expansion. The typical velocities of the redshifted and blueshifted knots are $\sim$7800 and $\sim$5000 km s^{-1}, respectively. Recently, [101] presented velocities of 59 metal-rich ejecta knots, based on Chandra HETG observations. As a result, the distribution of space velocities throughout the remnant suggests that the southeast quadrant generally expands faster than the rest of the SNR. Also, blueshifted knots are projected more in the northern shell, while redshifted knots are more in the southern shell, suggesting asymmetries in the CSM along the line of sight.

13.5 Circumstellar Medium

Core-collapse SNe and some of Type Ia SNe occur in a dense environment, i.e., CSM, created by the mass-loss of massive progenitor stars. After the SN explosion, the SN ejecta interact with the surrounding CSM, giving rise to a variety of intense radiation. This emission provides us with an excellent opportunity to study the nature of the progenitor star. Below, we will briefly summarise results from high-resolution X-ray spectroscopy of the shock-heated CSM.

The elemental abundance of the CSM has vital information to infer the evolutionary state of the progenitor star just before the SN explosion. In this regard, the relative abundances of C, N, and O are particularly important, because the C/N/O ratio in the CSM changes significantly with the degree of the CNO processing that takes place in the H-rich envelope. The relative C/N/O abundances of the stellar surface significantly vary with the evolutionary stage: the solar ratio (1/0.25/3) is expected at the main sequence phase, and then a mild N over-abundance and O depletion are expected at a red supergiant phase, and finally an extreme N over-abundance and O depletion (approaching to 1/30/0.5 at CNO equilibrium) are expected at a luminous blue variable or Wolf-Rayet (WR) phase [89]. Detection of the CSM with the CNO-equilibrium abundance immediately tells us that the progenitor star was a LBV/WR star at the time of the explosion. On the other hand, detection of the RSG-like abundance pattern does not necessarily means that the progenitor star was a RSG, given that the CSM ejected during the RSG phase are naturally expected at a distance of a few pc around LBV/WR stars.

There is growing evidence for N and O K lines arising from CSM associated with SNRs. The X-ray measured N/O ratios of the CSM are summarised in Table 13.2. There are two patterns of the N/O ratio, i.e., mildly enhanced and approximately solar values. Not surprisingly, the majority of N/O ratios seen in core-collapse SNRs are mildly elevated from the solar value, suggesting that the X-ray emitting plasmas are the stellar winds from RSG stars. Of these, the CSM ring associated with SN 1987A has been exceptionally well studied. Reference [136] reported an interesting finding that the metal abundances of N, O, Ne, and Mg significantly declined in the past few years. This may be the result that the blast wave left the dense CSM ring and the relative contribution from the H II region compared with the dense CSM became larger than before. On the other hand, the unusually low (near solar) N/O ratio measured at the equatorial belt in G292.0 + 1.8 could be explained if the CSM was created by the stellar wind ejected before CNO-processed material was dredged up to the surface of the progenitor star. It may be possible that a companion star is the main contributor to the equatorial belt, especially because the ring-like CSM feature such as the equatorial belt in G292.0 + 1.8 may have originated from binary interactions [104].

In the case of Type Ia SNRs, the CSM detection itself is important, because it strongly suggests the presence of a companion star, favoring the single-degenerate scenario rather than the double-degenerate scenario for the progenitor system. It is reasonable to assume that the CSM within SNRs must have originated from a donor star, based on simple timescale estimates; the CSM must have been blown some 10^4 yr ago from a non-degenerate (e.g., asymptotic giant branch or red giant) star, whereas it takes $\sim 10^6$ yr for a white dwarf to reach the Chandrasekhar mass by accretion from a donor star, so that the exploding star should have been a white dwarf when the CSM was ejected. In this context, the CSM associated with SNRs allows us to study the donor star. Kepler's SNR and N103B in the LMC are prototypical Ia SNRs with CSM. The N/O ratios in the CSM were measured in both optical and X-ray wavelengths. Given that the N/O ratio on the surface of stars with masses below $8\,M_\odot$ generally increases with increasing stellar mass [69], we can constrain the mass of the donor star from the N/O ratio. The CSM observed in Kepler's SNR exhibits an over-abundance of N, which suggests a relatively massive donor star, possibly an ABG star. In this case, a surviving companion star is expected to remain very bright ($\sim 10^3 L_\odot$) for at least 10^5 yr after the SN explosion [95]. However, no such bright surviving companion star has been found near the center of the remnant, posing a challenge to the single-degenerate scenario. On the other hand, [10, 11, 37] have shown that the N enhancements in the CSM knots of Kepler's SNR are actually consistent with those expected in the local ISM at its location only $\sim$3 kpc away from the Galactic center. Moreover, a recent RGS observation [70] estimated the N abundance of the CSM to be only 1–2 solar, which conflicts with past X-ray (and optical) results, indicating an interesting possibility that the metal abundance of the CSM is even lower than the surrounding ISM. If true, a massive AGB donor star may not be required. In this context, it is important to revisit the CSM abundance with the upcoming XRISM [140]. Lastly, the low N/O ratio measured in N103B is

Table 13.2 Summary of N/O abundances of the CSM detected in SNRs

Name	SN type	N/O (solar)	Refs.
SN 1987A	CC	~8	[2, 136]
SN 1978K	CC	~12	[28]
RX J1713.7-3946	CC	~7	[141]
G296.1-0.5	CC	~4	[25, 139]
G292.0+1.8	CC	~1	[152]
Kepler's SNR	Ia	1–6	[10, 70, 75]
N103B	Ia	~0.5	[11, 170]

(also) consistent with the local ISM of the LMC [11, 170]. Reference [160] argued that the CSM originated from a mass loss prior to the first dredge-up of material in a donor star.

13.6 Absorption

The Crab Nebula, the remnant of SN 1054, provides us with a unique opportunity to measure column densities of intervening material through an occultation technique. The Crab Nebula would be the best object for this study, because it is one of the brightest sources in the whole X-ray sky with its flux being relatively steady, and it has a simple power-law spectrum. Given that X-rays see only atoms and cannot distinguish between monoatomic gas and molecules, the X-ray occultation method allows us to measure the total atomic number densities along the line of sight.

[66] performed high-resolution X-ray spectroscopy for the Crab Nebula using the XMM-Newton RGS, following earlier works with Einstein [130] and Chandra [156]. As shown in Fig. 13.9, the RGS successfully detected strong $1s - 2p$ absorption lines of neutral O, absorption K-shell edges of neutral O and Ne, and L-shell edges of neutral Fe. These absorption features are due to the ISM. The column densities (and abundances relative to H) for N, O, Ne, Mg, and Fe were reliably obtained. It was found that N/H and O/H abundances fully agree with the solar abundances [91], and Mg/H and Fe/H abundances are slightly lower than the solar values. On the other hand, the Ne/H abundance is elevated by a factor of 1.7, which is equivalent to the Ne/O number ratio of 0.26. Interestingly, this Ne/O ratio is in excellent agreement with those obtained in the Orion Nebula and of B-stars in that nebula. In fact, the Ne

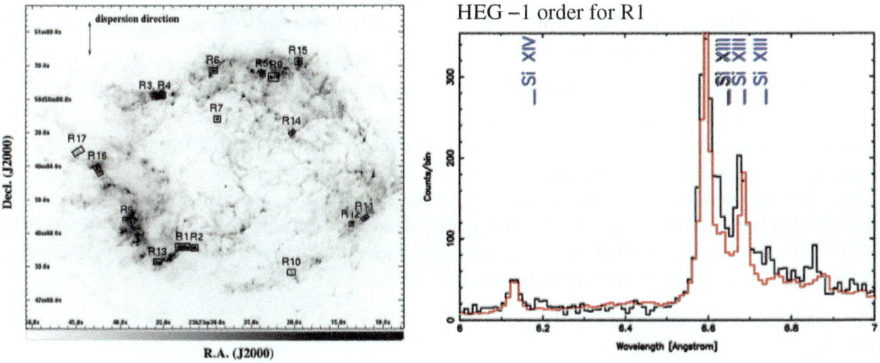

Fig. 13.9 RGS spectrum from the Crab Nebula. RGS1 and RGS2 spectra are separately shown in dark and light colors, respectively. Two narrow regions near the O K-edge and Fe L-edge (indicated by vertical lines) are excluded in the fitting. The best-fit model, which consists of an absorbed (pure gas), curvature-corrected power-law component, is shown as a solid line. Lower panel shows the residuals. This figure is taken from [66]. Reproduced with permission © ESO

abundance of the Sun has been subject to relatively large uncertainty, because Ne lacks detectable photospheric lines in cool stars like the Sun. A more recent standard solar abundance [92] provides the Ne/O abundance ratio of 0.21, which is still smaller than, but closer to the value obtained for the ISM toward the Crab Nebula.

Later, [157] analysed long-exposure data with the Chandra LETG and the most recent response functions. The superb angular resolution of Chandra enabled distinguishing the Crab pulsar from the surrounding nebula, making the spectral analysis simpler. The line spacing of the LETG is however as large as 10,000 Å, which is 25 times larger than that of the RGS, resulting in substantial spectral blurring due to the spatial extent of the Crab Nebula along the dispersion axis. Nonetheless, the Chandra LETG successfully detected the O K-edge. The O/H abundance of the ISM toward the Crab Nebula was estimated to be $(5.28 \pm 0.28) \times 10^{-4}$, which is consistent with the RGS measurement [66]. The dust scattering depth, i.e., the effect of flux scattering out of the image by interstellar grains, was also obtained to be $\tau = 0.147 \pm 0.043$ at 1 keV.

Another intriguing application of the occultation technique is to measure vertical density profiles of the atmosphere of solar planets. The atmospheric thickness of Titan (i.e., Saturn's largest satellite) was measured from the transit of the Crab Nebula on January 05, 2003 [103]. An atmospheric density profile at Mars was measured by occultations of $\sim$10 keV X-rays from Scorpius X-1, using the SEP instrument on the MAVEN spacecraft [115]. In addition, vertical density profiles of the Earth's upper atmosphere (altitude 60–200 km) have been also measured in X-rays. [32, 72, 73, 169, 172] analysed Earth's atmospheric occultations of the Crab Nebula with various X-ray astronomy satellites. The most recent measurements of the vertical density profiles [169] are in good agreement with the latest version of the atmospheric model, i.e., NRLMSIS 2.0 [39]. Also, [73] recently showed that the density of the upper atmosphere has been gradually decreasing possibly due to the increasing greenhouse gases. Future prospects of this study look bright. As shown in [72], the unprecedented spectral resolution of $\sim$5 eV in 0.3–12 keV of the microcalorimeter aboard XRISM will allow us to measure not only total densities but also atmospheric composition from absorption edges, just like the case of the ISM absorption as shown in Fig. 13.9.

Acknowledgements I would like to thank Dr. Hiroyuki Uchida for a long-term collaboration on high-resolution X-ray spectroscopy of SNRs and for providing me with the idea for Fig. 13.1. This work was supported by the Grants-in-Aid for Scientific Research from the Japanese Ministry of Education, Culture, Sports, Science, and Technology (MEXT) of Japan, No. JP20K20935, JP20H00174, and JP21H01121. I also deeply appreciate the Observational Astrophysics Institute at Saitama University for supporting the research fund.

References

1. H. Akamatsu, W.B. Doriese, J.A.B. Mates, B.D. Jackson, Handbook of X-ray and Gamma-ray Astrophysics, Chapter XX (2022)
2. D. Alp, J. Larsson, C. Fransson, ApJ **916**(2), 76 (2021). https://doi.org/10.3847/1538-4357/ac052d
3. Y. Amano, H. Uchida, T. Tanaka, L. Gu, T.G. Tsuru, ApJ **897**(1), 12 (2020). https://doi.org/10.3847/1538-4357/ab90fc
4. F.E. Bauer, V.V. Dwarkadas, W.N. Brandt, S. Immler, S. Smartt, N. Bartel, M.F. Bietenholz, ApJ **688**(2), 1210 (2008). https://doi.org/10.1086/589761
5. W. Becker, P. Tobias, P.F. Winkler, R. Petre, ApJ **755**(2), 141 (2012). https://doi.org/10.1088/0004-637X/755/2/141
6. E. Behar, A.P. Rasmussen, R.G. Griffiths, K. Dennerl, M. Audard, B. Aschenbach, A.C. Brinkman, A & A **365**, L242 (2001). https://doi.org/10.1051/0004-6361:20000082
7. E.M. Berkhuijsen, A. Fletcher, MNRAS-L **390**(1), L19 (2008). https://doi.org/10.1111/j.1745-3933.2008.00526.x
8. J. Bhalerao, S. Park, D. Dewey, J.P. Hughes, K. Mori, J.-J. Lee, ApJ **800**(1), 65 (2015). https://doi.org/10.1088/0004-637X/800/1/65
9. M.F. Bietenholz, N. Bartel, MNRAS **386**(3), 1411 (2008). https://doi.org/10.1111/j.1365-2966.2008.13058.x
10. W.P. Blair, P. Ghavamian, K.S. Long, B.J. Williams, K.J. Borkowski, S.P. Reynolds, R. Sankrit, ApJ **662**(2), 998 (2007). https://doi.org/10.1086/518414
11. W.P. Blair, P. Ghavamian, J.C. Raymond, B.J. Williams, R. Sankrit, K.S. Long, P.F. Winkler, N. Pirzkal, I.R. Seitenzahl, ApJ **902**(2), 153 (2020). https://doi.org/10.3847/1538-4357/abb3c7
12. W.P. Blair, K.S. Long, O. Vancura, ApJ **366**, 484 (1991). https://doi.org/10.1086/169583
13. R. Blandford, R. Bühler, *Handbook of Supernovae*, vol. 83 (2017) ISBN:978-3-319-21845-8
14. J.A.M. Bleeker, R. Willingale, K. van der Heyden, K. Dennerl, J.S. Kaastra, B. Aschenbach, J. Vink, A & A **365**, L225 (2001). https://doi.org/10.1051/0004-6361:20000048
15. E. Bray, D.N. Burrows, S. Park, A.P. Ravi, ApJ **899**(1), 21 (2020). https://doi.org/10.3847/1538-4357/ab9c9e
16. A.C. Brinkman et al., ApJ **530**(2), L111 (2000). https://doi.org/10.1086/312504
17. A.C. Brinkman, J.J. van Rooijen, J.A.M. Bleeker, J.H. Dijkstra, J. Heise, P.A.J. de Korte, R. Mewe, F. Paerels, ApL & C **26**, 73 (1987)
18. S. Broersen, A. Chiotellis, J. Vink, A. Bamba, MNRAS **441**(4), 3040 (2014). https://doi.org/10.1093/mnras/stu667
19. S. Broersen, J. Vink, J. Kaastra, J. Raymond, A & A **535**, A11 (2011). https://doi.org/10.1051/0004-6361/201117390
20. S. Broersen, J. Vink, M. Miceli, F. Bocchino, G. Maurin, A. Decourchelle, A & A **552**, A9 (2013). https://doi.org/10.1051/0004-6361/201220526
21. C.R. Canizares, G.W. Clark, D. Bardas, T. Markert, SPIE **106**, 154 (1977). https://doi.org/10.1117/12.955467
22. C.R. Canizares, P.F. Winkler, ApJ **246**, L33 (1981). https://doi.org/10.1086/183547
23. C.R. Canizares, J.E. Davis, D. Dewey, K.A. Flanagan, E.B. Galton, D.P. Huenemoerder, K. Ishibashi, T.H. Markert, H.L. Marshall, M. McGuirk, M.L. Schattenburg, N.S. Schulz, H.I. Smith, M. Wise, PASP **117**(836), 1144 (2005). https://doi.org/10.1086/432898
24. C.R. Canizares, K.A. Flanagan, D.S. Davis, D. Dewey, J.C. Houck, in *ASP Conference Proceedings*, vol. 234 (2001), p. 173. ISBN:1-58381-071-4
25. D. Castro, P.O. Slane, B.M. Gaensler, J.P. Hughes, D.J. Patnaude, ApJ **734**(2), 86 (2011). https://doi.org/10.1088/0004-637X/734/2/86
26. P. Chandra, SSR **214**(1), 27 (2018). https://doi.org/10.1007/s11214-017-0461-6
27. R.A. Chevalier, ApJ **619**(2), 839 (2005). https://doi.org/10.1086/426584
28. Y. Chiba, S. Katsuda, T. Yoshida, K. Takahashi, H. Umeda, PASJ **72**(2), 25 (2020). https://doi.org/10.1093/pasj/psz148

29. A. Decourchelle, J.L. Sauvageot, M. Audard, B. Aschenbach, S. Sembay, R. Rothenflug, J. Ballet, T. Stadlbauer, R.G. West, A & A **365**, L218 (2001). https://doi.org/10.1051/0004-6361:20000115
30. A. Decourchelle, *Handbook of Supernovae*, vol. 117 (2017) ISBN:978-3-319-21845-8
31. J.W. den Herder et al., A & A **365**, L7 (2001). https://doi.org/10.1051/0004-6361:20000058
32. J.R. Determan, S.A. Budzien, M.P. Kowalski, M.N. Lovellette, P.S. Ray, M.T. Wolff, K.S. Wood, L. Titarchuk, R. Bandyopadhyay, JGR: Space Phys. **112**(A6), A06323 (2007) https://doi.org/10.1029/2006JA012014
33. D. Dewey, SSR **157**(1–4), 229 (2010). https://doi.org/10.1007/s11214-010-9718-z
34. D. Dewey, F.E. Bauer, V.V. Dwarkadas, AIPC **1358**, 289 (2011). https://doi.org/10.1063/1.3621791
35. D. Dewey, V.V. Dwarkadas, F. Haberl, R. Sturm, C.R. Canizares, ApJ **752**(2) (2012). https://doi.org/10.1088/0004-637X/752/2/103
36. D. Dewey, S.A. Zhekov, R. McCray, C.R. Canizares, ApJ **676**(2), L131 (2008). https://doi.org/10.1086/587549
37. M.A. Dopita, I.R. Seitenzahl, D.C. Nicholls, F.P.A. Vogt, P. Gavamian, A.J. Ruiter, **157**(2), 50 (2019) https://doi.org/10.3847/1538-3881/aaf235
38. V.V. Dwarkadas, D. Dewey, F. Bauer, MNRAS **407**(2), 812 (2010). https://doi.org/10.1111/j.1365-2966.2010.16966.x
39. J.T. Emmert et al., ESS **8**(3), e01321 (2021). https://doi.org/10.1029/2020EA001321
40. Y. Ezoe, T. Ohashi, K. Mitsuda, Rev. Mod. Plasma Phys. **5**(1), 4 (2021). https://doi.org/10.1007/s41614-021-00052-2
41. R.A. Fesen, K.E. Weil, I.A. Cisneros, W.P. Blair, J.C. Raymond, MNRAS **481**(2), 1786 (2018). https://doi.org/10.1093/mnras/sty2370
42. S.L. Finkelstein et al., ApJ **641**(2), 919 (2006). https://doi.org/10.1086/500570
43. K.A. Flanagan, C.R. Canizares, D. Dewey, J.C. Houck, A.C. Fredericks, M.L. Schattenburg, T.H. Markert, D.S. Davis, ApJ **605**(1), 230 (2004). https://doi.org/10.1086/382145
44. M. Fraser, D. Boubert, ApJ **871**(1), 92 (2019). https://doi.org/10.3847/1538-4357/aaf6b8
45. B.M. Gaensler, S.P. Hendrick, S.P. Reynolds, K.J. Borkowski, ApJ **594**(2), L111 (2003). https://doi.org/10.1086/378687
46. B.M. Gaensler, B.J. Wallace, ApJ **594**(1), 326 (2003). https://doi.org/10.1086/376861
47. P. Ghavamian, C.E. Rakowski, J.P. Hughes, T.B. Williams, ApJ **590**(2), 833 (2003). https://doi.org/10.1086/375161
48. D. Graczyk et al., ApJ **904**(1), 13 (2020). https://doi.org/10.3847/1538-4357/abbb2b
49. E. Greco, et al., ApJ **931**(2) (2022). https://doi.org/10.3847/1538-4357/ac679d
50. R. Giacconi et al., ApJ **230**, 540 (1979). https://doi.org/10.1086/157110
51. F. Haberl, U. Geppert, B. Aschenbach, G. Hasinger, A & A **460**(3), 811 (2006). https://doi.org/10.1051/0004-6361:20066198
52. K. Heng, F. Haberl, B. Aschenbach, G. Hasinger, ApJ **676**(1), 361 (2008). https://doi.org/10.1086/526517
53. F. Haberl, M.D. Filipović, L.M. Bozzetto, E.J. Crawford, S.D. Points, W. Pietsch, A.Y. De Horta, N. Tothill, J.L. Payne, M. Sasaki, A & A **543**, A154 (2012). https://doi.org/10.1051/0004-6361/201218971
54. Hitomi Collaboration, PASJ **70**(2), 16 (2017). https://doi.org/10.1093/pasj/psx151
55. Hitomi Collaboration, PASJ **70**(2), 14 (2018). https://doi.org/10.1093/pasj/psx072
56. Hitomi Collaboration, PASJ **70**(3), 38 (2018). https://doi.org/10.1093/pasj/psy027
57. S.S. Holt, E.V. Gotthelf, H. Tsunemi, H. Negoro, PASJ **46**(46), L151 (1994)
58. L. Hovey, J.P. Hughes, K. Eriksen, **809**(2), 119 (2015). https://doi.org/10.1088/0004-637X/809/2/119
59. J.P. Hughes, M. Rafelski, J.S. Warren, C. akowski, P. Slane, D. Burrows, J. Nousek, ApJ **645**(2), L117 (2006). https://doi.org/10.1086/506323
60. U. Hwang, S.S. Holt, R. Petre, ApJ **537**(2), L119 (2000). https://doi.org/10.1086/312776
61. U. Hwang, J.P. Hughes, C.R. Canizares, T.H. Markert, ApJ **414**, 219 (1993). https://doi.org/10.1086/173070

62. U. Hwang, T.H. Markert, ApJ **431**, 819 (1994). https://doi.org/10.1086/174532
63. U. Hwang, R. Petre, K.A. Flanagan, ApJ **767**(1), 378 (2008). https://doi.org/10.1086/528925
64. U. Hwang, A.E. Szymkowiak, R. Petre, S.S. Holt, ApJ **560**(2), L175 (2001). https://doi.org/10.1086/324177
65. F. Jansen, D. Lumb, B. Altieri, J. Clavel, M. Ehle, C. Erd, C. Gabriel, M. Guainazzi, P. Gondoin, R. Much, R.R. Munoz, M. Santos, N. Schartel, D. Texier, G. Vacanti, A & A **365**, L1 (2001). https://doi.org/10.1051/0004-6361:20000036
66. J.S. Kaastra, R. Mewe, H. Nieuwenhuijzen, in *Proceedings of 11th Colloquium on UV and X-Ray Spectroscopy of Astrophysical and Laboratory Plasmas*, vol. 411 (1996)
67. J.S. Kaastra, R. Mewe, A & A **302**, L13 (1995)
68. J.S. Kaastra, R. Mewe, A & A **302**, L13 (1995)
69. A.I. Karakas, M. Lugaro, ApJ **825**(1), 26 (2016). https://doi.org/10.3847/0004-637X/825/1/26
70. T. Kasuga, J. Vink, S. Katsuda, H. Uchida, A. Bamba, T. Sato, J.P. Hughes, ApJ **915**(1), 42 (2021). https://doi.org/10.3847/1538-4357/abff4f
71. S. Katsuda, *Handbook of Supernovae*, vol. 63 (2017). https://doi.org/10.1007/978-3-319-21846-5_45
72. S. Katsuda, H. Fujiwara, Y. Ishisaki, Y. Maeda, K. Mori, Y. Motizuki, K. Sato, M.S. Tashiro, Y. Terada, JGR: Space Phys. **126**(4), e28886 (2021). https://doi.org/10.1029/2020JA028886
73. S. Katsuda, T. Enoto, A.N. Lonmen, K. Mori, Y. Motizuki, M. Nakajima, N.C. Ruhl, K. Sato, G. Stober, M.S. Tashiro, Y. Terada, K.S. Wood, JGR: Space Phys. (202X). https://doi.org/10.1002/essoar.10511820.1
74. S. Katsuda, U. Hwang, R. Petre, S. Park, K. Mori, H. Tsunemi, ApJ **714**(2), 1725 (2010). https://doi.org/10.1088/0004-637X/714/2/1725
75. S. Katsuda, K. Mori, K. Maeda, M. Tanaka, K. Koyama, H. Tsunemi, H. Nakajima, Y. Maeda, M. Ozaki, R. Petre, ApJ **808**(1), 49 (2015). https://doi.org/10.1088/0004-637X/808/1/49
76. S. Katsuda, K. Mori, H. Tsunemi, S. Park, U. Hwang, D.N. Burrows, J.P. Hughes, P.O. Slane, ApJ **678**(1), 297 (2008). https://doi.org/10.1086/586891
77. S. Katsuda, Y. Ohira, K. Mori, H. Tsunemi, H. Uchida, K. Koyama, T. Tamagawa, ApJ **768**(2), 182 (2013). https://doi.org/10.1088/0004-637X/768/2/182
78. S. Katsuda, H. Tsunemi, K. Mori, H. Uchida, H. Kosugi, M. Kimura, H. Nakajima, S. Takakura, R. Petre, J.W. Hewitt, H. Yamaguchi, ApJ **730**(1), 24 (2011). https://doi.org/10.1088/0004-637X/730/1/24
79. S. Katsuda, H. Tsunemi, K. Mori, H. Uchida, R. Petre, S. Yamada, H. Akamatsu, S. Konami, T. Tamagawa, ApJ **756**(1), 49 (2012). https://doi.org/10.1088/0004-637X/756/1/49
80. R.L. Kelley et al., SPIE **9905**, 99050V (2016). https://doi.org/10.1117/12.2232509
81. B.-C. Koo, C. Park, *Handbook of Supernovae*, vol. 161 (2017) ISBN:978-3-319-21845-8
82. B.S. Koribalski, et al., AJ **128**(1), 16 (2004). https://doi.org/10.1086/421744
83. D. Kosenko, E.A. Helder, J. Vink, A & A **519**, A11 (2010). https://doi.org/10.1051/0004-6361/200913903
84. D. Kosenko, W. Hillebrandt, M. Kromer, S.I. Blinnikov, R. Pakmor, J.S. Kaastra, MNRAS **449**(2), 1441 (2015). https://doi.org/10.1093/mnras/stv348
85. D. Kosenko, J. Vink, S. Blinnikov, A. Rasmussen, A & A **490**(1), 223 (2008). https://doi.org/10.1051/0004-6361:200809495
86. Y. Koshiba, H. Uchida, T. Tanaka, Y. Amano, H. Sano, T.G. Tsuru, PASJ **74**(4), 757 (2022). https://doi.org/10.1093/pasj/psac033
87. O. Krause, M. Tanaka, T. Usuda, T. Hattori, M. Goto, S. Birkmann, K. Nomoto, Nature **456**(7222), 617 (2008). https://doi.org/10.1038/nature07608
88. R. Lallemen, A & A **422**, 391 (2004). https://doi.org/10.1051/0004-6361:20035625
89. H.J.G.L.M. Lamers, A. Nota, N. Panagia, L.J. Smith, N. Langer, ApJ **551**(2), 764 (2001). https://doi.org/10.1086/320229
90. J.S. Lazendic, D. Dewey, N.S. Schulz, C.R. Canizares, ApJ **651**(1), 250 (2006). https://doi.org/10.1086/507481
91. K. Lodders, ApJ **591**(2), 1220 (2003). https://doi.org/10.1086/375492

92. K. Lodders, H. Palme, H.-P. Gail, Sol. Syst. **4B**, 712 (2009). https://doi.org/10.1007/978-3-540-88055-4_34
93. K.S. Lum, C.R. Canizares, G.W. Clark, J.M. Coyne, T.H. Markert, P.J. Saez, M.L. Schattenburg, P.F. Winkler, ApJS **78**, 423 (1992). https://doi.org/10.1086/191635
94. P. Lundqvist, N. Lundqvist, Y.A. Shibanov, A & A **658**, A30 (2022). https://doi.org/10.1051/0004-6361/202141931
95. E. Marietta, A. Burrows, B. Fryxell, ApJS **128**(615) (2000). https://doi.org/10.1086/313392
96. T.H. Markert, C.R. Canizares, G.W. Clark, P.F. Winkler, ApJ **268**, 134 (1983). https://doi.org/10.1086/160939
97. R. McCray, C. Fransson, ARA & A **54**, 19 (2016). https://doi.org/10.1146/annurev-astro-082615-105405
98. M. Miceli, S. Orlando, D.N. Burrows, K.A. Frank, C. Argiroffi, F. Reale, G. Peres, O. Petruk, F. Bocchino, Nat. Astron. **3**, 236 (2019). https://doi.org/10.1038/s41550-018-0677-8
99. E. Michael, S. Zhekov, R. McCray, U. Hwang, D.N. Burrows, S. Park, G.P. Garmire, S.S. Holt, G. Hasinger, ApJ **574**(1), 166 (2002). https://doi.org/10.1086/340591
100. M.J. Millard, J. Bhalerao, S. Park, T. Sato, J.P. Hughes, P. Slane, D. Patnaude, D.N. Burrows, C. Badenes, ApJ **893**(2), 98 (2020). https://doi.org/10.3847/1538-4357/ab7db1
101. M.J. Millard, S. Park, T. Sato, J.P. Hughes, P. Slane, D. Patnaude, D.N. Burrows, C. Badenes, ApJ **937**(2), 121 (2022). https://doi.org/10.3847/1538-4357/ac8f30
102. E. Miyata, H. Tsunemi, R. Pisarski, S.E. Kissel, PASJ **46**, L101 (1994)
103. K. Mori, H. Tsunemi, H. Katayama, D.N. Burrows, G.P. Garmire, A.E. Metzger, ApJ **607**(2), 1065 (2004). https://doi.org/10.1086/383521
104. T. Morris, P. Podsiadlowski, MNRAS **399**(2), 515 (2009). https://doi.org/10.1111/j.1365-2966.2009.15114.x
105. M. Ozawa, K. Koyama, H. Yamaguchi, K. Masai, T. Tamagawa, ApJL **706**(1), L71 (2009). https://doi.org/10.1088/0004-637X/706/1/L71
106. J. Quirola-Vásquez, F.E. Bauer, V.V. Dwarkadas, C. Badenes, W.N. Brandt, T. Nymark, D. Walton, MNRAS **490**(4), 4536 (2019). https://doi.org/10.1093/mnras/stz2858
107. F. Paerels, SSR **157**(1–4), 15 (2010). https://doi.org/10.1007/s11214-010-9719-y
108. F.B.S. Paerels, S.M. Kahn, ARA & A **41**, 291 (2003). https://doi.org/10.1146/annurev.astro.41.071601.165952
109. S. Park, J.P. Hughes, D.N. Burrows, P.O. Slane, J.A. Nousek, G.P. Garmire, ApJ **598**(2), L95 (2003). https://doi.org/10.1086/380599
110. S. Park, J.P. Hughes, P.O. Slane, D.N. Burrows, J.-J. Lee, K. Mori, ApJ **748**(2), 117 (2012). https://doi.org/10.1088/0004-637X/748/2/117
111. G. Pietrzyński, et al., Nature **567**(7747), 200 (2019). https://doi.org/10.1038/s41586-019-0999-4
112. P.P. Plucinsky, A.P. Beardmore, A. Foster, F. Haberl, E.D. Miller, A.M.T. Pollock, S. Sembay, A & A **597**, A35 (2017). https://doi.org/10.1051/0004-6361/201628824
113. D. Porquet, J. Dubau, N. Grosso, SSR **157**(1–4), 103 (2010). https://doi.org/10.1007/s11214-010-9731-2
114. G. Qing, W. Wang, J.-F. Liu, P. Yoachim, ApJ **799**(1), 19 (2015). https://doi.org/10.1088/0004-637X/799/1/19
115. A. Rahmati, D.E. Larson, T.E. Cravens, R.J. Lillis, C.O. Lee, P.A. Dunn, GRL **47**(21), e88927 (2020). https://doi.org/10.1029/2020GL088927
116. S. Ranasinghe, D.A. Leahy, AJ **155**(5), 204 (2018). https://doi.org/10.3847/1538-3881/aab9be
117. S. Ranasinghe, D. Leahy. arXiv:2209.04570 TBD
118. A.P. Rasmussen, E. Behar, S.M. Kahn, J.W. den Herder, K. van der Heyden, A & A **365**, L231 (2001). https://doi.org/10.1051/0004-6361:20000231
119. A.P. Ravi, S. Park, S.A. Zhekov, M. Miceli, S. Orlando, K.A. Frank, D.N. Burrows, ApJ **922**(2), 140 (2021). https://doi.org/10.3847/1538-4357/ac249a
120. A. Rest et al., Nature **438**(7071), 1132 (2005). https://doi.org/10.1038/nature04365

121. E.M. Reynoso, S. Cichowolski, A.J. Walsh, MNRAS **464**(3), 3029 (2017). https://doi.org/10.1093/mnras/stw2219
122. M. Ross, V.V. Dwarkadas, ApJ **153**(6), 246 (2017). https://doi.org/10.3847/1538-3881/aa6d50
123. P. Ruiz-Lapuente, J.I. González Hernández, R. Mor, M. Romero-Gómez, N. Miret-Roig, F. Figueras, L.R. Bedin, R. Canal, J. Méndez, ApJ **870**(2), 135 (2019) https://doi.org/10.3847/1538-4357/aaf1c1
124. J. Rutherford, D. Dewey, E. Figueroa-Feliciano, S.N.T. Heine, F.A. Bastien, S. Kosuke, C.R. Canizares, ApJ **769**(1), 64 (2013). https://doi.org/10.1088/0004-637X/769/1/64
125. S. Ryder, L. Staveley-Smith, M. Dopita, R. Petre, E. Colbert, D. Malin, E. Schlegel, ApJ **416**, 167 (1993). https://doi.org/10.1086/173223
126. T. Sako, J.P. Hughes, ApJ **840**(2), 112 (2017). https://doi.org/10.3847/1538-4357/aa6f60
127. T. Sako, J.P. Hughes, ApJ **845**(2), 167 (2017). https://doi.org/10.3847/1538-4357/aa8305
128. K. Sato, Y. Uchida, K. Ishikawa, *This Book* (2023)
129. M.L. Schattenburg, C.R. Canizares, C.J. Berg, G.W. Clark, T.H. Markert, P.F. Winkler, ApJ **241**, L151 (1980). https://doi.org/10.1086/183380
130. M.L. Schattenburg, C.R. Canizares, ApJ **301**, 759 (1986). https://doi.org/10.1086/163942
131. F.D. Seward, A.R. Foster, R.K. Smith, S.D. Points, ApJ **909**(1), 13 (2021). https://doi.org/10.3847/1538-4357/abd561
132. F.D. Seward, R.K. Smith, P.O. Slane, S.S. Murray, S.D. Points, A.J.R. Gordon, J.R. Dickel, ApJ **861**(2), 154 (2018). https://doi.org/10.3847/1538-4357/aabf43
133. S.-S. Shan, H. Zhu, W.-W. Tian, H.-Y. Zhang, A.-Y. Yang, M.-F. Zhang, RAA **19**(7), 92 (2019). https://doi.org/10.1088/1674-4527/19/7/92
134. R. Smith, et al., *ASTRO-H White Paper* (2014). arXiv:1412.1172
135. R. Sturm, F. Haberl, B. Aschenbach, G. Hasinger, A & A **515**, A5 (2010). https://doi.org/10.1051/0004-6361/200913317
136. L. Sun, J. Vink, Y. Chen, P. Zhou, D. Prokhorov, G. Pühlhofer, D. Malyshev, ApJ **916**(1), 41 (2021). https://doi.org/10.3847/1538-4357/ac033d
137. H. Suzuki, H. Yamaguchi, M. Ishida, H. Uchida, P.P. Plucinsky, A.R. Foster, E.D. Miller, ApJ **900**(1), 39 (2020). https://doi.org/10.3847/1538-4357/aba524
138. T. Takahashi et al., JATIS **4**, 021402 (2018). https://doi.org/10.1117/1.JATIS.4.2.021402
139. Y. Tanaka, H. Uchida, T. Tanaka, Y. Amano, Y. Koshiba, T.G. Tsuru, H. Sano, Y. Fukui, ApJ **933**(1), 101 (2022). https://doi.org/10.3847/1538-4357/ac738f
140. M.S. Tashiro et al., SPIE **11444**, 1144422 (2020). https://doi.org/10.1117/12.2565812
141. D. Tateishi, S. Katsuda, Y. Terada, F. Acero, T. Yoshida, S. Fujimoto, H. Sano, ApJ **923**(2), 187 (2021). https://doi.org/10.3847/1538-4357/ac2c00
142. H. Tsunemi, K. Yamashita, K. Masai, S. Hayakawa, K. Koyama, ApJ **306**, 248 (1986). https://doi.org/10.1086/164337
143. H. Uchida, S. Katsuda, H. Tsunemi, K. Mori, L. Gu, R.S. Cumbee, R. Petre, T. Tanaka, ApJ **871**(2), 234 (2019). https://doi.org/10.3847/1538-4357/aaf8a6
144. K.J. van der Heyden, E. Behar, J. Vink, J.A.P. Rasmussen, J.S. Kaastra, J.A.M. Bleeker, S.M. Kahn, R. Mewe, A & A **392**, 955 (2002). https://doi.org/10.1051/0004-6361:20020963
145. K.J. van der Heyden, J.A.M. Bleeker, J.S. Kaastra, A & A **421**, 1031 (2004). https://doi.org/10.1051/0004-6361:20034156
146. K.J. van der Heyden, J.A.M. Bleeker, J.S. Kaastra, J. Vink, A & A **406**, 141 (2003). https://doi.org/10.1051/0004-6361:20030658
147. K.J. van der Heyden, F. Paerels, J. Cottam, J.S. Kaastra, G. Branduardi-Raymont, A & A **365**, L254 (2001). https://doi.org/10.1051/0004-6361:20000113
148. P.W. Vedder, C.R. Canizares, T.H. Markert, A.K. Pradhan, ApJ **307**, 269 (1986). https://doi.org/10.1086/164413
149. J. Vink, AIPC **774**, 241 (2005). https://doi.org/10.1063/1.1960935
150. J. Vink, A & ARv **20**, 49 (2012). https://doi.org/10.1007/s00159-011-0049-1
151. J. Vink, *Handbook of Supernovae*, vol. 139 (2017). ISBN:978-3-319-21845-8

152. J. Vink, J. Bleeker, J.S. Kaastra, A. Rasmussen, Nuc. Phys. B Proc. Suppl. **132**, 62 (2004). https://doi.org/10.1016/j.nuclphysbps.2004.04.009
153. J. Vink, J.M. Laming, M.F. Gu, A. Rasmussen, J.S. Kaastra, ApJ **587**(1), L31 (2003). https://doi.org/10.1086/375125
154. F. Vogt, M.A. Dopita, ApSS **331**(2), 521 (2011). https://doi.org/10.1007/s10509-010-0479-7
155. M.C. Weisskopf, B. Brinkman, B.C. Canizares, G. Garmire, S. Murray, L.P. Van Speybroeck, PASP **114**(791), 1 (2002). https://doi.org/10.1086/338108
156. M.C. Weisskopf, S.L. O'Dell, F. Paerels, R.F. Elsner, W. Becker, A.F. Tennant, D.A. Swartz, ApJ **601**(2), 1050 (2004). https://doi.org/10.1086/380600
157. M.C. Weisskopf, A.F. Tennant, D.G. Yakovloev, A. Harding, V.E. Zavlin, S.L. O'Dell, R.F. Ronald, W. Becker, ApJ **743**(2), 139 (2011). https://doi.org/10.1088/0004-637X/743/2/139
158. B.J. Williams, W.P. Blair, K.J. Borkowski, P. Ghavamian, S.P. Hendrick, K.S. Long, R. Petre, J.C. Raymond, A. Rest, S.P. Reynolds, R. Sankrit, I.R. Seitenzahl, P.F. Winkler, ApJL **865**(2), L13 (2018). https://doi.org/10.3847/2041-8213/aae08d
159. B.J. Williams, K.J. Borkowski, S.P. Reynolds, P. Ghavamian, J.C. Raymond, K.S. Long, W.P. Blair, R. Sankrit, R.C. Smith, S. Points, P.F. Winkler, S.P. Hendrick, ApJ **729**(1), 65 (2011). https://doi.org/10.1088/0004-637X/729/1/65
160. B.J. Williams, K.J. Borkowski, S.P. Reynolds, P. Ghavamian, J.C. Raymond, K.S. Long, W.P. Blair, R. Sankrit, P.F. Winkler, S.P. Hendrick, ApJ **790**(2), 139 (2014). https://doi.org/10.1088/0004-637X/790/2/139
161. B.J. Williams, P. Ghavamian, I.R. Seitenzahl, S.P. Reynolds, K.J. Borkowski, R. Petre, ApJ **935**(2), 78 (2022). https://doi.org/10.3847/1538-4357/ac81ca
162. B.J. Williams, S. Katsuda, R. Cumbee, R. Petre, J.C. Raymond, H. Uchida, ApJL **898**(2), L51 (2020). https://doi.org/10.3847/2041-8213/aba7c1
163. R. Willingale, J.A.M. Bleeker, K.J. van der Heyden, J.S. Kaastra, J. Vink, A & A **381**, 1039 (2002). https://doi.org/10.1051/0004-6361:20011614
164. P.F. Winkler, G.W. Clark, T.H. Markert, K. Kalata, H.M. Schnopper, C.R. Canizares, ApJ **246**, L27 (1981). https://doi.org/10.1086/183546
165. P.F. Winkler, G.W. Clark, T.H. Markert, R. Petre, C.R. Canizares, ApJ **245**, 574 (1981). https://doi.org/10.1086/158832
166. P.F. Winkler, K. Twelker, C.N. Reith, K.S. Long, ApJ **692**(2), 1489 (2009). https://doi.org/10.1088/0004-637X/692/2/1489
167. M.W. Wise, C.L. Sarazin, ApJ **345**, 384 (1989). https://doi.org/10.1086/167913
168. L. Xi, T.J. Gaetz, P.P. Plucinsky, J.P. Hughes, D.J. Patnaude, ApJ **874**(1), 14 (2019). https://doi.org/10.3847/1538-4357/ab09ea
169. W. Xue, et al., ApJS (2022) (in press)
170. H. Yamaguchi, F. Acero, C.-J. Li, Y.-H. Chu, ApJL **910**(2), L24 (2021). https://doi.org/10.3847/2041-8213/abee8a
171. H. Yamaguchi, M. Ozawa, K. Koyama, K. Masai, J.S. Hiraga, M. Ozaki, D. Yonetoku, ApJL **705**(1), L6 (2009). https://doi.org/10.1088/0004-637X/705/1/L6
172. D. Yu, H. Li, B. Li, M. Ge, Y. Tuo, X. Li, W. Xue, Y. Liu, A. Wang, Y. Zhu, L. Bingxian, AMT **15**(10), 3141 (2022). https://doi.org/10.5194/amt-15-3141-2022
173. S.A. Zhekov, R. McCray, K.J. Borkowski, D.N. Burrows, S. Park, ApJ **628**(2), L127 (2005). https://doi.org/10.1086/432794
174. S.A. Zhekov, R. McCray, K.J. Borkowski, D.N. Burrows, S. Park, ApJ **645**(1), 293 (2006). https://doi.org/10.1086/504285
175. S.A. Zhekov, R. McCray, D. Dewey, C.R. Canizares, K.J. Borkowski, D.N. Burrows, S. Park, ApJ **692**(2), 1190 (2009). https://doi.org/10.1088/0004-637X/692/2/1190